Application of Analytical Chemistry to Foods and Food Technology

Application of Analytical Chemistry to Foods and Food Technology

Editors

Daniele Naviglio
Monica Gallo

MDPI • Basel • Beijing • Wuhan • Barcelona • Belgrade • Manchester • Tokyo • Cluj • Tianjin

Editors
Daniele Naviglio
University of Naples Federico II
Italy

Monica Gallo
University of Naples Federico II I
taly

Editorial Office
MDPI
St. Alban-Anlage 66
4052 Basel, Switzerland

This is a reprint of articles from the Special Issue published online in the open access journal *Foods* (ISSN 2304-8158) (available at: https://www.mdpi.com/journal/foods/special_issues/Application_Analytical_Chemistry_Foods_Technology).

For citation purposes, cite each article independently as indicated on the article page online and as indicated below:

LastName, A.A.; LastName, B.B.; LastName, C.C. Article Title. *Journal Name* **Year**, *Article Number*, Page Range.

ISBN 978-3-03943-460-2 (Hbk)
ISBN 978-3-03943-461-9 (PDF)

Cover image courtesy of Francesco Filippelli.

Contents

About the Editors

Daniele Naviglio currently works at the Department of Chemical Sciences and is an expert in using classical analytical procedures and new instrumental methods of analysis such as chromatographic techniques. Prof. Daniele Naviglio is the owner of an industrial invention patent entitled "Rapid and dynamic solid-liquid extractor working at high pressures and low temperatures for obtaining in short times solutions containing substances that initially were in solid matrixes insoluble in extracting liquid". The extractor was patented as Extractor Naviglio®. He is co-author of more than 100 publications in international journals with impact factors.

Monica Gallo currently works at the Department of Molecular Medicine and Medical Biotechnology, University of Naples Federico II. Her current research interest is oriented towards health sciences, with a deep interest in biomedical sciences; biopharmaceutical sciences; biomedicine and natural products; conventional and innovative solid–liquid extraction techniques; circular economy; green chemistry; agri-food waste recovery; purification and characterization of bioactive compounds; antioxidants; functional foods; and in vitro, in vivo, and in silico models.

Editorial

Application of Analytical Chemistry to Foods and Food Technology

Daniele Naviglio [1,*] and Monica Gallo [2,*]

[1] Department of Chemical Sciences, University of Naples Federico II, via Cintia, 21, 80126 Naples, Italy
[2] Department of Molecular Medicine and Medical Biotechnology, University of Naples Federico II, via Pansini, 5, 80131 Naples, Italy
* Correspondence: naviglio@unina.it (D.N.); mongallo@unina.it (M.G.); Tel.: +39-081-674063 (D.N.); +39-081-7463117 (M.G.)

Received: 1 September 2020; Accepted: 12 September 2020; Published: 15 September 2020

Abstract: Foods are a mixture of substances capable of supplying the human body with nutrients, which, once metabolized, are used mainly for the production of energy, heat, replenishment, and growth material for organs and tissues, ensuring the normal performance of vital functions necessary for growth of the human body. Therefore, the study of the chemical composition of foods and the properties of their constituents helps to define their nutritional and commodity values. Furthermore, it allows for evaluation of the chemical modifications that the constituents of the food undergo following the treatments (Food Technology) to which they are subjected. Analytical chemistry is the branch of chemistry based on the qualitative and quantitative determination of compounds present in a sample under examination. Therefore, through its application, it is possible to determine the quality of a product and/or its nutritional value, reveal adulterations, identify the presence of xenobiotic substances potentially harmful to human health (heavy metals, IPA, pharmaceuticals, etc.). Furthermore, some foods, in particular those of plant origin, contain numerous substances, secondary metabolites, with huge beneficial effects for human health. These functional components can be taken both through a correct diet, but also obtained from different food matrices by technological or biotechnological processes for the formulation of both functional foods and/or nutraceutical products. This Special Issue brings together 10 original studies and two comprehensive reviews on the above topics, in particular: (i) processes of extraction, identification, and characterization of biologically active compounds from different food matrices, (ii) overview of the main techniques applied for the determination of food colors, (iii) newer and greener solid-liquid extraction techniques.

Keywords: antioxidants; bioactive compounds; functional foods; gas chromatography; health effects; liquid chromatography (HPLC); mass spectrometry; nutraceuticals; phytochemicals; solid-liquid extraction techniques

---

Two centuries ago, the application of Analytical Chemistry to the study of the composition of food gave rise to a new science called Bromatology (from the Greek βρῶμα, which means food). This science, currently referred to as Food Chemistry, can be considered as a branch of chemistry that deals with the study of food, deepening the aspects related to the qualitative and quantitative characterization of its main components (lipids, proteins, carbohydrates, vitamins, and minerals). On the other hand, food products can be consumed as such or subjected to treatments including conservation and transformation technologies, with all the resulting consequences due to the potential lowering in quality of final products [1]. Therefore, the enormous growth of the food industry over the last fifty years has broadened the scope of analytical chemistry not only to food, but also to food technology, which is fundamental for increasing the production of a large gamma of foods.

Furthermore, various scientific evidence has now definitively demonstrated the positive role that some nutritional factors, such as vegetable fibers, antioxidant compounds, particular classes of lipids, bioactive peptides, and so on, present in food matrices, especially of vegetable origin, can play in the prevention of widespread chronic and degenerative diseases (cardiovascular diseases, neoplasms, metabolic syndrome). More specifically, a high consumption of fruit and vegetables has been associated with a lower risk of cardiovascular disease, as well as of some types of neoplasms [2].

In addition, in recent times, even in the food sector, the concept of acircular economy is becoming more and more widespread, i.e., a system based on the ability to Reuse, Recover and Recycle (Three R) waste materials from the various production phases, or even on preventing them. As is known, the production of waste contributes to air, water, and soil pollution, as well as to climate change and the loss of biodiversity. However, it must be considered that some agri-food waste, even after use, retains a mixture of substances that are not removed. Therefore, the current green concept foresees the creation and research of new ways of extracting valuable compounds from plants, herbs, algae, other organisms, but also from waste material, in order to promote sustainable growth of the world population. Consequently, through appropriate technologies, it is possible to recover bioactive compounds to be used in various sectors [3,4].

Starting from the above premises, in recent years, numerous researches have been carried out concerning the identification of substances with beneficial action from various matrices, especially of vegetable origin, in order to evaluate their use in various sectors, such as pharmaceutical, cosmetic, herbal, and food. These studies foresee that the matrices under examination are subjected to extraction processes, with subsequent identification and characterization of molecules of nutraceutical interest. The structural characterization of these molecules is obtained using classical biochemical methods, which, more recently, are usually integrated with advanced techniques, by means of proteomic and metabolomic approaches based on chromatographic, electrophoretic, mass spectrometry, and nuclear magnetic resonance technologies. The same extraction, identification, and characterization procedures are also applied to processed products and/or other systems of interest.

Below is a brief review of the articles made by international research groups that have allowed the realization of this Special Issue. The research areas range from studies on biologically active compounds obtained from different food matrices to the most recent extraction techniques, from food technologies proper up to quality control and food safety, using some of the most innovative approaches available today. In particular, they concern the extraction, identification and characterization of molecules of nutraceutical interest, especially of plant origin by means of proteomic and metabolomic approaches based on chromatographic, electrophoretic, and spectroscopic technologies.

Curcuma is a perennial herb of Indonesian origin. The noble part of the plant is the rhizome. The latter contains curcumin and curcuminoids, active ingredients of turmeric, much studied for their beneficial effects. Binello et al. (2020) describe the extraction of curcuminoids from *Curcuma longa* L. rhizomes by ultrasound-assisted extraction (UAE). The results show, for the first time, an analytical evaluation of the stability of curcuminoids under sonication in different solvents [5].

Kobbah is an oriental dish consisting of ground bulgur (grain-based food) mixed with ground beef. Typically, the dish is prepared in the form of balls filled with cooked minced meat, onions, nuts, and spices, cooked by frying. Al-Asmar et al. (2020) investigated the influence of different hydrocolloid-based coatings (containing pectin and chickpea flour prepared in the presence or absence of nanoparticles and/or transglutaminase) on the content of acrylamide, water, oil, digestibility and color of fried kobbah. The physico-chemical properties of different coating solutions were also evaluated. The results show that the best coating solution that significantly reduced acrylamide was the one made from pectin, confirming that increasing the water content within the fried food by coating is an effective way to mitigate the formation of acrylamide and the oil content [6].

In a paper by Wang et al. (2020), quantitative analysis of spectinomycin and lincomycin in poultry egg samples was performed by accelerated solvent extraction (ASE) coupled with gas chromatography tandem mass spectrometry (GC-MS/MS). The results show that the proposed method is faster, requires fewer reagents and more samples can be processed at a time, compared to conventional extraction methods [7].

A paper by Hirondart et al. (2020) reports the comparison of two extraction techniques, such as conventional Soxhlet extraction and pressurized liquid extraction (PLE) to determine the initial composition of the key antioxidants contained in rosemary leaves. The data obtained show that there are no significant differences between the two procedures in terms of extraction, but PLE is a quick, clean, and environmentally friendly extraction technique [8].

Colchicum triphyllum is a little-known Turkish cultivar belonging to the *Colchicaceae*. Senizza et al. (2020) evaluated the antioxidant and enzyme inhibitory effects in vitro of extracts of flowers, tubers and leaves of this cultivar, obtained by different extraction methods, such as maceration, infusion, and Soxhlet. The interesting data obtained show the potential of *C. triphhyllum* extracts in food and pharmaceutical applications [9].

The presence of lead in the environment and in the food chain represents a serious pollution problem. In fact, multiple health effects are associated with exposure to heavy metals, with different degrees of severity and conditions: kidney and bone problems, neurobehavioral and developmental disorders, high blood pressure and, potentially, even lung cancer. In this context, immunoassays for the quantitative measurement of environmental heavy metals offer numerous advantages over other traditional methods. Therefore, the methods developed by Xu et al. (2020) through the use of ELISA and chemiluminescent enzyme immunoassay allowed the determination of trace lead (II) in various samples with high sensitivity, simplicity, and accuracy [10].

Liu et al. (2020) investigated the characteristic aroma components of five Chinese mango varieties using headspace solid phase microextraction (HS-SPME) coupled with gas chromatography-mass spectrometry-gas chromatography-olfactometry (GC-MS-O) techniques. Based on these techniques, five main types of substances have been detected, including alcohols, terpenes, esters, aldehydes, and ketones, which are responsible for their special flavor [11].

Literature data report that the fruits of the oleaster (*Elaeagnus angustifolia*) as well as being used as food, contain components with antinociceptive and anti-inflammatory effects. However, narrow-leaved olive fruits have different geographic origins that vary in chemical and physical properties and differ in their nutritional and commercial values. In a study by Gao et al. (2019), near-infrared hyperspectral imaging was used to identify the geographic origins of dry narrow-leaved olive fruits with machine learning methods. The overall results illustrated that this approach could be used to trace the geographical origins of oleaster fruits [12].

Tsochatzis et al. (2019) developed and validated an Ultra High Performance Hydrophilic Liquid Chromatography (UHPLC) (HILIC) tandem mass spectrometry (MS) method for the quantification of amino acids in organic and conventional flour samples with different extraction rates. The results showed significant differences in the amino acid profiles of the flours studied [13].

Hsian-tsao (*Platostoma palustre* Blume) is a species of plant belonging to the genus Platostoma of the mint family. In a paper by Kung et al. (2019) analysis of volatile components present in eight varieties of Hsian-tsao was performed using headspace solid phase microextraction (HS-SPME) and simultaneous distillation-extraction (SDE) coupled with gas chromatography (GC) and gas chromatography-mass spectrometry (GC/MS). The analysis of the results obtained made it possible to identify 120 volatile components. In particular, SDE was able to detect more components, while the HS-SPME analysis was more convenient [14].

Food colorants are natural or artificial substances that give color to a food or restore its original color. Therefore, they are classifiable as food additives and are widely used in the food industry. However, in some cases, their presence can pose risks to human health. Consequently, their determination is extremely important. In a review by Ntrallou et al. (2020) an overview of the main techniques applied for the determination of food colorants and sample preparation procedures that strongly depend on the food matrix is presented over the last 10 years [15].

The extraction process represents a very important step in obtaining compounds of interest from a certain matrix. Alongside the conventional extraction techniques still used, there are currently so-called green techniques. In fact, in order for the extraction processes to be defined sustainable, it is important to make use of technologies with high energy efficiency, but based on low environmental impact solvents. In the review by Naviglio et al. (2020) a comparison was made between the most recent solid liquid extraction techniques with the rapid dynamic solid liquid extraction (RSLDE) as a valid alternative to those examined. In particular, in the RSLDE, the extraction takes place for the generation of a negative pressure gradient from the inside to the outside of the solid matrix, so it can be carried out at room temperature or even sub-environment [16].

In conclusion, the results obtained from the various studies presented can contribute significantly to the identification of new bioactive molecules and to the description of the structural and functional properties of the various bioactive compounds, in order to consider their possible use in various sectors. On the other hand, the research areas range from studies on the extraction, identification, and characterization of bioactive compounds, from food technologies proper up to quality control and food safety, using some of the most innovative approaches available today. Therefore, the multidisciplinary approach contained in this Special Issue, in terms of both knowledge and technologies, can constitute a reference point for the study and evaluation of the effect of foods, natural substances, and nutraceuticals on social wellness.

Author Contributions: M.G. and D.N. conceived and wrote this editorial. All authors have read and agreed to the published version of the manuscript.

Funding: This research received no external funding.

Acknowledgments: The authors thank all the eminent scientists who contributed to the realization of this Special Issue.

Conflicts of Interest: The authors declare no conflict of interest.

References

1. Gallo, M.; Ferrara, L.; Calogero, A.; Montesano, D.; Naviglio, D. Relationships between food and diseases: What to know to ensure food safety. *Food Res. Int.* **2020**, *137*, 109414. [CrossRef]

2. Ortega, A.M.M.; Campos, M.R.S. Bioactive Compounds as Therapeutic Alternatives. In *Bioactive Compounds*; Campos, M.R.S., Ed.; Woodhead Publishing: Cambridge, UK, 2019; pp. 247–264.

3. Salvatore, M.M.; Ciaravolo, M.; Cirino, P.; Toscano, A.; Salvatore, F.; Gallo, M.; Naviglio, D.; Andolfi, A. Fatty acids from *Paracentrotus lividus* sea urchin shells obtained via Rapid Solid Liquid Dynamic Extraction (RSLDE). *Separations* **2019**, *6*, 50. [CrossRef]

4. Gallo, M.; Formato, A.; Ciaravolo, M.; Formato, G.; Naviglio, D. Study of the Kinetics of Extraction Process for The Production of Hemp Inflorescences Extracts by Means of Conventional Maceration (CM) and Rapid Solid-Liquid Dynamic Extraction (RSLDE). *Separations* **2020**, *7*, 20. [CrossRef]

5. Binello, A.; Grillo, G.; Barge, A.; Allegrini, P.; Ciceri, D.; Cravotto, G.A. Cross-Flow Ultrasound-Assisted Extraction of Curcuminoids from *Curcuma longa* L.: Process Design to Avoid Degradation. *Foods* **2020**, *9*, 743. [CrossRef] [PubMed]

6. Al-Asmar, A.; Giosafatto, C.V.L.; Sabbah, M.; Mariniello, L. Hydrocolloid-Based Coatings with Nanoparticles and Transglutaminase Crosslinker as Innovative Strategy to Produce Healthier Fried Kobbah. *Foods* **2020**, *9*, 698. [CrossRef] [PubMed]

7. Wang, B.; Wang, Y.; Xie, X.; Diao, Z.; Xie, K.; Zhang, G.; Zhang, T.; Dai, G. Quantitative Analysis of Spectinomycin and Lincomycin in Poultry Eggs by Accelerated Solvent Extraction Coupled with Gas Chromatography Tandem Mass Spectrometry. *Foods* **2020**, *9*, 651. [CrossRef] [PubMed]

8. Hirondart, M.; Rombaut, N.; Fabiano-Tixier, A.S.; Bily, A.; Chemat, F. Comparison between Pressurized Liquid Extraction and Conventional Soxhlet Extraction for Rosemary Antioxidants, Yield, Composition, and Environmental Footprint. *Foods* **2020**, *9*, 584. [CrossRef] [PubMed]

9. Senizza, B.; Rocchetti, G.; Okur, M.A.; Zengin, G.; Yıldıztugay, E.; Ak, G.; Montesano, D.; Lucini, L. Phytochemical Profile and Biological Properties of *Colchicum triphyllum* (Meadow Saffron). *Foods* **2020**, *9*, 457. [CrossRef] [PubMed]

10. Xu, L.; Suo, X.Y.; Zhang, Q.; Li, X.P.; Chen, C.; Zhang, X.Y. ELISA and Chemiluminescent Enzyme Immunoassay for Sensitive and Specific Determination of Lead (II) in Water, Food and Feed Samples. *Foods* **2020**, *9*, 305. [CrossRef] [PubMed]

11. Liu, H.; An, K.; Su, S.; Yu, Y.; Wu, J.; Xiao, G.; Xu, Y. Aromatic Characterization of Mangoes (Mangifera indica L.) Using Solid Phase Extraction Coupled with Gas Chromatography–Mass Spectrometry and Olfactometry and Sensory Analyses. *Foods* **2020**, *9*, 75. [CrossRef] [PubMed]

12. Gao, P.; Xu, W.; Yan, T.; Zhang, C.; Lv, X.; He, Y. Application of near-infrared hyperspectral imaging with machine learning methods to identify geographical origins of dry narrow-leaved oleaster (*Elaeagnus angustifolia*) fruits. *Foods* **2019**, *8*, 620. [CrossRef] [PubMed]

13. Tsochatzis, E.; Papageorgiou, M.; Kalogiannis, S. Validation of a HILIC UHPLC-MS/MS Method for Amino Acid Profiling in Triticum Species Wheat Flours. *Foods* **2019**, *8*, 514. [CrossRef] [PubMed]

14. Kung, T.L.; Chen, Y.J.; Chao, L.K.; Wu, C.S.; Lin, L.Y.; Chen, H.C. Analysis of Volatile Constituents in Platostoma palustre (Blume) Using Headspace Solid-Phase Microextraction and Simultaneous Distillation-Extraction. *Foods* **2019**, *8*, 415. [CrossRef] [PubMed]

15. Ntrallou, K.; Gika, H.; Tsochatzis, E. Analytical and sample preparation techniques for the determination of food colorants in food matrices. *Foods* **2020**, *9*, 58. [CrossRef] [PubMed]

16. Naviglio, D.; Scarano, P.; Ciaravolo, M.; Gallo, M. Rapid Solid-Liquid Dynamic Extraction (RSLDE): A powerful and greener alternative to the latest solid-liquid extraction techniques. *Foods* **2019**, *8*, 245. [CrossRef] [PubMed]

Article

A Cross-Flow Ultrasound-Assisted Extraction of Curcuminoids from *Curcuma longa* L.: Process Design to Avoid Degradation

Arianna Binello [1], Giorgio Grillo [1], Alessandro Barge [1], Pietro Allegrini [2], Daniele Ciceri [2] and Giancarlo Cravotto [1,*]

[1] Dipartimento di Scienza e Tecnologia del Farmaco, University of Turin, Via P. Giuria 9, 10125 Turin, Italy; arianna.binello@unito.it (A.B.); giorgio.grillo@unito.it (G.G.); alessandro.barge@unito.it (A.B.)
[2] INDENA S.pA., Viale Ortles, 12, 20139 Milan, Italy; pietro.allegrini@indena.com (P.A.); daniele.ciceri@indena.com (D.C.)
*   Correspondence: giancarlo.cravotto@unito.it; Tel.: +39-011-6707183; Fax: +39-011-6707162

Received: 9 May 2020; Accepted: 2 June 2020; Published: 4 June 2020

Abstract: Rhizomes of *Curcuma longa* L. are well known for their content of curcuminoids, which are compounds with interesting biological activity against various inflammatory states and diseases. Curcuminoids can degrade during processing. This piece of work investigates fast, efficient and cost-effective metabolite recovery from turmeric under ultrasound-assisted extraction (UAE). An analytical evaluation of curcuminoid stability under sonication in different solvents is reported for the first time. HPLC and quantitative ^{1}H-NMR were used. Under the applied conditions, EtOAc was found to be the optimal extraction medium, rather than EtOH, due to its lower radical generation, which facilitates better curcuminoid stability. Kinetic characterization, by means of the Peleg equation, was applied for single-step UAE on two different rhizome granulometries. Over a time of 90 min, maximum extraction yields were 25.63% and 47.56% for 6 and 2 mm matrix powders, respectively. However, it was observed that the largest portion of curcuminoid recovery was achieved in the first 30 min. Model outcomes were used as the basis for the design of a suitable multi-step cross-flow approach that supports and emphasizes the disruptive role of cavitation. The maximum curcuminoid yield was achieved over three steps (92.10%) and four steps (80.04%), for lower and higher granulometries, respectively. Finally, the central role of the solvent was further confirmed by turmeric oleoresin purification. The EtOAc extract was purified via crystallization, and a 95% pure curcuminoid product was isolated without any chromatographic procedure. No suitable crystallization was observed for the EtOH extract.

Keywords: *Curcuma longa* L.; curcuminoid stability; multi-step extraction; ultrasound-assisted extraction; extraction kinetic

---

1. Introduction

In recent years, non-synthetic and biologically active compounds from vegetal sources have gained increasing interest because of their important role in health-care systems worldwide; pigments have found use as additives or supplements in food, pharmaceutical and cosmetic industries. One of the most widely studied sources of natural pigment plants is *Curcuma longa* L., a perennial rhizomatous shrub belonging to the *Zingiberaceae* family. Also known as turmeric, this plant is commonly used as a coloring and flavoring agent in the food industry and, in particular, it is known as the 'golden spice of life', and constitutes the main component of curry [1]. However, it is also appreciated in traditional medicine for its biological properties, which are mainly related to its curcuminoids; chemical components that include curcumin (CUR), demethoxycurcumin (DMC) and bisdemethoxycurcumin (BDMC) (Figure 1) [2].

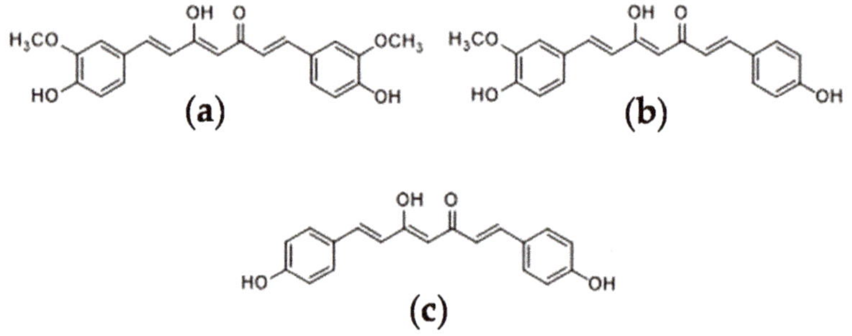

Figure 1. Chemical structures of curcuminoids: (**a**) curcumin (CUR), (**b**) demethoxycurcumin (DMC), (**c**) bisdemethoxycurcumin (BDMC).

Although these substances can be chemically synthesized, is worth noting that the Joint FAO/WHO Expert Committee on Food Additives (JECFA) specifications only allow curcuminoids extracted from natural source material to be used as food additives [3].

Turmeric has curcuminoid contents of 2–9% depending on its growing conditions and origin. The main biological activities exhibited by these compounds are antioxidant, anti-inflammatory, antibacterial, antiviral, antifungal, anticancer, immune-stimulatory and neuroprotective [4–7].

The most common techniques to obtain curcuminoids from turmeric involve solvent extraction followed by column chromatography. However, the choice of the solvent and the conditions applied must take into account the safety of the final use to which these compounds are intended, for example, their acceptability in the food industry. Soxhlet, ultrasonic-assisted extraction (UAE) and microwave-assisted extraction (MAE) are currently the most commonly used methods, although other techniques, such as pulsed-ultrasonic and supercritical-fluid extraction, are also reported to be efficient processes [8–10]. It is worth noting that UAE is now widely used in vegetal-matrix extraction thanks to its efficiency, which allows it to: (1) enhance extraction yields and rates; (2) make use of alternative solvents; (3) reduce costs and required extraction times; and (4) preserve heat-sensitive compounds. Several cavitation devices are available, including bath, probe and flow systems, and can provide huge process-type flexibility, such as scalability, and the use of counter-current or co-current systems [11]. In particular, the co-current cross-flow approach is a relatively simple sequential-step protocol that is used in solid/liquid extraction for different purposes, such as decontamination, leaching and food washing. Essentially, the solid to be extracted is mixed with fresh solvent, then recovered and drained. The extracted solid then meets fresh solvent and is recovered and drained again. The higher the number of steps, the more the matrix approaches depletion [12].

A great deal of metabolite-extraction research has investigated this mechanism, and it has been described using Fick's second law of diffusion [13,14]. However, a deeper modeling of the kinetics underlying the solid–liquid extraction processes is not common in literature; mathematical assets, such as useful engineering tools, are needed to cast light on method applicability and optimization. Desorption processes, on the other hand, are commonly interpreted by mathematical descriptors in literature, and Peleg's Model is one of the best known [15]. The common principles shared by dehydration/rehydration phenomena and metabolite extraction allow the ability of this model to render UAE kinetics to be investigated [16,17].

An equation that describes the time-dependent trend of a system can ensure the understanding necessary to define the fastest extraction rates and the best compromises between metabolite yield and time consumption. This approach is fundamental from an industrial point of view, where the main focus is the simultaneous maximization of productivity and the minimization of energy consumption and costs.

The international authorities (FAO/WHO and European Commission), which monitor the development and commercialization of food additives, have decreed that a limited number of solvents are permitted for use in the preparation of curcuminoid-based products [18]. The permitted polar and non-polar organic solvents include acetone, methanol, ethanol, iso-propanol, hexane, ethyl acetate and supercritical carbon dioxide, which have been accepted for use in the extraction of CUR and its analogues from turmeric.

S.R. Shirsath et al. [17] have thoroughly explored, using kinetic investigations, CUR extraction from *Curcuma amada*, and have screened parameters, such as time, temperature, solvent and granulometry. Ethanol was confirmed as the solvent of choice (vs. methanol, acetone and ethyl acetate), as it was able to give 72% of total metabolites. A direct comparison confirmed the efficiency of UAE, compared to silent conditions which provided an average yield decrease of 10%. A broader screening can be performed to consider other aspects that are involved in the industrial processing of vegetal matrixes, such as easier-to-handle granulometries (less prone to filter clogging), final product stability and purification steps.

In fact, although ethanol is the preferred solvent, the choice of extraction solvent must also be evaluated in relation to the curcuminoid purification steps. On the laboratory scale, chromatography is the technique most commonly employed to isolate curcuminoids, for example using silica as the stationary phase and various organic-solvent mixtures (e.g., chloroform:methanol) as eluents [19]. Crystallization can be included among the purification steps. Ukrainczyk et al. have reported the influence that process conditions have on the purification of crude curcumin via successive cooling crystallizations with isopropanol. [20]. Concentrated oleoresin was selected for the formation of crystals via the slow addition of petroleum ether, water and hexane. The best crystal quality was found when petroleum ether was used, whereas the crystals prepared using the other two solvents were sticky in nature [21]. A non-classical crystallization pathway for curcumin particles was found by Alpana et al. when this process is carried out in sonochemical conditions with or without stabilizers [22].

The quantification of curcuminoid content in turmeric extracts is commonly performed using chromatographic techniques (HPLC, UPLC and capillary electrophoresis). However, in recent years, the quali-quantitative control of herbal products have also been carried out with spectroscopic fingerprinting [23,24]. In particular, Gad and Bouzabata [25] have recently investigated the use of UV, FT-IR and ^{1}H NMR for the quality control of *Curcuma longa* by comparing data with those obtained from HPLC analyses. They observed that NMR shows good potential efficiency.

Although curcuminoids possess health-promoting factors, they have found limited application in the food and pharmaceuticals industries because of their low water solubility, poor bioavailability and poor stability in in-vivo and in-vitro environment [26]. These compounds undergo degradation by acidic or alkaline hydrolysis, oxidation, photo-degradation and, moreover, are sensitive to light. The autoxidative degradation of CUR at physiological pH gives bicyclopentandione as the major product, while vanillin, ferulic acid and are reported to be minor compounds [27,28]. The same compounds have been identified as curcuminoid photodegradation products [29].

A systematic stability study of curcuminoids has been carried out by Peram et al. using a RP-HPLC method that analyzes their behavior under different stress degradation conditions (i.e., acidic, alkaline, oxidative, photolytic, and thermal degradation) [30]. The authors reported that the order of stability of curcuminoids was: BDMC, followed by DMC and CUR. This proves that these compounds possess a precise synergistic stabilizing mechanism when present in a mixture, as compared to their pure forms.

The processing conditions that are applied to obtain purified bioactives from turmeric, can affect the labile stability and the biological activity of curcuminoids. Hence, it is of the utmost importance that degradation behavior be considered. Valuable information can be obtained from stability studies in order to understand how the health-promoting effects can be retained. Although mechanistic studies have been published on curcuminoid degradation under different stress conditions [30,31], the kinetics under US treatment, both when curcuminoids are present in standard solutions and when they are found in turmeric extracts, have yet to be reported to the best of our knowledge.

In our work, the rapid and exhaustive UAE of turmeric has been defined using multi step extractions and kinetic studies in ethyl acetate media. Solvent choice was subordinated to bioactive stability and the effectiveness of purification by crystallization.

For the sake of comparison, curcuminoid stability under US irradiation in ethanol and ethyl acetate systems has been tested both on optimized extracts and in standard mixtures.

NMR has been applied as both a qualitative and quantitative analytical method to monitor the stability of curcuminoids under sonication treatment. The degradation products were also detected using HPLC and UPLC-MS analyses.

2. Materials and Methods

2.1. Chemicals

Ethyl acetate (ACS grade, ≥99%) (Sigma-Aldrich, Milan, Italy) was used in the extraction procedures. Acetonitrile CHROMASOLV® (gradient grade, for HPLC, ≥99.9%) for HPLC analysis was purchased from Sigma-Aldrich, while Milli-Q H_2O was obtained in the laboratory using a Milli-Q Reference A+System (Merck Millipore). Standards of Curcumin (87.02% Curcumin, 12.98% other curcuminoids) were purchased from Sigma Aldrich.

2.2. Curcuma longa L. Matrix

Curcuma longa L. mother rhizomes (India), which were cured and sun dried, were kindly provided by Indena SpA (Milano, Italy), in two different average granulometries, 6 and 2 mm. Biomass was ground in a hammer mill. The biomass was stored at room temperature (RT) in a dry and dark environment to avoid metabolite degradation.

2.3. Curcuminoid Stability Tests under US

The lability of curcuminoids under US irradiation was evaluated in EtOAc, and compared with the most common GRAS solvent for *Curcuma longa* L. extraction, namely EtOH. In order to evaluate the stabilizing effects of co-extracted molecules, both the dry extract (see conventional extraction, Section 2.5) and the curcumin standard were subjected to sonication prior to solubilization. A total of 2 mL of each solvent were used to dissolve 5 mg of the chosen sample in an analytical tube. The solutions were sonicated in a cup-horn PEX 3 Sonifier (24 kHz, 200W, REUS, Contes, France) for either 30, 60, 90 or 120 min. In order to preserve cavitation efficiency, an average temperature of 40 °C was maintained during the tests thanks to a double layered mantle that was crossed by cooling tap water. After treatment, the samples were dried under vacuum for HPLC analyses.

2.4. Ultrasound-Assisted Extraction (UAE)

Curcuma longa L. rhizome powder (10 g) was transferred into a 100 mL glass vessel. Ethyl acetate was added, based on a previous screening, to maintain a solid/liquid (S/L) ratio of 1:5. Extraction was performed in two US devices: a probe system equipped with a titanium horn (20.5 kHz, 350-500W, HNG-20500-SP, Hainertec Suzhou, China), and a cup-horn PEX 3 Sonifier (see Section 2.3). The US horn requires an ice bath to control the temperature in the medium. In both systems, an average temperature of 40 °C was maintained to preserve cavitation efficiency while a time screening was performed.

Extractions were conducted in one or multiple sequential steps (cross-flow), with the aim of depleting the matrix.

A technique comparison was performed with 6 mm rhizome powder and the optimized parameters were then transposed to a 2 mm matrix in order to evaluate the incisiveness of mass transport and the physical effects of US treatment. The crude extract was filtered on sintered glass and dried under vacuum.

2.5. Conventional Extraction

Classical extractions were carried out as a benchmark for technique screening and for overall curcuminoid yield definition. For the sake of comparison, parameters that were used for optimized UAE, such as temperature, solvent (EtOAc), extraction steps and time, were transposed to the 2 mm matrix in a conventional magnetic stirred system (silent conditions). The maximum curcuminoid yield, which corresponded to matrix depletion, was defined in accordance to the rhizome-powder extraction conditions (S/L ratio 1:5), with homogenization pre-treatment (5 min, OV5 rotor-stator homogenizer Velp Scientifica, Usmate Velat, Italy) and 4 sequential extraction steps (4h each) for a total of 16 h, at RT and under protection from light. For the sake of comparison, the same conditions were used with EtOH as a substitute for EtOAc.

2.6. Curcuminoid Determination

Total curcuminoids (TC) were determined by HPLC using the external standard method. Analyses were performed on a Waters 1525 binary pump equipped with a 2998 PDA, and a Phenomenex Kinetex® Column (5 μm C18 100 Å, 250 × 4.6 mm). Data acquisition was accomplished using Empower PRO (Waters Associates, Milford, MA). A CH_3CN 5% acetic acid aqueous solution was used as the mobile phase. Chromatographic separation was performed in isocratic (50:50 *v/v*) at 25 °C, and a flow rate of 1 mL/min. The injection volume was 10 μL, while sample detection was carried out at 425 nm. Before injection, all samples were dissolved in MeOH, giving concentrations of between 1 and 2 mg/mL. All the samples were passed through 0.2-μm membrane filters before injection into the HPLC apparatus. The calibration curve was obtained using curcumin standard solutions (from 0.02 to 2 mg/mL); a linear regression with $R^2 > 0.999$ was obtained. The reported results express both TC (expressed as curcumin equivalents, percentage yields and mg/mL amounts) and curcumin concentration. Relative limit of detection (LOD) and limit of quantification (LOQ) were determined as 0.005 mg/mL and 0.02 mg/mL, respectively.

2.7. NMR Quali and Quantitative Analyses

The quali-quantitative determination of curcumin was also carried out by ^{1}H-NMR. NMR spectra were acquired on a Jeol ECZR 600 spectrometer, operating at a 14 T magnetic field strength and equipped with a Jeol Royal standard probe. Signal acquisition and FID processing were carried out using Jeol DELTA software. Qualitative evaluation, via spectrum analysis and a comparison with the spectra of standards, allowed the different curcuminoids that were present in the whole extract to be identified. The quantitation of curcumin was achieved by acquiring ^{1}H-NMR spectra in MeOD using a 90° pulse, ^{13}C decoupling and a repetition time that was longer than 7-times the longest T1 (typically 40–60 s). FIDs were processed with a zero-filling that was double that of the experimental point (64K+64K), an exponential apodization function with 0.1Hz width and final interactive baseline correction. Specific peak area at 7.20 ppm was compared with the area of potassium terephthalate (analytical standard grade) D_2O solution (with a precisely known concentration) signals, which were obtained in the same receiver gain conditions as the sample spectra. A set of three different concentrations of standard curcumin was used to verify the correspondence between real curcumin concentration and curcumin concentration as determined by NMR quantitation (using an external standard). Relative LOD and LOQ were calculated as 0.5 mg/mL and 1 mg/mL, respectively.

2.8. Statistic Treatment

To validate reproducibility and give soundness to the experimental section, every procedure (Sections 2.3–2.5) was performed in triplicate and percentual standard deviation was consequently calculated. The results are expressed as the mean ± %SD in Tables as well as in Appendix A. Degradation percentage SDs are graphically depicted in Figures 4–6. Moreover, upper and lower SD envelopes have been extrapolated by means of linear regression, describing deviation trends where possible.

2.9. Kinetic Model

The hyperbolic model of Peleg (see Equation (1)) was used to evaluate the extraction kinetics and to determine the point of maximum extraction rate by means of the related constants.

$$C(t) = C_0 + \frac{t}{k_1 + k_2 t} \tag{1}$$

$C(t)$ is the concentration of the extract after extraction time t, whilst C_0 is equal to 0 at the beginning of the process. The *Peleg Initial Extraction Rate* (k_1) is correlated to the starting extraction rate (B_0, Equation (2)) and can be exploited to calculate the relative extraction rate in each moment of the extraction (B_t).

$$B_0 = \frac{1}{k_1} \tag{2}$$

This parameter can be used to calculate the instant of maximum extraction speed, the critical point for cross-flow and counter-current extractions, which is fundamental for an industrial transposition of the process.

The *Peleg Capacity Constant* (k_2) is correlated to the highest extraction yield at the steady state (Y_s, Equation (3)). For an ideal process, it can be used to evaluate the number of sequential extraction steps necessary to deplete the matrix. From a graphical point of view, this parameter represents an horizontal asymptote.

$$c_0 = c_{eq} = Y_s = \frac{1}{k_2} \tag{3}$$

Two extraction sets were conducted using a probe system, according to Section 2.4, on the 2 and 6 mm Curcuma powder. Sampling was performed at 2, 5, 7.5, 15, 30, 60, 90 and 120 min, by collecting 1 mL of solution. The crude extract was filtered on sintered glass and dried under vacuum for HPLC analyses.

Equation (1) can be conveniently linearized in Equation (4), thus providing a fast and easy way to extrapolate k_1 and k_2 as the intercept and slope, respectively. Hence, the kinetic constants can be calculated by linear interpolation of the experimental yields at different extraction times (see Table A1 and Figure A1 for 6 mm and Table A2 and Figure A2 for 2 mm), and then inputted into general Equation (1).

$$\frac{t}{C(t)} = k_1 + k_2 t \tag{4}$$

The obtained hyperbolic curve describes a time-dependent extraction trend. This model is a useful means to display the horizontal asymptote of Y_s and the extraction rates (slope of the curve). Furthermore, the knee-point can be exploited to determine the best trade-off between productivity and process extent.

2.10. Crystallization

The ethyl acetate extracts were concentrated under vacuum to approximately an S/L ratio of 1:1. The mixture was stirred at 20–25 °C for 24 h, then the suspension was filtered under vacuum. The wet solid was ground twice at 20–25 °C for 1 h with 3 volumes of isopropanol 90% (3 mL per g of wet solid). The wet solid was finally dried at 50 °C under vacuum.

3. Results

3.1. Curcuminoid Stability Test

In order to investigate the effect of the US irradiation of EtOH and EtOAc on curcumin, a known amount of analytical standard was subjected to prolonged sonication in both solvents. Due to the small volume of the samples, the cup-horn was thought to be the most suitable device for the degradation treatments. As shown in Table 1, the effective stability of the pure curcumin was determined by monitoring its concentration every 30 min using HPLC. Total treatment lasted two hours, and the results are reported in Table 1 as curcumin amounts and degradation percentages.

Table 1. Curcumin standard US degradation tests.

Irradiation (min)	EtOH			EtOAc		
	Curcumin conc. ± %SD (mg/mL)		Degradation (%)	Curcumin conc. ± %SD (mg/mL)		Degradation (%)
0	4.21	± 0.42	-	4.21	± 0.39	-
30	3.70	± 0.61	12.11	3.89	± 0.46	7.60
60	3.35	± 0.54	20.43	3.73	± 0.65	11.40
90	3.17	± 0.73	24.70	3.52	± 0.63	16.39
120	2.96	± 0.68	29.69	3.48	± 0.70	17.34

Curcumin concentration quantified by HPLC.

In order to evaluate the degradation that was only related to solvent sonolysis, the degradation test was repeated in EtOAc, in closed vials. The curcumin concentration was determined by quantitative NMR. Although its sensitivity is much lower than HPLC-UV or HPLC-MS, NMR quantitation offers the opportunity to determine the analyte concentration using an external standard whose spectrum is acquired once. The external standard can be a different molecule to the analyte as the only requirement is that its concentration is exactly known, and that its NMR spectrum is acquired following the quantitative protocol. Table 2 reports the comparison between the two behaviors.

Table 2. Sonolysis: the influence of air on curcumin degradation.

Irradiation (min)	Closed Vessel	
	Degradation (%)	± SD
30	4.00	± 0.58
60	8.54	± 0.72
90	10.72	± 1.14
120	11.27	± 1.48

Solvent: EtOAc; NMR quantifications.

Many studies have indicated that curcuminoid stability increases when all of the components are studied together, rather than in their pure form [30]. In order to verify whether this statement can also be applied to US-mediated degradation, the screening that was performed on the analytical standard was repeated on the extract obtained from the conventional procedure (see Section 2.5). The results for EtOH and EtOAc are reported in Table 3.

Table 3. Curcuminoid extract US degradation tests.

Irradiation (min)	EtOH			EtOAc		
	Curcumin conc. ± %SD (mg/mL)		Degradation (%)	Curcumin conc. ± %SD (mg/mL)		Degradation (%)
0	3.83	± 0.39	-	4.79	± 0.51	-
30	3.71	± 0.61	3.13	4.74	± 0.70	1.05
60	3.54	± 0.57	7.57	4.61	± 0.38	3.76
90	3.16	± 0.49	17.49	4.48	± 0.57	6.47
120	3.08	± 0.67	19.58	4.25	± 0.44	11.27

HPLC quantification of curcuminoids, expressed as curcumin equivalents.

3.2. Total Curcuminoid Content Determination

The quantification of total curcuminoids in plant rhizomes was obtained by performing sequential extractions with EtOAc under ultraturrax$^{®}$ treatment (as described in Section 2.5), with the aim of achieving full matrix depletion.

The triplicate average gives a total amount of 67.14 mg/g$_{Matrix}$ curcuminoids (HPLC determination, expressed as curcumin equivalents), which corresponds to 6.71% of raw material and 40.49 mg/g$_{Matrix}$ (4.05%) of curcumin. The same extraction protocol was performed with EtOH as the solvent and revealed a negligible difference of +0.84% compared to the previous test. This evidence indicates that no difference subsists between the two solvent systems in terms of total matrix depletion, and that 67.14 mg/g$_{Matrix}$ is admissible as the maximum metabolite yield.

3.3. US-Assisted Extraction

This work investigated the UAE of *Curcuma longa* L. and did so by varying US equipment and process parameters. A multi-step approach, which depended on single-step kinetics, was finally defined and investigated for two different granulometries.

Firstly, the effects of acoustic cavitation were studied on the coarse particle size (6 mm), using two different US reactors: an immersion horn (two power intensity, 350 W and 500 W) and a cup-horn (200 W). In order to emphasize the difference between each run, sequential UAE was adopted, and this allowed a preventive evaluation of multi-step feasibility to be performed. Solvent volumes were minimized to an S/L ratio of 1:5, in order to move towards a more sustainable process. Hence, three sequential extractions of 30 min were chosen for each step, in order to avoid the overheating of the system due to solution consistency.

3.3.1. Kinetic Model—Single-Step UAE

Once the extraction device was selected, a better understanding of the kinetic was required. For this purpose, 6 mm rhizomes (treated in Table 4) were considered together with smaller particles (size of 2 mm) for extraction modeling. Samples were gathered after a suitable time-span of 120 min of UAE.

Table 4. US technology screening.

US Technology	Dry Extract (%)	Curcuminoid conc. $\pm$ %SD (%)	
Cup-horn [a]	19.19	72.07	$\pm$ 0.22
Horn [b]	16.31	68.43	$\pm$ 0.38
Horn [c]	18.17	72.68	$\pm$ 0.29

Matrix 6 mm, 3 steps of 30 min, S/L ratio 1:5. [a] 200W; [b] 350W; [c] 500W. Curcuminoid yields, calculated as percentage recovery of total curcuminoid-content. HPLC quantification expressed as curcumin equivalents.

Detected yields were processed according to the Peleg Model, as described in Section 2.9. Using linearization (see Equation (4)), it was possible to extrapolate the kinetic constants by building a general equation for the system that expressed a curve with a physical meaning that describes curcuminoid recovery in a time-dependent function.

The plots (see Figure 2 for 6 mm and Figure 3 for 2 mm) and relative equations (see Equation (5) for 6 mm and Equation (6) for 2 mm) are reported below. Linearization gave impressive interpolation of the experimental values, with R^2 being over 0.99 (see Figures A1 and A2). It is possible to observe how the theoretical curves match the experimental points.

Extraction rates (B_t) can be expressed as the inverse of k_1, which is computable for every instant (t). Bt expresses the extraction efficiency at any precise moment and is mathematically depicted as the slope of a line tangent to the model curve at time t (see Tables A1 and A2).

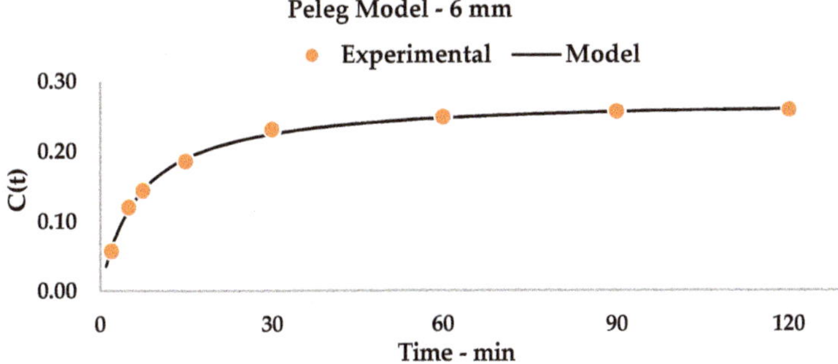

Figure 2. Kinetic Model, 6 mm matrix. Statistical values are reported in Table A1.

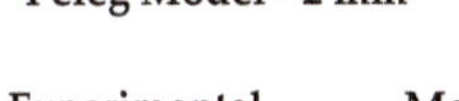

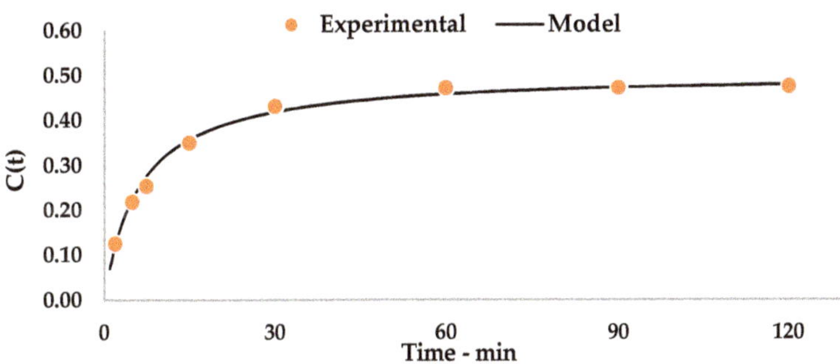

Figure 3. Kinetic Model, 2 mm matrix. Statistical values are reported in Table A2.

$$C(t) = \frac{t}{24.5350 + 3.6412\,t} \tag{5}$$

$$C(t) = \frac{t}{12.2240 + 1.9831\,t} \tag{6}$$

3.3.2. Cross-Flow UAE

The kinetic model gave 30 min as the best extraction time, as it allowed a good compromise between curcuminoid yield and process duration to be achieved. This discovery was developed and sequential UAE was tested. Biomass was recovered run-by-run and was then submitted to a next stage, feeding fresh EtOAc. Both particle sizes were used in order to define granulometry dependency. Multi-step yields and yield increases are reported in Table 5.

Table 5. Cross-flow UAE, step screening with 6 and 2 mm matrix.

Step	6 mm Matrix				2 mm Matrix		
	Dry Extract %	Curcuminoid Yield ± %SD %		Yield Increase %	Dry Extract %	Curcuminoid Yield ± %SD %	Yield Increase %
1	5.44	23.10	± 0.37	-	11.19	43.10 ± 0.22	-
2	12.24	51.97	± 0.33	28.87	22.56	88.54 ± 0.19	45.44
3	18.17	72.68	± 0.29	20.71	23.77	92.10 ± 0.30	3.56
4	20.12	80.04	± 0.31	7.36	-	-	-

30 min per step, S/L ratio 1:5, Horn 500 W; Curcuminoid yields, calculated as percentage recovery of total curcuminoid content. HPLC quantification expressed as curcumin equivalents.

The results depict, as expected, that the extraction rate of the 2 mm rhizome powder was clearly higher than that of the 6 mm analogue. Furthermore, the maximum extraction yield was reached with one extraction step of difference.

3.4. Crystallization

The reported procedure (Section 3.1) was applied to both the EtOH and EtOAc extracts. It was found that the alcoholic turmeric oleoresin could not be purified via crystallization, whilst EtOAc successfully led to suitable precipitate formation. HPLC quantification confirmed an average curcuminoid purity of 95%, expressed as curcumin equivalents.

4. Discussion

4.1. Curcuminoid Stability Test

The fact that radicals are formed as a consequence of acoustic cavitation has been well described in the literature [32]. Although the production of $OH^{\bullet}$ and $H^{\bullet}$ in aqueous media is something that researchers have robustly proven using different approaches [33], very few demonstrations of radical production in organic solvents (as methanol) are available [34]. Radical concentration is frequency-dependent in linear mode, but it is also possible to produce traces of $R^{\bullet}$ or $RO^{\bullet}$ at low frequencies [35]. The extreme reactivity of curcumin-like compounds means that the presence of radicals in the extraction media can affect extraction yields, triggering the degradation and the accumulation of oxidized compounds, which will reduce general shelf-life and product activity.

These considerations can inform the choice of solvent for curcuminoid extractions. Ethanolic solutions are, according to the literature, the best performing media, as they achieve high yields in relatively short times. It is reasonable to think that ethanol is subjected to the formation of $CH_3CH_2^{\bullet}$ and $OH^{\bullet}$ radicals, due to his polarity, as these reactive species can also be generated from MeOH and H_2O [34]. On the other hand, EtOAc (a GRAS approved solvent) is less likely to undergo the same fate.

In the case of pure curcumin, the onset of degradation after 30 min supports the low stability of the compound in both solvent systems (Figure 4, related to Table 1). Tests were performed using a common US set-up, in order to be a closer fit for the usual extraction conditions. This approach cannot avoid oxidation due to atmospheric O_2, which overlaps with the other degradation pathways. In any case, it is possible to state that EtOH media display higher sensitivity than EtOAc media. Furthermore, the time increase leads to a stronger lowering of curcumin concentration in alcoholic systems, reaching a maximum degradation of 29.76% vs. 17.28%.

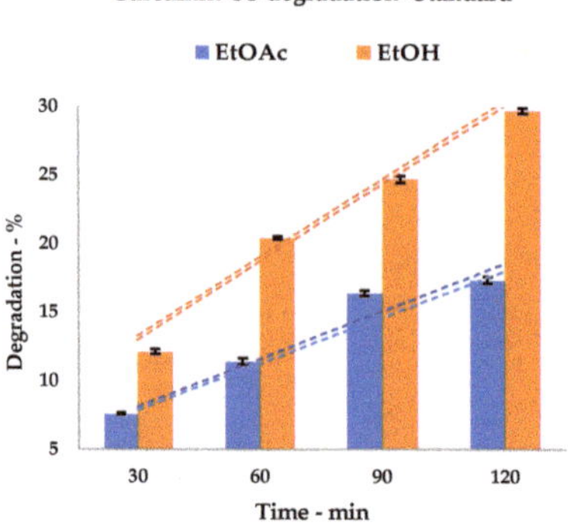

Figure 4. Curcumin standard degradation trend; dashed lines represent upper and lower SD envelopes.

Tests were repeated in sealed vials to investigate the correlation between atmospheric oxygen and overall degradation. With strongly reduced O_2 presence, stability should be principally affected by solvent sonolysis.

The concentration of curcumin decreases over sonication time, but its decrease is lower than when measured in open air. The graph depicted in Figure 5 shows behavior over time.

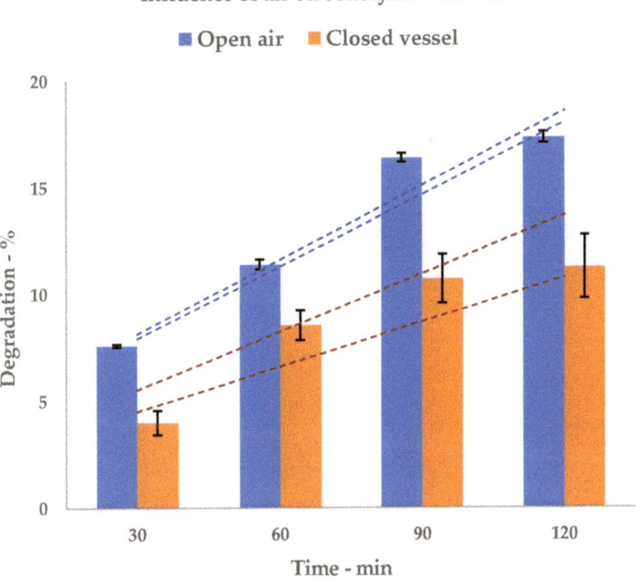

Figure 5. Sonolysis: influence of air. HPLC (open air) and NMR (closed vessel) quantification; dashed lines represent upper and lower SD envelopes.

According to the literature, the stability of curcuminoids increases if all three components (CUR, DMC, BDMC) are mixed together [29].

The stability screenings were therefore repeated with the analytical standard of curcumin being replaced with a conventional extract (see procedure in Section 2.5). US extracts were not used as the starting material to ensure that the matrix was not affected by previous irradiation.

The results fit into the abovementioned studies; general stability was increased by the presence of curcuminoids and co-extracts, shifting from 29.76% to 19.56% and from 17.28% to 11.14%, for EtOH and EtOAc media, respectively. It was possible to observe a sudden increase in the degradation of the alcoholic solvent between 60 and 90 min. This is a retardant effect that is possibly due to the presence of the co-extracts (see Figure 6), as this behavior is not visible in pure curcumin (see Figure 4).

These results demonstrate that different solvent media can give rise to different radical concentrations, which is critical for labile molecules, such as curcumin-like compounds, that are well known for their low shelf-life and stability. EtOAc appeared to provide milder radical generation, which paves the way for a suitable multistep extraction protocol that reduces the degradation that is caused by prolonged treatments. These tests are fundamental for a preliminary characterization of a multistep extraction protocol, and for the design of an efficient procedure for the complete depletion of turmeric rhizome powder.

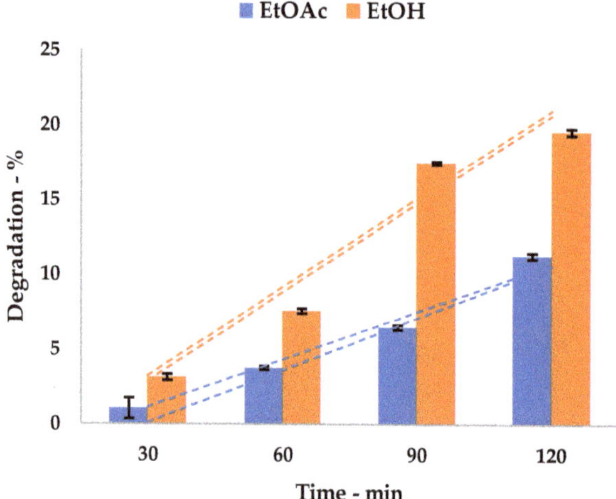

Figure 6. Curcumin extract degradation trend; A: ethanol; B: ethyl acetate; dashed lines represent upper and lower SD envelopes.

4.2. US-Assisted Extraction

In the early stages, the work was focused on 6 mm turmeric powder and finding the best US device for extraction. This approach was a useful means to properly evaluate the physical effects of sonication upon a roughly milled biomass. Bulky powders require better extraction performance for successful cellular/particle cleavage and convenient mass-transfer enhancement to occur. In summary, this type of matrix requires harsher conditions and is a challenge for UAE intensification.

The first approach to UAE was the screening of several US devices, and, in particular, two different set-ups: the immersion horn and cup-horn. The results indicate that a slightly better result was obtained for the horn at 500 W (see Table 4). Technology comparisons shed light on handling and process suitability, and thus influence the practical applications. The cup-horn results were promising even if the power was lower, although temperature control appeared to be very challenging. The sedimentation of heavier particles on the cavitating surface, causing overheated spots, is a possible explanation for this. The media was not easily cooled by the water jacket because of the consequent poorer mass-transfer. The immersion horn, on the other hand, appeared to efficiently transmit energy to the media, despite the radiating surface being smaller. The free tip maintained high mass transfer and easier heat transfer towards the cooling bath. Hence, this first screening facilitated the choice of the 500 W immersion horn as the elective device for the subsequent investigations.

4.2.1. Kinetic Model—Single Step UAE

After the selection of the US device, and before performing the cross-flow extraction design, it was essential to characterize a single-step process. Thus, UAE was approached from a kinetic point of view in order to shed light on the dependency of yield on treatment time. The influence of granulometry was considered for this screening, including the smaller particle size of 2 mm.

When comparing the two kinetic systems (Figures 2 and 3), some considerations can be pointed out. Firstly, the k_2 constants describe, as their inverse values, the curcuminoid total yield in the steady state, and are translated by a horizontal asymptote on the graph. This quantity is, as expected, higher for the 2 mm matrix than for the 6 mm one, as it describes, for the first, a matrix that is more prone to extraction (Y_S: 50.43% vs. 27.46%). Interestingly, the chosen step-length for the initial optimization

(30 min, Table 4) appears to give results that are near the maximum yield for single-step extractions: 23.10% vs. 27.46% for 6 mm and 43.10% vs. 50.43% for 2 mm, respectively. This information allowed us to focus our attention on a dedicated multistep cross-flow UAE design.

4.2.2. Cross-Flow UAE

Sequential UAE was based on a value of 30 min as it was the best productivity trade-off for single-step extractions.

The highest curcuminoid yield for the 6 mm matrix was obtained with four-step cross-flow, and a progressive reduction in extraction rate was observed (Table 5). As expected, the curcuminoid output steeply increased with the cross-flows and was more pronounced in the 2 mm than in the 6 mm matrix; the yield was almost double at the first extraction (43.10% vs. 23.10%, also see Figure 7).

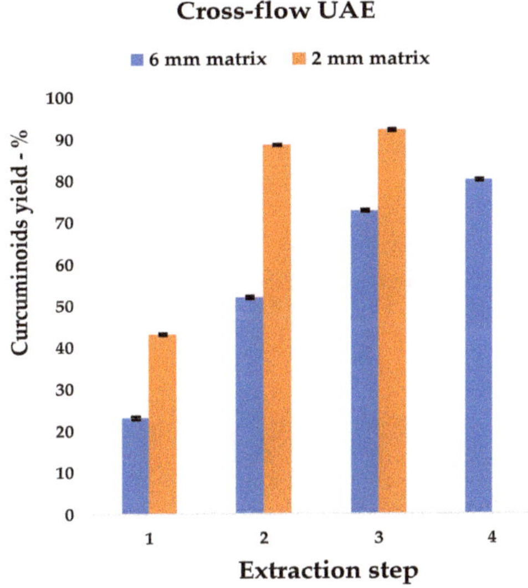

Figure 7. Cross-flow trend for 6 and 2 mm particle sizes.

It is interesting to observe that the maximum production achieved for the coarse material after 4 steps, is overcome by 2 mm turmeric with two UAE steps (80.04% vs. 88.54%, respectively). A fourth extraction was not suitable with the finer Curcuma powder because of the poor quantity of the curcuminoids left in the rhizomes (less than 8%); an additional step was considered not to be cost-effective.

Due to its nature, it is possible to describe a cross-flow protocol (fresh solvent on recycled biomass), as a progressive sum of single extraction stages. Starting from this approximation, experimental yields can be compared step-by-step to the Peleg Model as extrapolated by a reiteration of every stage. In Table 6, it is possible to observe the progress of the model experiments in relation to their step number.

The prediction of sequential extractions using model reiterations clearly shows dramatic discrepancies from the experimental data. This behavior can be explained by the constant biomass modification produced by acoustic cavitation. A likely explanation may be the capacity of US to physically disaggregate vegetal materials [36]. UAE reduces particle size during extraction, enhancing the mass-transfer and modifying the kinetic profile, and thus explains the underestimation of the theoretical approach.

Table 6. Model fitting to cross-flow screening.

Step	6 mm Curcuminoid Yield		2 mm Curcuminoid Yield	
	Model %	Experimental %	Model %	Experimental %
1	22.92	23.10	41.06	43.10
2	40.59	51.97	65.26	88.54
3	54.20	72.68	79.52	92.10
4	64.70	80.04	87.93	-

Thirty min per step, S/L ratio 1:5, Horn 500 W; curcuminoid yields, calculated as the percentage recovery of total curcuminoid content. HPLC quantification expressed as curcumin equivalents.

Moreover, it is possible to state that the observed trend for sequential extraction changes when approaching matrix depletion. This is possibly due to the increase in the magnitude of the degradation mechanisms in light of the accumulation of the treatment periods. Indeed, according to Section 3.1, the total degradation for curcuminoid extracts in EtOAc, is 6.45% after 90 and 11.14% after 120 min (equal to 3 and 4 steps). As a matter of fact, the peculiarities of UAE, such as particle comminution and mass-transfer enhancement, helped to provide a constant yield increase, thus balancing and overcoming curcuminoid degradation.

The granulometric variation can also be investigated by means of B_t values (Equation (2), Tables A1 and A2). As the most relevant evolution of this parameter occurs in the initial moments of extraction (see Figure 8), the first three points sampled for Peleg's Model are taken as being descriptive.

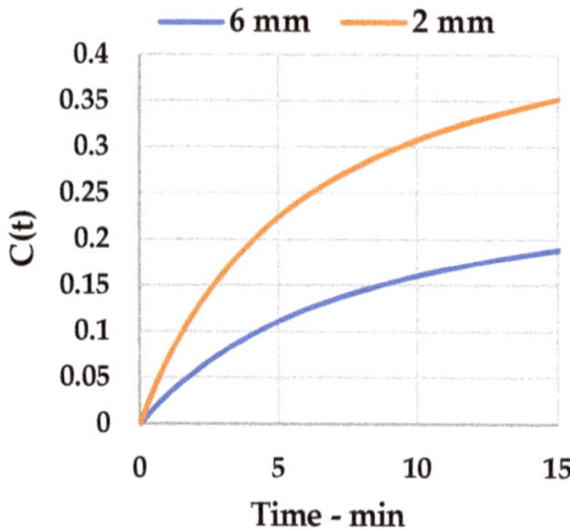

Figure 8. Granulometry comparison: slope variation during initial extraction phase, Peleg model.

Hence, it is possible to appreciate the slope divergence by plotting the *extraction rates* (Figure 9) for 2, 5 and 7.5 min for every particle size. The reverse proportionality of B_t to extraction time was confirmed. It is important to underline that the dashed bars represent the average *extraction rate*, which results from the accumulation of all the rates and is extrapolated from the kinetic constant k_1, of the systems (B_0). Therefore, although cavitation can progressively reduce particle size during extraction, the early stages, in which disaggregation is negligible, have a prevailing influence on the overall process.

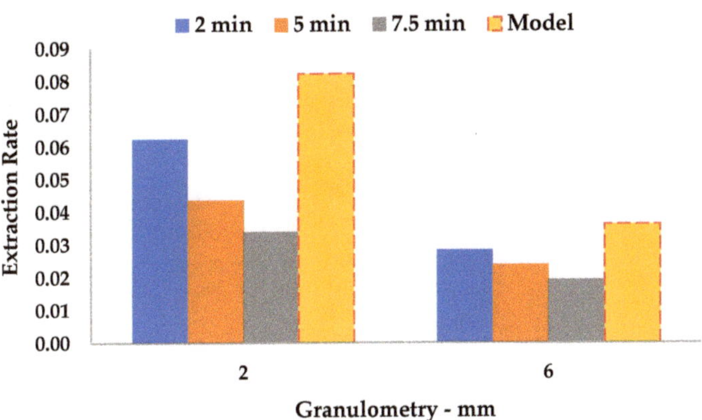

Figure 9. Extraction-rate variation with granulometry. B_t of 2, 5 and 7.5 min compared to overall B_0.

4.3. Crystallization

The features and the advantages of using EtOAc as the extraction solvent were also investigated in terms of metabolite purification via crystallization. In addition to the better stability against curcuminoid radical degradation, the acetate showed an interesting predisposition to crystal generation, which was not observed in the EtOH extracts, under the tested conditions. The efficiency of this purification step is crucial if complex and expensive protocols, such as chromatography, are to be avoided. The final detected purity of the crystallized curcuminoids is 95% on average, as determined by HPLC and expressed as curcumin equivalents.

5. Conclusions

A thorough investigation into curcuminoid stability under US interactions has been performed with the elective solvent and its derived ester (EtOH vs. EtOAc). The latter was expected to be less prone to sonolysis. For the sake of comparison, both pure curcumin and a curcuminoid mixture were tested, and HPLC-UV and quantitative ^{1}H-NMR were used to determine metabolite degradation. The main advantage of this method is the mere requirement of one analysis for the defined external reference, and no specific and expensive standard for each target analyte. Stability tests have shown that solvents can lead to different radical concentrations, which is crucial for labile molecules, such as curcumin-like compounds. US irradiation caused milder radical generation in the presence of EtOAc than in EtOH, even in relation to the presence of oxygen. In particular, the use of a sealed vessel decreased curcumin degradation by 6.01% after 2 h of US irradiation in the aprotic solvent. The synergistic stabilizing mechanism of curcuminoids when present in a mixed form was confirmed for both the systems, as reported in the literature. In detail, degradation behavior was strongly prevalent in EtOH, with 12.48% and 8.42% decreases being observed for curcumin and the curcuminoids mix, respectively, after 2 h.

These preliminary tests have designated EtOAc as the elective solvent for efficient US-assisted extraction procedures that aim for curcuminoid stability and maximum yields. The use of UAE was first evaluated in a single-step extraction of *Curcuma longa* L., and the impact of different granulometries, namely 6 and 2 mm, was screened. Particular attention was paid to the kinetic features. UAE equilibria were reached before rhizome depletion (25.63% and 47.56% for 6 and 2 mm, respectively), proving the need for a multi-step protocol. Stability tests supported the use of EtOAc, even for sequential extractions, due to its capacity to minimize degradation. Hence, the multi-step co-current cross-flow approach was exploited to maximize yield, and the influence of granulometry is to be highlighted.

Maximum curcuminoid recovery was achieved in three steps (92.10%) and four steps (80.04%), respectively, for the 2 and 6 mm rhizome powders.

Comparisons between single extraction kinetics and the cross-flow trend highlight the disaggregation power of US, which is able to accelerate extraction through sequential stages. EtOAc has shown interesting results and applicability in the UAE of *Curcuma longa* L. but also has a crucial role in the final purification steps. In this work, EtOAc produced an extract that is prone to the crystallization process, and an average of 95% pure product was achieved, whilst ethanol prevented crystal formation from the oleoresin extract.

In conclusion, the effect of US on curcumin and curcuminoid degradation has been screened with EtOH and EtOAc for the first time herein, to the best of our knowledge. The results were then exploited to define a UAE process for *Curcuma longa* L. A Peleg kinetic model was used to describe the single-step extraction, and the optimized process time was the groundwork upon which the multi-step cross-flow process was designed. A small particle size was crucial to the obtaining of improved final yields. Finally, the efficacy of the applied solvent was then confirmed by means of suitable curcuminoid purification via crystallization.

Author Contributions: Conceptualization, D.C., A.B. (Arianna Binello) and G.C.; methodology, G.G. and A.B. (Alessandro Barge); validation, A.B. (Alessandro Barge); formal analysis, A.B. (Arianna Binello); investigation, G.G.; data curation, D.C., P.A.; writing—original draft preparation, A.B. (Arianna Binello) and G.G.; writing—review and editing, G.C. and D.C.; supervision, P.A. and G.C.; All authors have read and agreed to the published version of the manuscript.

Funding: This work was supported by the University of Turin (ricerca locale 2019).

Acknowledgments: G.G. acknowledges Indena SpA for their research scholarship.

Conflicts of Interest: The authors declare no conflicts of interest.

Appendix A

Table A1. Experimental UAE of 6 mm *Curcuma longa* L.

t min	Curcuminoid Yield ± %SD %		B_t [a]	Linearization t/C_t [b]
0	-		-	-
2	0.057	± 0.23	0.0285	35.0877
5	0.12	± 0.21	0.024	41.6667
7.5	0.144	± 0.32	0.0192	52.0833
15	0.1857	± 0.29	0.0124	80.7754
30	0.231	± 0.37	0.0077	129.8701
60	0.2482	± 0.33	0.0041	241.7405
90	0.2563	± 0.24	0.0028	351.1510
120	0.2588	± 0.24	0.0022	463.6785

[a] Extraction rates; [b] linearization values.

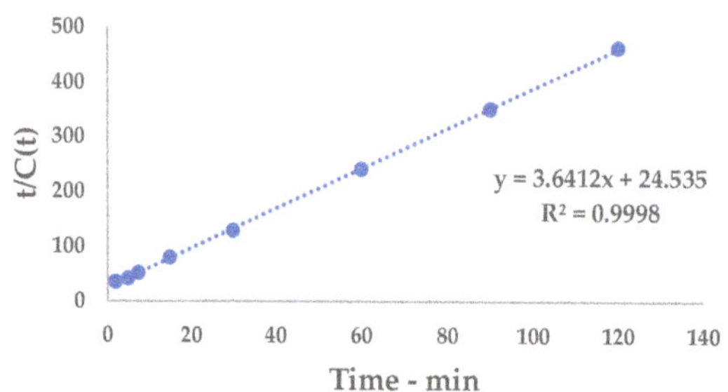

Figure A1. Peleg Model linearization, 6 mm.

Table A2. Experimental UAE of 2 mm *Curcuma longa* L.

t min	Curcuminoid Yield ± %SD %		B_t [a]	Linearization t/C_t [b]
0	-		-	-
2	0.1252	± 0.33	0.0625	15.9744
5	0.2188	± 0.25	0.0438	22.8519
7.5	0.2543	± 0.26	0.0339	29.4927
15	0.3501	± 0.30	0.0233	42.8449
30	0.4310	± 0.22	0.0144	69.6056
60	0.4710	± 0.31	0.0079	127.3885
90	0.4718	± 0.28	0.0052	190.7588
120	0.4756	± 0.29	0.004	252.3129

[a] Extraction rates; [b] linearization values.

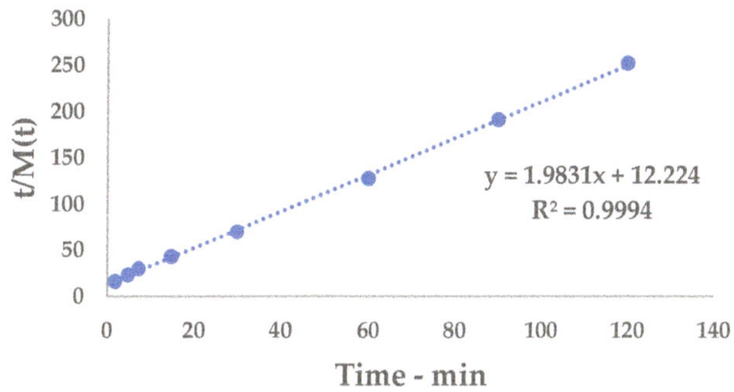

Figure A2. Peleg Model linearization, 2 mm.

References

1. Ravindran, P.N. *Turmeric—The golden spice of life, In Turmeric: The Genus Curcuma*; Ravindran, P.N., Babu, K.N., Sivaraman, K., Eds.; CRC Press: Boca Raton, FL, USA, 2007; Volume 1, pp. 1–14.

2. Dao, T.T.; Nguyen, P.H.; Won, H.K.; Kim, E.H.; Park, J.; Won, B.Y.; Oh, W.K. Curcuminoids from *Curcuma longa* and their inhibitory activities on influenza a neuraminidases. *Food Chem.* **2012**, *134*, 21–28. [CrossRef]

3. Nguyen, T.T.H.; Si, J.; Kang, C.; Chung, B.; Chung, D.; Kim, D. Facile preparation of water soluble curcuminoids extracted from turmeric (*Curcuma longa* L.) powder by using steviol glucosides. *Food Chem.* **2017**, *214*, 366–373. [CrossRef] [PubMed]

4. Sueth-Santiago, V.; Mendes-Silva, G.P.; Decoté-Ricardo, D.; de Lima, M.E.F. Curcumin, the golden powder from turmeric: Insights into chemical and biological activities. *Quim. Nova.* **2015**, *38*, 538–552. [CrossRef]

5. Esatbeyoglu, T.; Huebbe, P.; Ernst, I.M.; Chin, D.; Wagner, A.E.; Rimbach, G. Curcumin—From molecule to biological function. *Angew. Chem. Int. Ed.* **2012**, *51*, 5308–5332. [CrossRef] [PubMed]

6. Niranjan, A.; Prakash, D. Chemical constituents and biological activities of turmeric (*Curcuma longa* L.)—A review. *J. Food Sci. Technol.* **2008**, *45*, 109–116.

7. Bessone, F.; Argenziano, M.; Grillo, G.; Ferrara, B.; Pizzimenti, S.; Barrera, G.; Cravotto, G.; Guiot, C.; Stura, I.; Cavalli, R.; et al. Low-dose curcuminoid-loaded in dextran nanobubbles can prevent metastatic spreading in prostate cancer cells. *Nanotechnology* **2019**, *30*, 214004–214016. [CrossRef] [PubMed]

8. Priyadarsini, K.I. The chemistry of curcumin: From extraction to therapeutic agent. *Molecules* **2014**, *19*, 20091–20112. [CrossRef] [PubMed]

9. Patra, J.K.; Das, G.; Lee, S.; Kang, S.-S.; Shin, H.-S. Selected commercial plants: A review of extraction and isolation of bioactive compounds and their pharmacological market value. *Trends Food Sci. Technol.* **2018**, *82*, 89–109. [CrossRef]

10. Chemat, F.; Vian, M.A.; Fabiano-Tixier, A.-S.; Nutrizio, M.; Jambrak, A.R.; Munekata, P.E.S.; Lorenzo, J.M.; Barba, F.J.; Binello, A.; Cravotto, G. A review of sustainable and intensified techniques for extraction of food and natural products. *Green Chem.* **2020**, *22*, 2325–2353. [CrossRef]

11. Grillo, G.; Boffa, L.; Binello, A.; Mantegna, S.; Cravotto, G.; Chemat, F.; Dizhbite, T.; Lauberte, L.; Telysheva, G. Cocoa bean shell waste valorization; extraction from lab to pilot-scale cavitational reactors. *Food Res. Int.* **2019**, *115*, 200–208. [CrossRef]

12. Berk, Z. *Extraction, In Food Process Engineering and Technology*, 3rd ed.; Taylor, S.L., Ed.; Elsevier Inc.: Amsterdam, The Netherlands, 2018; Volume 1, pp. 289–310.

13. Cacace, J.E.; Mazza, G. Optimization of extraction of anthocyanins from black currants with aqueous ethanol. *J. Food Sci.* **2003**, *68*, 240–248. [CrossRef]

14. Herodez, S.S.; Hadolin, M.; Skerget, M.; Knez, Z. Solvent extraction study of antioxidants from Balm (Melissa officinalis L.) leaves. *Food Chem.* **2003**, *80*, 275–282. [CrossRef]

15. Peleg, M. An empirical model for the description of moisture sorption curves. *J. Food Sci.* **1988**, *54*, 1216–1217. [CrossRef]

16. Bucić-Kojić, A.; Planinić, M.; Tomas, S.; Bilić, M.; Velić, D. Study of solid—Liquid extraction kinetics of total polyphenols from grape seeds. *J. Food Eng.* **2007**, *81*, 236–242. [CrossRef]

17. Shirsath, S.R.; Sable, S.S.; Gaikwad, S.G.; Sonawane, S.H.; Saini, D.R.; Gogate, P.R. Intensification of extraction of curcumin from *Curcuma amada* using ultrasound assisted approach: Effect of different operating parameters. *Ultrason. Sonochem.* **2017**, *38*, 437–445. [CrossRef]

18. Aguilar, F.; Dusemund, B.; Galtier, P.; Gilbert, J.; Gott, D.M.; Grilli, S.; Gürtler, R.; König, J.; Lambré, C.; Larsen, J.-C.; et al. Scientific opinion on the re-evaluation of curcumin (E 100) as a food additive. *EFSA J.* **2010**, *8*, 1–46. [CrossRef]

19. Revanthy, S.; Elumalai, S.; Benny, M.; Antony, B.J. Isolation, purification and identification of curcuminoids from turmeric (*Curcuma longa* L.) by column chromatography. *Exp. Sci.* **2011**, *2*, 21–25.

20. Ukrainczyk, M.; Hodnett, B.K.; Rasmuson, A.C. Process parameters in the purification of curcumin by cooling crystallization. *Org. Process Res. Dev.* **2016**, *20*, 1593–1602. [CrossRef]

21. Joshi, P.; Jain, S.; Sharma, V. Turmeric (*Curcuma longa*) a natural source of edible yellow colour. *Int. J. Food Sci. Technol.* **2009**, *44*, 2402–2406. [CrossRef]

22. Thorat, A.A.; Dalvi, S.V. Particle formation pathways and polymorphism of curcumin induced by ultrasound and additives during liquid antisolvent precipitation. *CrystEngComm* **2014**, *16*, 11102–11114. [CrossRef]

23. Kuhnen, S.; Bernardi Ogliari, J.; Dias, P.F.; da Santos, M.S.; Ferreira, A.N.G.; Bonham, C.C.; Maraschin, M. Metabolic fingerprint of brazilian maize landraces silk (stigma/styles) using NMR spectroscopy and chemometric methods. *J. Agric. Food Chem.* **2010**, *58*, 2194–2200. [CrossRef] [PubMed]

24. Yang, S.-O.; Shin, Y.-S.; Hyun, S.-H.; Cho, S.; Bang, K.-H.; Lee, D.; Choi, H.-K. NMR-based metabolic profiling and differentiation of ginseng roots according to cultivation ages. *J. Pharm. Biomed. Anal.* **2012**, *58*, 19–26. [CrossRef] [PubMed]

25. Gad, H.A.; Bouzabata, A. Application of chemometrics in quality control of Turmeric (*Curcuma longa*) based on Ultra-violet, Fourier transform-infrared and ^{1}H NMR spectroscopy. *Food Chem.* **2017**, *237*, 857–864. [CrossRef] [PubMed]

26. Li, B.; Konecke, S.; Wegiel, L.A.; Taylor, L.S.; Edga, K.J. Both solubility and chemical stability of curcumin are enhanced by solid dispersion in cellulose derivative matrices. *Carbohydr. Polym.* **2013**, *98*, 1108–1116. [CrossRef]

27. Gordon, O.N.; Luis, P.B.; Sintim, H.O.; Schneider, C. Unraveling curcumin degradation: Autoxidation proceeds through spiroepoxide and vinylether intermediates en route to the main bicyclopentadione. *J. Biol. Chem.* **2015**, *290*, 4817–4828. [CrossRef]

28. Schneider, C.; Gordon, O.N.; Edwards, R.L.; Luis, P.B. Degradation of curcumin: From mechanism to biological implications. *J. Agric. Food Chem.* **2015**, *63*, 7606–7614. [CrossRef] [PubMed]

29. Khurana, A.; Ho, C.T. High performance liquid chromatographic analysis of curcuminoids and their photo-oxidative decomposition compounds in *Curcuma longa* L. *J. Liq. Chromatogr.* **1988**, *11*, 2295–2304. [CrossRef]

30. Peram, M.R.; Jalalpure, S.S.; Palkar, M.B.; Diwan, P.V. Stability studies of pure and mixture form of curcuminoids by reverse phase-HPLC method under various experimental stress conditions. *Food Sci. Biotechnol.* **2017**, *26*, 591–602. [CrossRef]

31. Price, L.C.; Buescher, R.W. Kinetics of alkaline degradation of the food pigments curcumin and curcuminoids. *J. Food Sci.* **1997**, *62*, 267–269. [CrossRef]
32. Didenko, Y.T.; Suslick, K.S. The energy efficiency of formation of photons, radicals and ions during single-bubble cavitation. *Nature* **2002**, *418*, 394–407. [CrossRef]
33. Riesz, P.; Kondo, T. Free radical formation induced by ultrasound and its biological implications. *Free Radic. Biol. Med.* **1992**, *13*, 247–270. [CrossRef]
34. Riesz, P.; Berdahl, D.; Christman, C.L. Free radical generation by ultrasound in aqueous and nonaqueous solutions. *Environ. Heal. Perspect.* **1985**, *64*, 233–252. [CrossRef] [PubMed]
35. Riesz, P.; Berdahl, D.; Christman, C.L. Dissolved gas and ultrasonic cavitation—A review. *Ultrason. Sonochem.* **2013**, *20*, 1–11. [CrossRef]
36. Toma, M.; Vinatoru, M.; Paniwnyk, L.; Mason, T.J. Investigation of the effects of ultrasound on vegetal tissues during solvent extraction. *Ultrason. Sonochem.* **2001**, *8*, 137–142. [CrossRef]

Article

Hydrocolloid-Based Coatings with Nanoparticles and Transglutaminase Crosslinker as Innovative Strategy to Produce Healthier Fried Kobbah

Asmaa Al-Asmar [1,2], Concetta Valeria L. Giosafatto [1], Mohammed Sabbah [3] and Loredana Mariniello [1,*]

1   Department of Chemical Sciences, University of Naples "Federico II", 80126 Naples, Italy; a.alasmar@najah.edu (A.A.-A.); giosafat@unina.it (C.V.L.G.)
2   Analysis, Poison Control and Calibration Center (APCC), An-Najah National University, P.O. Box 7 Nablus, Palestine
3   Department of Nutrition and Food Technology, An-Najah National University, P.O. Box 7 Nablus, Palestine; m.sabbah@najah.edu
*   Correspondence: loredana.mariniello@unina.it; Tel.: +39-081-2539470

Received: 3 May 2020; Accepted: 22 May 2020; Published: 1 June 2020

Abstract: This study addresses the effect of coating solutions on fried kobbah. Coating solutions were made of pectin (PEC) and grass pea flour (GPF), treated or not with transglutaminase (TGase) and nanoparticles (NPs)—namely mesoporous silica NPs (MSN) or chitosan NPs (CH–NPs). Acrylamide content (ACR), water, oil content and color of uncoated (control) and coated kobbah were investigated. Zeta potential, Z-average and in vitro digestion experiments were carried out. Zeta potential of CH–NPs was stable from pH 2.0 to pH 6.0 around + 35 mV but decreasing at pH > 6.0. However, the Z-average of CH–NPs increased by increasing the pH. All coating solutions were prepared at pH 6.0. ACR of the coated kobbah with TGase-treated GPF in the presence nanoparticles (MSN or CH–NPs) was reduced by 41.0% and 47.5%, respectively. However, the PEC containing CH–NPs showed the higher reduction of the ACR by 78.0%. Water content was higher in kobbah coated by PEC + CH–NPs solutions, while the oil content was lower. The color analysis indicated that kobbah with lower browning index containing lower ACR. Finally, in vitro digestion studies of both coating solutions and coated kobbah, demonstrated that the coating solutions and kobbah made by means of TGase or nanoparticles were efficiently digested.

Keywords: acrylamide; kobbah; transglutaminase; pectin; chitosan-nanoparticles; coatings; mesoporous silica nanoparticles; grass pea; HPLC-RP

---

1. Introduction

Kibbeh, kibbe, kobbah (also kubbeh, kubbah, kubbi) (pronunciation varies with region) is Eastern dish made of a ground bulgur (wheat-based food) mixed with minced beef meat formed as balls stuffed with cooked ground meat, onions, nuts and spices. They are usually cooked by deep frying for 8–10 min at high temperatures (160–180 °C), thus they have rough crust and thoroughly browned. They are home-made and consumed fresh or they are sold frozen in the super-markets and consumers can fry them at home [1].

Hydrocolloid materials are used for food protection, as well as for separating the different part of a food [2,3]. Coatings represent a thin layer of edible molecules that are laid on the surface of a food product and can be used to protect high perishable aliments. Pectin (PEC) is a polysaccharide present in the plant cell wall containing mainly galacturonic acid, but highly variable in composition, structure and molecular weight [4]. PEC is known as food additive (E440), useful for thickening mainly jam and marmalades and other products, since is provided with gelling properties [5]. Yossef [6], found that strawberry fruits dipped in PEC-based solutions retained physico-chemical properties as the same fruit protected by other hydrocolloid molecules, such as soy proteins, gluten or starch. Moreover, protein-based such as grass pea flour (GPF) was used for its high content in proteins. Grass pea (*Lathyrus sativus* L.) belongs to the leguminous family and is quite important in many Asian and African countries where it is cultivated either for animal feeding or human use. It is characterized by resistance to both abiotic and biotic stresses [7]. GPF containing proteins were able to act as transglutaminase substrates, giving arise to novel bioplastics endowed with improved technological properties than the ones cast without the enzyme [8–10].

Microbial transglutaminase (TGase) belongs to a family of enzymes (E.C. 2.3.2.13) (widely distributed in nature from microbes to animals and plants) capable of catalyzing iso–peptides bonds between endo-glutamine and endo-lysine residues belonging to proteins of different nature, giving arises to intra- and inter-molecular crosslinks [2,11]. TGase is widely used in the food industry as technological aid. Recently, Sabbah et al. [12], demonstrated that the proteins of *Nigella sativa* defatted cakes are act as TGase substrates and are responsible to enhance physico-chemical properties of the obtained films.

Using nanoparticles for developing of nanocomposite coatings is a way to improve their features and is of interest also for producing active packaging [13]. Mesoporous silica nanoparticles MSNs (Type MCM-41) are a kind of SiO_2–based nanoparticles that are promising materials for application in numerous aspects of biomedical and food purposes [14–16]. Recently, Fernandez-Bats et al. [17]; Giosafatto et al. [18]; Al-Asmar et al. [3], prepared and characterized the active protein, pectin and chitosan edible films grafted with MSNs, and they concluded that MSNs significantly influence the mechanical and permeability properties of the obtained materials. SiO_2 nanoparticles of different composition are labeled as E551, E554, E556 or E559 and used for instance as an anti-caking agent. The amount ingested daily is estimated to be 1.8 mg/kg (around 126 mg/day for a 70 kg person) [19]. Moreover, McCarthy [20] observed that SiO_2- based NPs with the size of 150 nm and 500 nm do not perform toxic effects on Calu-3 cells.

Chitosan nanoparticles (CH–NPs) are natural materials obtained from the marine byproducts, endowed with not able physico-chemical and antimicrobial characteristics, besides being sustainable and harmless for human health [21,22]. These properties suggest that CH–NPs can be used also as carrier for drug delivery. Lorevice et al. [23], obtained higher mechanical properties by adding CH–NPs to PEC films compared with control, allowing these novel materials to be an alternative to traditional food packaging production. Moreover, addition of small fractions of CH–NPs enhance mechanical and thermal stability of banana puree-based films [24]. Application of CH in foods is gaining interest, specifically after that shrimp chitin-derived CH has been recognized as Generally Recognized as Safe (GRAS) for common use in foods by the US Food and Drug Administration in 2011 [25,26].

Acrylamide (ACR) is a chemical that was discovered in foods in 2002 and it is present in a range of popular foods [27]. ACR is not present in raw foods, but it is formed from natural precursors during food treatment at high-temperature (>120 °C) following Maillard reaction [28]. ACR is formed during frying, baking roasting and toasting of the carbohydrate rich food and cereal products, as well as coffee. In particular, ACR occurs because of the reaction between the free amino acid asparagine and a carbonyl-containing compound. [29]. European Food Safety Authority (EFSA) scientists classified ACR is a 'probably carcinogenic to humans [30]. Hydrocolloid-based coatings recently become one of several strategies used to mitigate ACR formation and to improve the eating quality of different fried foods such as potato chips [31], bread [32] and fried banana [33]. Very recently, our research group demonstrated the effectiveness of hydrocolloid-based coatings prepared in the presence of TGase in reducing ACR content of French fries [34] and fried falafel, a typical Easter food [35]. In 2004, Al-Dmoor et al. [1] determined the ACR content in fried kobbah (food mostly eaten in Jordan, but also very popular in Palestine) finding values ranging from 2900 to 5300 μg kg^{-1}. Thus, kobbah contain ACR in values that impose attention to protect the health of consumers.

The aim of this study was to investigate the influence of different hydrocolloid-based coatings (containing PEC and GPF prepared in the presence or absence of MNS and/or TGase) on ACR, water content, oil content, digestibility and color of fried kobbah. The physicochemical properties of different coating solutions were also evaluated.

2. Materials and Methods

2.1. Materials

Methanol and ACR standard ≥99.8%, were purchased from Sigma–Aldrich Chemical Company (St. Louis, MO, USA). Acetonitrile HPLC analytical grade, n-hexane and formic acid were obtained from Carlo Erba reagents S.r.l. (Cornaredo, Milan, Italy). Oasis HLB 200 mg, 6 mL solid phase extraction (SPE) cartridges were from Waters (Milford, MA, USA). Syringe filters (0.45- and 0.22-μm PVDF) were from Alltech Associates (Deerfield, Italy). PEC of a low-methylated citrus peel (7%) (Aglupectin USP) was purchased from Silva Extracts s.r.l. (Gorle, Bergamo, Italy) and Activa®WM *Streptoverticillium* TGase was supplied by Ajinomoto Co (Tokyo, Japan). Sodium tripolyphosphate (TPP) and glycerol (GLY) was from Merck Chemical Company (Darmstadt, Germany). Chitosan (CH, mean molar mass of 3.7–104 g/mol) was procured from Professor R. Muzzarelli (University of Ancona, Ancona, Italy), with a degree of 9.0% N-acetylation. Grass pea seeds, corn oil, ground bulgur (wheat-based food), minced beef meat, onion, salt and spices were obtained from a local market in Naples (Italy).

2.2. Nanoparticles Preparation

MSN (MCM-41) were synthesized and characterized, as described in Fernandez-Bats et al. [17]. However, CH–NPs was prepared by using the ionic gelation method according to Chang et al. [36]. Briefly, adding the TPP 0.5% (*w/v*) dropwise to the CH solution (0.8%) by 40 min stirring the obtained suspension was centrifuged at 22,098× *g* for 10 min at 4 °C, then rinsed three times by Milli-Q water and dry at room temperature.

2.3. Kobbah Formulation and Manufacturing

Kobbah was made as described by Brazil et al. [37], soaking ground bulgur flour into hot water (80 °C) for 1 h, then the wheat flour, oil, salt and spices were added and mixed with the soaked ground bulgur. The dough was kept at the refrigerator for 1 h. Stuffing: onions, salt and spices were mixed with minced beef meat then cooked with olive oil. The dough was shaped into balls stuffed with cooked ground meat.

2.4. Preparation of the Coating Solutions

GPF was obtained according to Giosafatto et al. [8] and Al-Asmar et al. [34], in particular, the seeds were milled by a laboratory blender LB 20ES (Waring Commercial, Torrington, CT, USA) and the obtained flour was treated with a 425-μm stainless steel sieve (Octagon Digital Endecotts Limited, London, UK). A total of 83 g of GPF (containing 24% *w/w* proteins) were dissolved in 1 L Milli-Q water and the solutions were shacked for 1 h. the pH was brought to 9.0 with 1-M NaOH followed by centrifugation at 12,096× *g* for 10 min. After centrifugation, the supernatant was collected and used to prepare the GPF dipping solution. Nanoparticles, either MSN or CH–NP (1% *w/w* GPF proteins) were added to GPF at pH 6.0 than the solutions were mixed for 30 min at room temperature (GPF; GPF + MSN; GP + CH–NP). TGase (33 U/g of GPF proteins) was used to prepare the GPF + TGase; GPF + MSN + TGase and GPF + CH–NP + TGase, GLY was used as plasticizer (8% *w/w* GPF proteins) in all the GPF coating solutions, then incubated for 2 h at 37 °C. PEC-based solutions (1% *w/v*) prepared as described in Esposito et al. [38], were made from a PEC stock solution (2% *w/v*), then diluted with Milli-Q water. MSN or CH–NP (1% *w/w* PEC) were added to PEC and mixed for 30 min at room temperature (PEC; PEC + MSN; PEC + CH–NP). Each dipping solutions were adjusted to pH 6.0, then was used to coat kobbah before trying.

2.5. Dipping and Frying Method

Two hundred grams of kobbah (divided in 5 pieces) were immersed for 30 s into either in H_2O_d ("control" sample) or in one of these coating solutions: (1) GPF; (2) GPF reinforced with MSN (GPF + MSN); (3) GPF reinforced with CH–NP (GPF + CH–NP); (4) TGase-treated (GPF + TGase); (5) TGase-treated reinforced with MSN (GPF + MSN + TGase); (6) TGase-treated reinforced with CH–NP (GPF + CH–NP + TGase); (7) PEC; (8) PEC reinforced with MSN (PEC + MSN); and (9) PEC reinforced with CH–NP (PEC + CH–NP). Moreover, each sample was dripped for 2 min before frying to get rid of the excess of solutions. The frying conditions consisted in 2 L corn oil preheated (using a controlled temperature deep-fryer apparatus (GIRMI, Viterbo, Italy)) to the processing temperature (190 ± 5 °C), then the kobbah were deep fried for 4.5 min. The oil was replaced by new one for each different coating solutions. After frying, kobbah were drained for 2 min to remove oil excess [34,35,39].

2.6. Zeta Potential and Z-Average of Coating Solutions

The Zeta potential and particle size (Z-average) of the CH–NP solution (1 mg/mL) prepared at pH 2.0 were obtained by titration from pH 2.0 to pH 7.0, by means of Zetasizer Nano-ZSP (Malvern®, Worcestershire, UK) equipped with a He–Ne laser. All coating solutions used in this experiment were also tested for their Zeta potential and Z-average.

2.7. Oil and Water Content

The oil content was performed following frying and cooling of each processed samples around (3–5 g) in triplicate. The result was reported as a percentage on dry matter weight by n-hexane solvent extraction using the Soxhlet method [40].

Fried kobbah water content was obtained following the gravimetric method [41] in triplicate.

2.8. Acrylamide Detection

2.8.1. Preparation of Acrylamide Standard

ACR standard stock solution (1.0 mg/mL) was obtained as described in Al-Asmar et al. [34,35]. In particular 10 mg of the ACR standard were dissolved in 10 mL of Milli-Q water. From the stock solution, different concentrations of calibration standards (100, 250, 500, 1000, 2000, 3000, 4000 and 5000 µg/L), were prepared, respectively. All series of standard solutions were kept in glass dark bottles at 4 °C until used.

2.8.2. Acrylamide Extraction

ACR extraction was performed as described in Al-Asmar et al. [34,35]. Briefly, about 200 g of fried kobbah, were put in n-hexane for 30 min to get rid of the oil [42]. After that, n-hexane was let to evaporate under fume hood at room temperature, then samples were subjected for ACR extraction. The treated fried kobbah samples were milled at 1300 rpm for 1 min by means of a rotary mill Grindomix GM200, (Retsch GmbH, Haan, Germany). Freeze drying was used to dry the samples before ACR extraction that was carried out by following the procedure of Wang et al. [43]. Briefly, two different tubes were set up for each sample, one for detecting ACR formed in kobbah samples, and the second one to carry out the "Recovery test for ACR in all kobbah types (in each sample 150 µg/L of ACR standard were added". In both tubes, 1.0 g (dry weight) of sample, was placed in both tubes and only in the second one there was the ACR standard added. Carrez reagent potassium salt and Carrez reagent zinc salts (50 µL) were included in each sample. In each tube, 10 mL of HPLC water were finally added. The samples were put in an incubated shaker for 30 min at 25 °C and 170 rpm, then centrifuged at 7741× g for 10 min at 4 °C. The supernatant was filtered with 0.45-µm syringe filter for the clean-up of the Oasis HLB SPE cartridges. The SPE cartridge was preventively conditioned with 2.0 mL of methanol followed by washing with 2 mL of water before loading 2.0 mL of the filtered supernatant, the first 0.5 mL was discarded and the remaining elute collected (~1.5 mL; exact volume was measured by weight and converted by means of density). All extracts were kept in dark glass vials at 4 °C before analysis. The clean sample extracts were further filtered through 0.2-μm nylon syringe filters before HPLC-UV (ultra violet) analysis of fried kobbah [34]. Each determination was performed in triplicate.

2.8.3. HPLC-UV Analysis

HPLC-UV analysis was used to determine the ACR, by using the RP-HPLC method on Beckman Gold HPLC instrument equipped with a dual pump and a diode array detector [34]. The column Synergi™ 4-μm Hydro-RP 80 Å HPLC Column 250 × 3 mm (Phenomenex, Torrance, CA, USA) was used [44].

The operating conditions described also in Al-Asmar et al. [34] were the following: the wavelength detection was 210 nm, a gradient elution of 0.1% formic acid (*v/v*) in water: acetonitrile (97:3, *v/v*) was used. Solvent A was water and Solvent B was acetonitrile, both solvents containing 0.10% (*v/v*) formic acid; flow rate, 1 mL/min. The gradient elution program was applied as follows: 97% A (3% B) for 10 min, increased to 20% A (80% B) from 10 to 12 min and kept at 20% A (80% B) for 5.0 min, increased to 95% B (5% A) from 17 to 19 min and kept at 95% B for 5 min, increased to 97% A (3% B) from 24 to 26 min and kept for 4 min. The injection volume was equal to 20 μL. The total chromatographic runtime was 30 min for each sample and the temperature was kept at 30 °C (GECKO 2000 "HPLC column heater", Spectra Lab Scientific, Inc., Markham, ON, Canada) to ensure optimal separation. In all samples (ACR standard and fried kobbah-derived), the ACR retention time was 4.9 min.

2.9. Color Analysis

Color measurement of food products was considered as an indirect measure of other quality features such as flavor and contents of pigments [45]. Chroma Meter Konica Minolta CR-400 (Japan) was utilized to determine L*, a*, b* values of fried kobbah samples. L* a* b* is an international standard for color measurement adopted by the Commission Internationale d'Eclairage (CIE) in 1976. L* is the lightness component, which ranges from 0 to 100 and parameters a* (from green to red) and b* (from blue to yellow) are the two chromatic components, which range from −120 to 120 [46]. Total color difference to the control sample (ΔE) indicates the magnitude of color difference between coated kobbah and uncoated control kobbah and it was obtained by the following equation [33,45]:

$$\Delta E = \sqrt{(L* - L'*)^2 + (a* - a'*)^2 + (b* - b'*)} \tag{1}$$

where L'*, a'* and b'* are the parameters of treated kobbah and L*, a* and b* the ones of the control (uncoated fried kobbah).

The browning index (BI) allowed to define the overall changes in browning color [33,47]. BI of the fried kobbah was calculated by the following equation:

$$BI = \frac{100\,(x - 0.31)}{0.17} \tag{2}$$

where:

$$X = \frac{(a* + 1.75\,L*)}{(5.645\,L* + a* - 3.012b*)} \tag{3}$$

2.10. Sodium Dodecyl Sulfate Polyacrylamide Gel Electrophoresis (SDS-PAGE) and In Vitro Digestion

SDS-PAGE, performed as described in Lemmli [48], was carried out at a constant voltage (80 V for 2–3 h). The protein bands were stained with Coomassie Brilliant Blue R250 (Bio-Rad, Segrate, Milan, Italy). Bio-Rad Precision Protein Standards were used as molecular weight markers.

GPF-based FFSs and fried kobbah either treated or not by TGase (33 U/g protein) reinforced or not by MSN, were subjected to in vitro gastric digestion (IVD) by using an adult model [49–51]. Then,

100 mg of each sample was incubated with 4 mL of simulated salivary fluid (SSF, 150 mM of NaCl, 3 mM of urea, pH 6.9) containing 75 U of amylase enzyme/g protein for 5 min at 37 °C at 170 rpm [35]. The amylase activity was stopped by adjusting the pH at 2.5. Afterwards, the samples were subjected to IVD as described by Giosafatto et al. [49] and Al-Asmar et al. [35], with some modifications. Briefly, 100 µL of simulated gastric fluid (SGF, 0.15 M of NaCl, pH 2.5) were placed in 1.5 mL microcentrifuge tubes and added to 100 µL of oral phase and then incubated at 37 °C. Thereafter, 50 µL of pepsin (0.1 mg/mL dissolved in SGF) were added to initiate the digestion. At intervals of 1, 2, 5, 10, 20, 40 and 60 min, 40 µL of the 0.5 M of ammonium bicarbonate (NH_4HCO_3) were added to each vial to stop the pepsin reaction. The control was set up by incubating the sample for 60 min without the protease. The samples were then analyzed by SDS-PAGE (12%) procedure described above.

2.11. Statistical Analysis

The statistical analysis was performed by means of JMP software 10.0 (SAS Institute, Cary, NC, USA), Two-way ANOVA and the *t*-student test for mean comparisons were used. Differences were considered significant at $p < 0.05$

3. Results and Discussion

3.1. Chitosan Nanoparticles, Mesoporous Silica Nanoparticles and Film Forming Solutions Characterization

Zeta potential is an important value to indicate the stability of solutions. Moreover, Z-average shows the size of the particles. Figure 1 shows Zeta potential (panel A) and Z-average (panel B) of the CH–NP in the function of pH. The results indicate that Zeta potential of CH–NP was stable at +35 mV started from pH 2.0 to pH 6.0 and decreases to +20 mV when the pH is equal to 7.0. This finding is in accordance with other authors' results [23,52,53]. The Z-average of CH–NPs at pH 2.0 was around 99.5 d.nm and increased at higher pH to reach 800 d.nm at pH 7.0. The obtained results were in agreement with those from Ali et al. [52], who explained that at pH higher than 6.0 the protonated amino groups start to lose protons and the ionic bonds start decreasing. Thus, the rises of Z-average together with the reduction in Zeta potential at pH 6.0 is because of the particle aggregation at this pH, rather than the additional increase of individual particle size [52,54]. In addition, we synthesized the MSN according to Fernandez-Bats et al. [17], with the very similar Z-average. These authors have analyzed MSN also by TEM and evidenced an average size of 143 ± 26 nm. MSN were used to improve the physio-chemical of PEC and CH films the results reported in Giosafatto et al. [18].

The coating solutions used during this study were also characterized for their stability. Zeta potential and particle size results are reported in Table 1. The results showed that stability was significantly increased after treatment of GPF-based solutions (−13.7 mV) with MSN (−16.8 mV) or TGase (−19.8 mV), also when the enzymatic crosslinking was carried out in of GPF-solutions nanoreinforced either with MSN (−18.4 mV) or CH–NPs (−18.2 mV). However, no significant effect on Zeta potential were found by adding MSN or CH–NPs on PEC FFSs stability. On the contrary, the particle size of GPF solutions was 201.3 d.nm, but this value increased significantly when CH–NP were incorporated with or without TGase. No significant change on the Z-average of FFS after adding MSN on the GPF was observed and these results are similar to those published previously by Fernandez-Bats et al. [17]. Adding TGase as crosslinker to the GPF together with MSN or CH–NP showed a significant increasing on the Z-average of FFSs. In addition, PEC FFS Z-average was (3198 d.nm) and it rises significantly to (3421 d.nm) after the addition of CH–NPs.

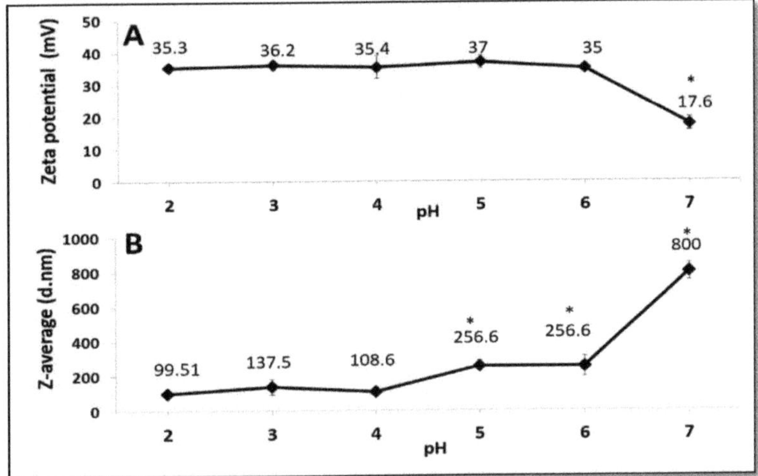

Figure 1. pH effect on Zeta potential (Panel **A**) and Z-average (Panel **B**) of 1 mg/mL chitosan nanoparticles (CH–NPs). Values marked with * were significantly different respect to the value at pH 2.0.

Table 1. Effect of 1% mesoporous silica nanoparticles (MSN) or 1% chitosan nanoparticles (CH–NPs) on Zeta potential and Z-average on either grass pea flour (GPF)-based (without or with transglutaminase (TGase) (33 U/g protein)) or pectin (PEC)-based film forming solutions at pH 6.

FFSs	Zeta Potential (mV)	Z-Average (d.nm)
GPF	−13.7 ± 0.6	201.3 ± 11.1
GPF + MSN	−16.8 ± 0.9 [a]	191.4 ± 14.2
GPF + CH–NP	−14.1 ± 0.8 [b]	385.6 ± 28.2 [a,b]
GPF + TGase	−19.8 ± 1.2 [a,b]	240.9 ± 14.4 [a,b]
GPF + MSN + TGase	−18.4 ± 0.5 [a]	333.0 ± 22.3 [a,b,c]
GPF + CH–NP + TGase	−18.2 ± 0.9 [a]	507.7 ± 18.9 [a,b,c,d]
PEC	−33.7 ± 2.1	3198 ± 79
PEC + MSN	−31.8 ± 2.9	3110 ± 77
PEC + CH–NP	−32.4 ± 3.2	3421 ± 63 *

The value significantly different from GPF FFSs are indicated by "a", the value indicated by "b" were significantly different from GPF + MSN film forming solution (FFS), whereas the value indicated by "c" were significantly different from GPF + TGase FFS, the value indicated by "d" was significantly different from GPF + MSN + TGase FFSs, the value indicated by "*" was significantly different respect to the PEC and PEC + MSN FFSs. Data represent the average values of three repetitions using (2-way ANOVA, $p < 0.05$ for mean comparison). Additional details are reported in the main text.

3.2. Influence on Nanoreinforced and TGase-Crosslinked Hydrocolloid Coating Solutions on Acrylamide Content

Kobbah is an ethnic food consumed dispersed among all the world not only in the Arab region. The main aim of this study consisted in studying the effect of the different coating solutions to decrease the ACR content that is formed during frying. The ACR content was performed by RP-HPLC and the results reported in Figure 2. Two main different dipping solutions (GPF and PEC), reinforced by means of two different nanoparticles (1% MSN and 1% CH–NP (*w/w*)) were used to coat the kobbah prior to frying. The GPF was enzymatically crosslinked by means of TGase in the presence or absence of NPs. The control sample was the kobbah dipped into distilled water. Figure 2 shows that control exhibited the highest ACR content reaching a value of 3039.7 µg/kg. On the contrary, all used coating materials were able to significantly reduce ACR content. Kobbah dipped into GPF solution showed about 22.5% reduction in ACR content, while PEC-based coating solution reduced the ACR to 55.5%. The previous work about potato French Fries showed that PEC alone reduced ACR formation about 48% [34]. Coating solutions containing NPs (either MSN or CH–NP) in addition to GPF provoked slightly significant reduction of ACR comparing to the GPF-based coating sample not containing NPs. Higher significant reduction was observed when even MSN or CH–NP were mixed with PEC. The lowest ACR content was detected in the kobbah coated by PEC solutions containing CH–NP. In fact, in these samples the ACR content was 678 µg/kg with the 78.0% ACR reduction in comparison to the control. Recently, Mekawi et al. [31] discovered that the addition of pomegranate peel NP extracts, to the sunflower oil during deep fat frying is responsible for ACR reduction to about (54%) in potato chips. The addition of the enzyme (33U TGase/g GPF protein) into nanoreinforced GPF (even with MSN or CH–NP) reduced the ACR formation significantly (about 41.0% and 47.5%, respectively) in respect to the nanoreinforced GPF prepared without TGase (Figure 2). The obtained data may indicate a potential synergistic effect between NPs and TGase which reduces the Maillard reaction. The ACR recovery was between 93% and 108% (Table 2).

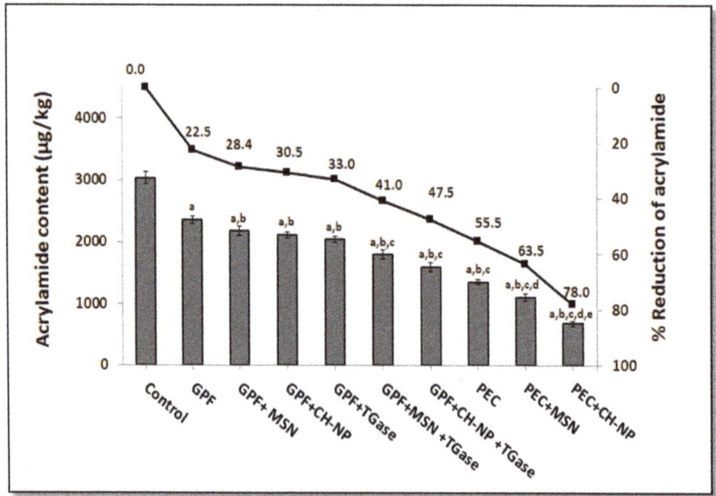

Figure 2. Influence of different hydrocolloid coatings on acrylamide content of fried kobbah (y-axis on the left based on fat-free dry matters (FFDM)) and% acrylamide reduction (y-axis on the right). "Control" represents kobbah sample dipped in distilled water. Columns "a" indicate values significantly different from the control sample; columns "b" indicate values significantly different from grass pea flour (GPF)-coated kobbah; columns "c" indicate values significantly different from GPF + mesoporous silica nanoparticles (MSN)-coated kobbah or GPF + transglutaminase (TGase)-coated kobbah; columns "d" indicate values significantly different from PEC-coated kobbah; columns "e" indicate values significantly different from pectin (PEC) + MSN-coated kobbah. Additional details are reported in the main text.

Table 2. Acrylamide content (ACR) recovery in all kobbah samples (in each sample 150 µg/L of ACR standard were used).

Kobbah Types	ACR Content in Spiked Sample (µg/Kg)	Recovery (%)
Control	3186 ± 61	98
Dipped in GPF	2511 ± 135 [a]	103
Dipped in GPF + MSN	2329 ± 103 [a,b]	101
Dipped in GPF + CH–NP	2255 ± 51 [a,b]	96
Dipped in GPF + TGase	2186 ± 48 [a,b]	98
Dipped in GPF + MSN + TGase	1934 ± 70 [a,b,c]	93
Dipped in GPF + CH–NP + TGase	1744 ± 49 [a,b,c]	99
Dipped in PEC	1495 ± 39 [a,b,c]	95
Dipped in PEC + MSN	1250 ± 50 [a,b,c,d]	95
Dipped in PEC + CH–NP	841 ± 37 [a,b,c,d,e]	108

Values significantly different from those obtained for the controls are indicated by "a", the value signed with "b" were significantly different from kobbah coated only by grass pea flour (GPF), whereas the value indicated by "c" were significantly different from kobbah coated with GPF in the presence of nanoparticles or TGase alone, the value indicated by "d" were significantly different from kobbah coated only by pectin (PEC) and the value indicated by "e" was significantly different respect to the kobbah coated by PEC + mesoporous silica nanoparticles (MSN). Data reported are the average values of three repetitions using (2-way ANOVA, $p < 0.05$ for mean comparison). Additional details are reported in the main text.

3.3. Influence of Nanoreinforced and Crosslinked Hydrocolloid Coating Solutions on Water and Oil Content

Water content of the kobbah (coated or not) was evaluated and the results reported in Figure 3. The obtained data have shown that the water content significantly increases in kobbah coated with any of the different hydrocolloid solutions used in this research. In fact, the lowest water content was found in the control sample (equal to 18%), while water content in coated kobbah by PEC-based solutions was (32.0%), significantly higher compared to the kobbah coated by GPF-based solutions (21.0%). Nanoreinforcement by using either MSN or CH–NP in both GPF-based or PEC-based solutions, provokes the increasing in water content of the kobbah significantly higher in comparison to samples coated by solutions made of only GPF or PEC. Our findings are supported by Osheba et al. [55], that concluded that CH–NP rise the moisture content of fish fingers up to 52.7%, while the uncoated samples exhibits 34.6% moisture. Regarding the use of TGase, our results prove that the enzyme action in both GPF-based and GPF + NP-based solutions show a significant higher water content respect to the kobbah coated without TGase. Comparable effects were observed by Rossi Marquez et al. [39], where TGase-mediated cross-links are responsible of the reduction of the water evaporation during frying. Moreover, the results demonstrate that the water content of kobbah coated by GPF + CH–NP + TGase is significantly higher compared to the kobbah coated by only GPF + TGase and GPF + MSN + TGase (Figure 3). Recently, Castelo Branco Melo et al. [56] found out that CH–NP led the delaying of the ripening process of the grapes as evident from the decreased weight loss, soluble solids and increased moisture retention.

One of the main health problems is the highest oil content of fried foods. Several studies concluded that coating the fried foods before frying by hydrocolloids materials reduced the oil uptake during frying [39,57]. Figure 4 shows the oil content of kobbah just dipped into water (and used as control) or the ones coated with different solutions. Coating significantly reduces the oil content in comparison to the control, which shows the highest oil content (36.9%), whereas the lowest value was obtained in the fried kobbah coated by PEC + CH–NP (15.2%). There was not any significant difference between the GPF coated kobbah and the kobbah protected by GPF nanoreinforced with MSN or CH–NP. On the other hand, significantly difference in oil content of the fried kobbah were observed between PEC-coated samples and PEC + NP-coated samples. Enzymatically cross-linking of GPF, without and with NPs, demonstrated a significant oil uptake reduction in the coated fried kobbah compared to kobbah coated by GPF or in the presence of NPs (Figure 4). PEC-based coating materials containing NPs (either MSN or CH–NP) induced a significant reduction in oil content of the coated kobbah (18.1%

and 15.2%, respectively) compared to kobbah coated with PEC (20.8%). Moreover, using CH–NPs for coating the fish fingers, Osheba et al. [55] have demonstrated a significant reduction of oil uptake which changed from 16.4% in uncoated fish fingers to 4.5% in coated ones.

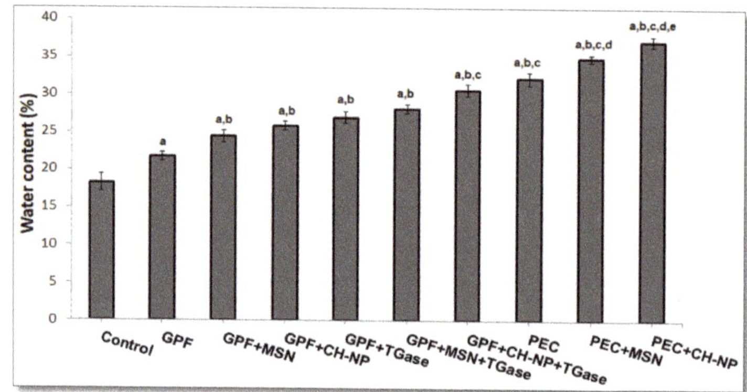

Figure 3. Effect of different hydrocolloid coatings on fried kobbah water content. "Control" represents the kobbah sample dipped in distilled water. Columns "a" indicate values significantly different from the control sample; columns "b" report values significantly different from grass pea flour (GPF)-coated kobbah; columns "c" indicate values significantly different from GPF + mesoporous silica nanoparticles (MSN)-coated kobbah or GPF + transglutaminase (TGase)-coated kobbah; columns "d" indicate values significantly different from pectin (PEC)-coated kobbah; columns "e" indicate values significantly different from PEC + MSN-coated kobbah. Additional details are reported in the main text.

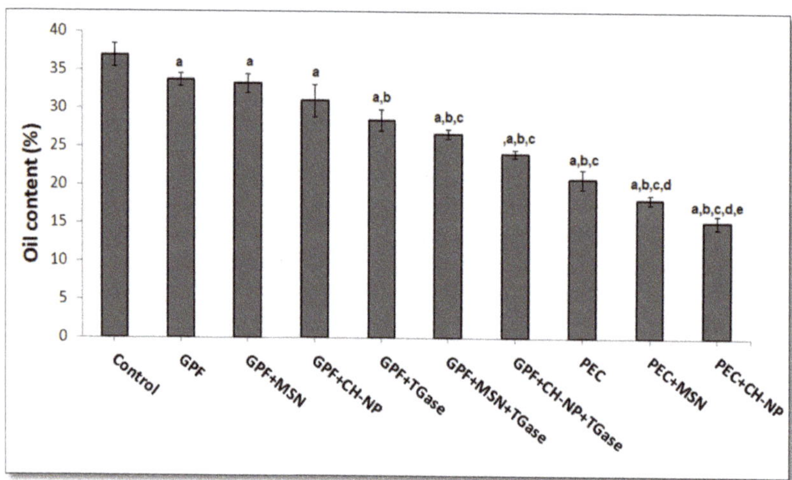

Figure 4. Influence of different hydrocolloid coatings on oil content of fried kobbah. Columns "a" indicate values significantly different from the control sample; columns "b" report values significantly different from grass pea flour (GPF)-coated kobbah; columns "c" indicate values significantly different from GPF + mesoporous silica nanoparticles (MSN)-coated kobbah or GPF + transglutaminase (TGase)-coated kobbah; columns "d" indicate values significantly different from pectin (PEC)-coated kobbah; columns "e" indicate values significantly different from PEC + MSN-coated kobbah. Additional details are reported in the main text.

3.4. Influence of Nanoreinforced and Crosslinked Hydrocolloid Coating Solutions on the Kobbah Color

Food color is important for the industries, as consumers are highly influenced by this feature. The color is dependent by several processes occurring during food processing [45]. Figure 5 shows the aspect of all the kobbah samples obtained in this study, while the results of color analysis are reported in Table 3, together with L*, a*, b* values and their derivatives, such as total color difference to control (ΔE) and Browning Index (BI).

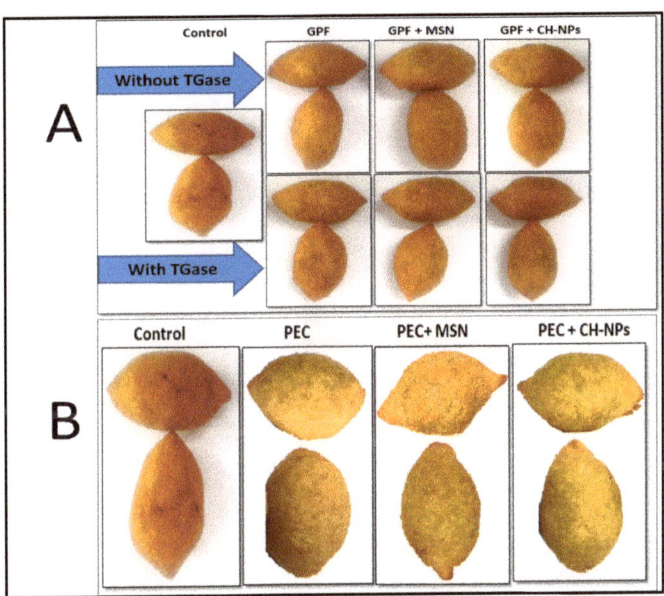

Figure 5. Images of kobbah samples coated by hydrocolloid coatings made of grass pea flour (GPF), GPF + mesoporous silica nanoparticles (MSN), GPF + chitosan nanoparticles (CH–NP), GPF + transglutaminase (TGase); GPF + MSN + TGase and GPF + CH–NP + TGase (Panel **A**); pectin (PEC), PEC + MSN and PEC + CH–NP (Panel **B**). "Control" represents the kobbah sample dipped in distilled water.

It was found that color of fried kobbah was influenced by coating which as a consequence could change the color of the final products. The lightness (L*) value showed that the lowest value was in the control samples (49.25 ± 0.68), that is uncoated and the highest value was founded in the kobbah coated by PEC + CH–NP (60.78 ± 1.02) and these results was conformed to Figure 5.

Moreover, kobbah coated with either PEC containing or not MSN or CH–NPs dipping solutions showed significant higher L* value comparing to the kobbah coated with GPF alone or with TGase or nanoparticles. The a* values showed a significant reduction after treated kobbah by different coating solutions the lowest value was in the kobbah coated with PEC + CH–NP (3.32 ± 1.35). GPF containing nanoparticles either with or without TGase showed significant reduction in the a* value comparing to the kobbah coated by only GPF. The b* value showed no significant different between the untreated and the treated kobbah except the coated kobbah by PEC + CH–NP was significant higher comparing to the kobbah coated by GPE.

Total color difference to control (ΔE) showed the highest value was in the kobbah coated by PEC–NP about (12.7 ± 0.98). However, the ΔE of the PEC coating solutions coating nanoparticles was significantly higher comparing to the kobbah coated with GPF solutions. The results indicated that the kobbah coated by the GPF containing CH–NP alone or with TGase was significantly higher than kobbah coated by only GPF (Table 3). In contrast, the highest BI was in the control kobbah

(110.09 ± 3.54) and it decreased significantly after coating the kobbah by different coating solutions and the lowest value was detected into the kobbah coated by PEC + CH–NP equal to 79.91 ± 1.72. This result is in correlation with the acrylamide results that indicated that the lowest ACR was in the kobbah coated by PEC–NPs. Jackson and Al-Taher. [58] and EFSA report [30], concluded that the surface color is highly correlated to acrylamide levels, where higher BI means higher ACR content. This demonstrates that the surface browning degree could be an indicator of ACR formation during cooking of kobbah product.

Table 3. Color properties of fried kobbah coated with different hydrocolloid-based solutions.

Kobbah Types	L*	a*	b*	ΔE	BI
Control	49.25 ± 0.68	8.28 ± 0.19	31.96 ± 0.76	0.0 ± 0.00	110.09 ± 3.54
Dipped in GPF	51.04 ± 0.34 *	7.77 ± 0.12	31.26 ± 0.57	2.02 ± 0.46 *	100.80 ± 2.99 *
Dipped in GPF + MSN	52.08 ± 1.01 *	6.69 ± 0.33 *,a	33.14 ± 0.18	3.48 ± 0.93 *	104.20 ± 3.36
Dipped in GPF CH–NP	53.23 ± 0.97 *,a	5.96 ± 0.16 *,a	31.88 ± 1.62	4.82 ± 0.76 *,a	95.00 ± 8.61 *
Dipped in GPF + TGase	54.93 ± 1.11 *,a	6.10 ± 0.82 *,a	33.01 ± 0.54	6.20 ± 1.30 *,a	95.16 ± 4.36 *,a
Dipped in GPF + MSN + TGase	55.35 ± 1.08 *,a	5.97 ± 0.31 *,a	31.55 ± 1.80	6.70 ± 1.10 *,a	88.68 ± 8.02 *,a
Dipped in GPF + CH–NP + TGase	55.70 ± 1.00 *,a	5.81 ± 0.42 *,a	32.06 ± 1.29	7.01 ± 0.84 *,a	89.44 ± 6.35 *,a
Dipped in PEC	56.56 ± 0.69 *,a,b	5.44 ± 0.31 *,a	32.58 ± 0.36	7.87 ± 0.75 *,a,b	88.75 ± 2.02 *,a
Dipped in PEC + MSN	58.03 ± 0.77 *,a,b	5.01 ± 0.25 *,a	32.99 ± 0.38	9.36 ± 0.77 *,a,b,c	86.78 ± 3.18 *,a
Dipped in PEC + CH–NP	60.78 ± 1.02 *,a,b,c	3.32 ± 1.35 *,a,c	33.34 ± 1.00 a	12.70 ± 0.98 *,a,b,c	79.91 ± 1.72 *,a,c

Columns significantly different from those obtained by analyzing the control are indicated by "*", the columns indicated by "a" were significantly different from kobbah coated only by grass pea flour (GPF), whereas the columns with "b" were significantly different from kobbah coated by GPF with transglutaminase (TGase) alone, the columns indicated by "c" were significantly different from kobbah coated only by pectin (PEC) or PEC in presence of mesoporous silica nanoparticles (MSN). The results represent the average values of three repetitions using (2-way ANOVA, $p < 0.05$ for mean comparison). Additional details are reported in the main text.

3.5. Effect of Nanoreinforced and Crosslinked Hydrocolloid Coating Solutions on the Digestibility of Fried Kobbah

In order to verify whether the coating composition could affect digestibility of the fried food, IVD experiments were performed by a protocol set up within the INFOGEST Cost Action [59]. According to INFOGEST protocol, IVD experiments were set up under physiological conditions, followed by SDS-PAGE (12%) analysis as shown in Figure 6. Samples "C" represents the control since such samples were treated with SGF prepared without pepsin. To study the digestibility rate two different kinds of bands were observed: 25 kDa band for samples containing GPF and GPF + MSN and 250 kDa band for samples set up in the presence of the enzyme (GPF + TGase and GPF + MSN + TGase). By visual inspection of the SDS-PAGE patterns of Figure 6, it is possible to assess that MSN do not affect digestibility (comparing Panel B to Panel A and Panel D to Panel C). However, looking at 250 kDa band present in TGase-treated samples (Figure 6, Panels C and D) it is not possible to note significative differences among different samples following pepsin treatment. Thus, densitometry analysis was performed, and the results reported in Figure 7, Panel A. It is possible to note that a significant rate of digestibility of 250 kDa band present in TGase-treated FFS samples, was observed after 10 min pepsin incubation. Similar results were obtained studying digestibility of GPF-based bioplastics crosslinked by means of TGase [8]. Densitometry analysis results of 25 kDa (present in FFS samples not treated with the enzyme) confirmed what was observed by visual inspection, namely an higher digestibility rate already after 1 min pepsin incubation (Figure 7, Panel A). Comparable data were reported by Romano et al. [9].

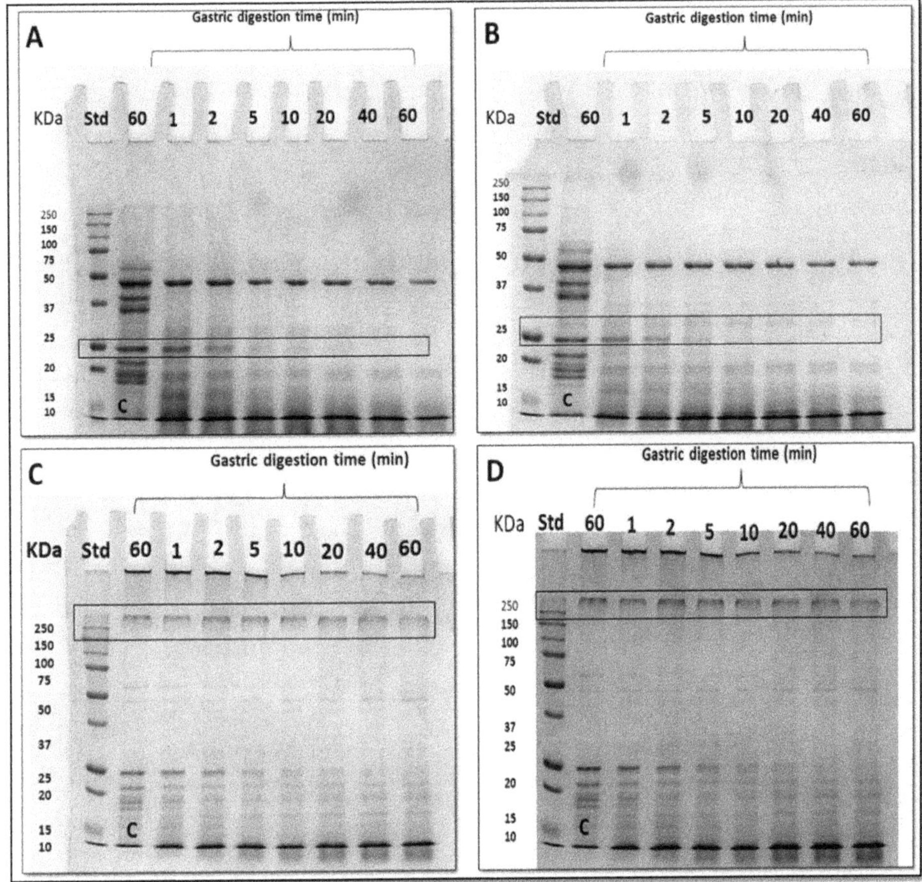

Figure 6. Sodium dodecyl sulfate polyacrylamide gel electrophoresis (SDS-PAGE) of grass pea flour (GPF) film forming solutions (FFSs) after in vitro digestion (IVD) experiments. (Panel **A**): GPF; (Panel **B**): GPF + mesoporous silica nanoparticles (MSN); (Panel **C**): GPF + transglutaminase (TGase 33 *U*/g); (Panel **D**): GPF + MSN + TGase (33 *U*/g). The bands in the rectangle are those chosen for densitometry analysis. C is control sample incubated with simulated gastric fluid (SGF) prepared without pepsin. Std, Molecular mass standards, Bio-Rad.

IVD experiments were performed also using kobbah dipped in GPF or GPF containing MSN FFSs-treated (Figure 8). IVD treatment was effective on protein component of kobbah, mainly proteins present in kobbah ingredients (i.e., mostly bulgur flour, beef meat). The ~45 kDa band of samples not treated with TGase was subjected to densitometry analysis, while in enzyme–treated samples the 250 kDa was analyzed.

Densitometry analysis of those bands are observed in Figure 7, Panel B. The digestion seems to be slower in the food coated by protein crosslinked by means of TGase enzyme. However, all the proteins were completely digested by pepsin at the longest incubation time in all different coated kobbah (Figure 7, Panel B).

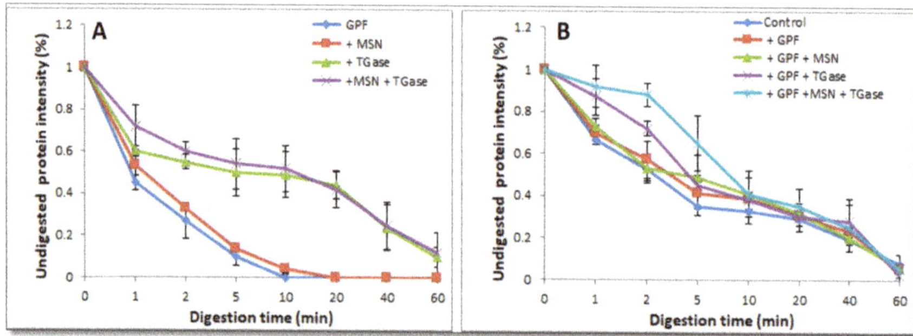

Figure 7. Intensity of the protein framed bands in gels of Figures 6 and 8, obtained after in vitro gastric digestion (IVD). Both grass pea flour (GPF)-based film forming solutions (FFSs) (Panel **A**) and fried kobbah coated with all GPF-based FFSs (Panel **B**), were subjected to densitometry analysis.

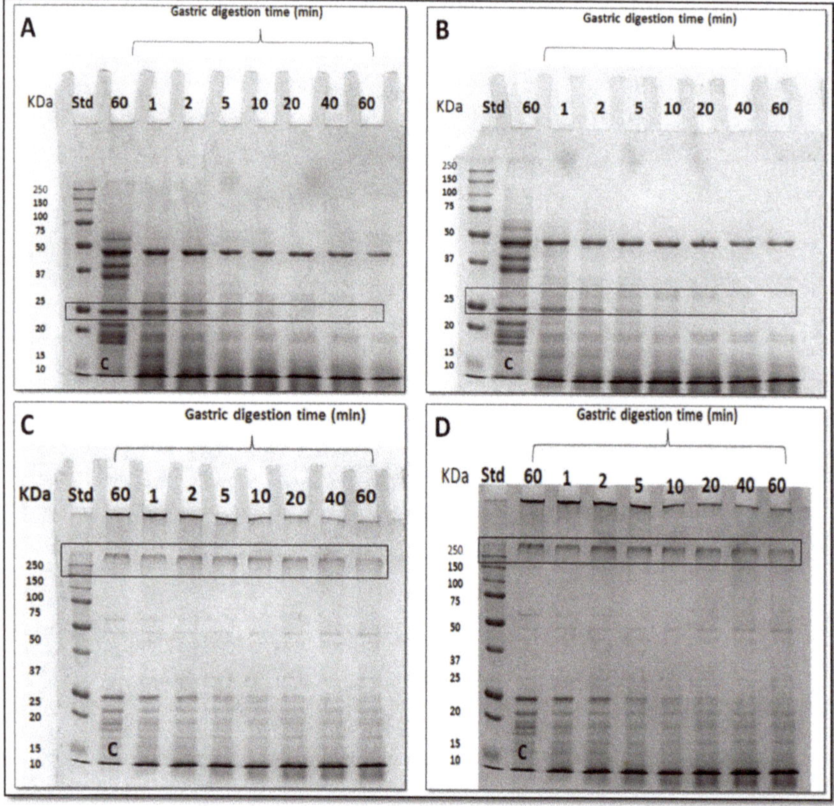

Figure 8. Sodium dodecyl sulfate polyacrylamide gel electrophoresis (SDS-PAGE) of fried kobbah digested by in vitro gastric digestion (IVD) experiments. (Panel **A**): kobbah dipped in water (control); (Panel **B**): kobbah dipped in grass pea flour (GPF); (Panel **C**): kobbah dipped in GPF + mesoporous silica nanoparticles (MSN); (Panel **D**): kobbah dipped in GPF + transglutaminase (TGase 33 *U*/g); (Panel **E**): kobbah dipped in GPF + MSN + TGase (33 *U*/g). The bands in the rectangle are those chosen for densitometry analysis. C is control sample incubated with simulated gastric fluid (SGF) prepared without pepsin. Std, Molecular mass standards, Bio-Rad.

4. Conclusions

Healthier fried kobbah was successfully obtained by dipping method using either GPF or PEC-based solutions. TGase-treated GPF in the presence of nanoparticles was demonstrated to have also important function to reduce ACR formation. The best coating solution that significantly reduced ACR was the one made of PEC nanoreinforced with CH–NP. From the obtained results we conclude that increasing water content inside the fried food by coating is an effective way to mitigate ACR formation and oil content. Reducing the browning index of the fried kobbah is a key indicator to the healthier kobbah. Moreover, the gastric digestion results showed that TGase-mediated modification fairly decreased the rate of digestion in both coating solutions and fried kobbah, even though protein component was completely digested at the end of the longest incubation time.

Author Contributions: Conceptualization, L.M. and A.A.-A.; methodology, A.A.-A., C.V.L.G. and M.S.; software, A.A.-A., C.V.L.G., M.S.; validation, A.A.-A., C.V.L.G. and L.M.; formal analysis, A.A.-A., C.V.L.G. and L.M.; investigation, A.A.-A. and L.M.; resources, L.M.; data curation, L.M.; writing—original draft preparation, A.A.-A.; writing—review and editing, L.M.; visualization, A.A.-A. and L.M.; supervision, L.M.; project administration, L.M.; funding acquisition, L.M. All authors have read and agreed to the published version of the manuscript.

Funding: This research received no external funding.

Acknowledgments: Authors would like to thank Mrs. Maria Fenderico for SDS-PAGE support.

Conflicts of Interest: The authors declare no conflict of interest.

References

1. Al-Dmoor, H.M.; Humeid, M.A.; Alawi, M.A. Investigation of acrylamide levels in selected fried and baked foods in Jordan. *Food Agric. Environ.* **2004**, *2*, 157–165.

2. Sabbah, M.; Giosafatto, C.V.L.; Esposito, M.; Di Pierro, P.; Mariniello, L.; Porta, R. Transglutaminase cross-linked edible films and coatings for food applications. In *Enzymes in Food Biotechnology*, 1st ed.; Kuddus, M., Ed.; Academic Press: New York, NY, USA, 2019; pp. 369–388.

3. Al-Asmar, A.; Giosafatto, C.V.L.; Sabbah, M.; Sanchez, A.; Villalonga Santana, R.; Mariniello, L. Effect of mesoporous silica nanoparticles on the physicochemical properties of pectin packaging material for strawberry wrapping. *Nanomaterials* **2020**, *10*, 52. [CrossRef] [PubMed]

4. Lara-Espinoza, C.; Carvajal-Millán, E.; Balandrán-Quintana, R.; López-Franco, Y.; Rascón-Chu, A. Pectin and pectin-based composite materials: Beyond food texture. *Molecules* **2018**, *23*, 942. [CrossRef] [PubMed]

5. Padmaja, N.; Bosco, S.J.D. Preservation of jujube fruits by edible Aloe *vera* gel coating to maintain quality and safety. *Indian J. Sci. Res. Technol.* **2014**, *3*, 79–88.

6. Yossef, M.A. Comparison of different edible coatings materials for improvement of quality and shelf life of perishable fruits. *Middle East J. Applied Sci.* **2014**, *4*, 416–424.

7. Campbell, C.G. *Grass Pea, Lathyrus sativus L.; Promoting the Conservation and Use of Underutilized and Neglected Crops*; IPGRI: Rome, Italy, 1997; Volume 18, pp. 1–91.

8. Giosafatto, C.V.L.; Al-Asmar, A.; D'Angelo, A.; Roviello, V.; Esposito, M.; Mariniello, L. Preparation and characterization of bioplastics from grass pea flour cast in the presence of microbial transglutaminase. *Coatings* **2018**, *8*, 435. [CrossRef]

9. Romano, A.; Giosafatto, C.V.L.; Al-Asmar, A.; Aponte, M.; Masi, P.; Mariniello, L. Grass pea (*Lathyrus sativus*) flour: Microstructure, physico-chemical properties and in vitro digestion. *Eur. Food Res. Technol.* **2019**, *245*, 191–198. [CrossRef]

10. Romano, A.; Giosafatto, C.V.L.; Al-Asmar, A.; Masi, P.; Romano, R.; Mariniello, L. Structure and in vitro digestibility of grass pea (*Lathyrus sativus* L.) flour following transglutaminase treatment. *Eur. Food Res. Technol.* **2019**, *245*, 1899–1905. [CrossRef]

11. Giosafatto, C.V.L.; Al-Asmar, A.; Mariniello, L. Transglutaminase protein substrates of food interest. In *Enzymes in Food Technology: Improvement and Innovation*; Kuddus, M., Ed.; Springer Nature; Singapore Pte Ltd.: Singapore, 2018; pp. 293–317.

12. Sabbah, M.; Altamimi, M.; Di Pierro, P.; Schiraldi, C.; Cammarota, M.; Porta, R. Black edible films from protein-containing defatted cake of *Nigella sativa* seeds. *Int. J. Mol. Sci.* **2020**, *21*, 832. [CrossRef]

13. Zink, J.; Wyrobnik, T.; Prinz, T.; Schmid, M. Physical, chemical and biochemical modifications of protein-based films and coatings: An extensive review. *Int. J. Mol. Sci.* **2016**, *17*, 1376. [CrossRef]

14. Liu, J.; He, D.; Liu, Q.; He, X.; Wang, K.; Yang, X.; Shangguan, J.; Tang, J.; Mao, Y. Vertically ordered mesoporous silica film-assisted label-free and universal electrochemiluminescence aptasensor platform. *Anal. Chem.* **2016**, *88*, 11707–11713. [CrossRef] [PubMed]

15. Slowing, I.I.; Vivero-Escoto, J.L.; Trewyn, B.G.; Lin, V.S.-Y. Mesoporous silica nanoparticles: Structural design and applications. *J. Mater. Chem.* **2010**, *20*, 7924–7937. [CrossRef]

16. Tyagi, V.; Sharma, A.; Gupta, R.K. Design and properties of polysaccharide based: Silica hybrid packaging material. *Int. J. Food Sci. Nutr.* **2017**, *2*, 140–150.

17. Fernandez-Bats, I.; Di Pierro, P.; Villalonga-Santana, R.; Garcia-Almendarez, B.; Porta, R. Bioactive mesoporous silica nanocomposite films obtained from native and transglutaminase-crosslinked bitter vetch proteins. *Food Hydrocolloid.* **2018**, *82*, 106–115. [CrossRef]

18. Giosafatto, C.V.L.; Sabbah, M.; Al-Asmar, A.; Esposito, M.; Sanchez, A.; Villalonga-Santana, R.; Cammarota, M.; Mariniello, L.; Di Pierro, P.; Porta, R. Effect of mesoporous silica nanoparticles on glycerol plasticized anionic and cationic polysaccharide edible films. *Coatings* **2019**, *9*, 172. [CrossRef]

19. Dekkers, S.; Krystek, P.; Peters, R.J.B.; Lankveld, D.P.K.; Bokkers, B.G.H.; van Hoeven-Arentzen, P.H.; Bouwmeester, H.; Oomen, A.G. Presence and risks of nanosilica in food products. *Nanotoxicology* **2011**, *5*, 393–405. [CrossRef]

20. McCarthy, J.; Inkielewicz-Stępniak, I.; Corbalan, J.J.; Radomski, M.W. Mechanisms of toxicity of amorphous silica nanoparticles on human lung submucosal cells in vitro: Protective Effects of Fisetin. *Chem. Res. Toxicol.* **2012**, *25*, 2227–2235. [CrossRef]

21. Malmiri, H.J.; Jahanian, M.A.G.; Berenjian, A. Potential applications of chitosan nanoparticles as novel support in enzyme immobilization. *Am. J. Biochem. Biotechnol.* **2012**, *8*, 203–219.

22. Divya, K.; Jisha, M.S. Chitosan nanoparticles preparation and applications. *Environ. Chem. Lett.* **2018**, *16*, 101–112. [CrossRef]

23. Lorevice, M.V.; Otoni, C.G.; de Moura, M.R.; Mattoso, L.H.C. Chitosan nanoparticles on the improvement of thermal, barrier, and mechanical properties of high-and low-methyl pectin films. *Food Hydrocoll.* **2016**, *52*, 732–740. [CrossRef]

24. Martelli, M.R.; Barros, T.T.; de Moura, M.R.; Mattoso, L.H.C.; Assis, O.B.G. Effect of chitosan nanoparticles and pectin content on mechanical properties and water vapor permeability of banana puree films. *J. Food Sci.* **2013**, *78*, N98–N104. [CrossRef] [PubMed]

25. Radhakrishnan, Y.; Gopal, G.; Lakshmanan, C.C.; Nandakumar, K.S. Chitosan nanoparticles for generating novel systems for better applications: A Review. *J. Mol. Genet. Med.* **2015**, S4-005. [CrossRef]

26. Hu, Z.; Gänzle, M.G. Challenges and opportunities related to the use of chitosan as a food preservative. *J. Appl. Microbiol.* **2018**, *126*, 1318–1331. [CrossRef] [PubMed]

27. Tareke, E.; Rydberg, P.; Karlsson, P.; Eriksson, S.; Törnqvist, M. Analysis of acrylamide, a carcinogen formed in heated foodstuffs. *J. Agric. Food Chem.* **2002**, *50*, 4998–5006. [CrossRef]

28. Mottram, D.S.; Wedzicha, B.L.; Dodson, A.T. Food chemistry: Acrylamide is formed in the Maillard reaction. *Nature* **2002**, *419*, 448–449. [CrossRef] [PubMed]

29. Zyzak, D.V.; Sanders, R.A.; Stojanovic, M.; Tallmadge, D.H.; Eberhart, B.L.; Ewald, D.K.; Gruber, D.C.; Morsch, T.R.; Strothers, M.A.; Rizzi, G.P.; et al. Acrylamide formation mechanism in heated foods. *J. Agric. Food Chem.* **2003**, *51*, 4782–4787. [CrossRef]

30. EFSA CONTAM Panel (EFSA Panel on Contaminants in the Food Chain). Scientific opinion on acrylamide in food. *EFSA J.* **2015**, *13*, 4104. [CrossRef]

31. Mekawi, E.M.; Sharoba, A.M.; Ramadan, M.F. Reduction of acrylamide formation in potato chips during deep-frying in sunflower oil using pomegranate peel nanoparticles extract. *J. Food Meas. Characteriz.* **2019**. [CrossRef]

32. Liu, J.; Liu, X.; Man, Y.; Liu, Y. Reduction of acrylamide content in bread crust by starch coating. *J. Sci. Food Agric.* **2017**, *98*, 336–345. [CrossRef]

33. Suyatma, N.E.; Ulfah, K.; Prangdimurti, E.; Ishikawa, Y. Effect of blanching and pectin coating as pre-frying treatments to reduce acrylamide formation in banana chips. *Int. Food. Res. J.* **2015**, *22*, 936–942.

34. Al-Asmar, A.; Naviglio, D.; Giosafatto, C.V.L.; Mariniello, L. Hydrocolloid-based coatings are effective at reducing acrylamide and oil content of French fries. *Coatings* **2018**, *8*, 147. [CrossRef]

35. Al-Asmar, A.; Giosafatto, C.V.L.; Panzella, L.; Mariniello, L. The effect of transglutaminase to improve the quality of either traditional or pectin-coated falafel (Fried Middle Eastern Food). *Coatings* **2019**, *9*, 331. [CrossRef]

36. Chang, P.R.; Jian, R.; Yu, J.; Ma, X. Fabrication and characterisation of chitosan nanoparticles/plasticised-starch composites. *Food Chem.* **2010**, *120*, 736–740. [CrossRef]

37. Brasil, T.A.; Capitani, C.D.; Takeuchi, K.P.; de Castro Ferreira, T.A.P. Physical, chemical and sensory properties of gluten-free kibbeh formulated with millet flour (*Pennisetum glaucum* (L.) R. Br.). *Food Sci. Technol.* **2015**, *35*, 361–367. [CrossRef]

38. Esposito, M.; Di Pierro, P.; Regalado-Gonzales, C.; Mariniello, L.; Giosafatto, C.V.L.; Porta, R. Polyamines as new cationic plasticizers for pectin-based edible films. *Carbohydr. Polym.* **2016**, *153*, 222–228. [CrossRef]

39. Rossi Marquez, G.; Di Pierro, P.; Esposito, M.; Mariniello, L.; Porta, R. Application of transglutaminase-crosslinked whey protein/pectin films as water barrier coatings in fried and baked foods. *Food Bioprocess. Technol.* **2013**, *7*, 447–455. [CrossRef]

40. *Association of Official Analytical Chemists (AOAC) Official Method 960.39 Fat (Crude) or Ether Extract in Meat First Action 1960 Final Action*; AOAC International: Arlington, MA, USA, 2006.

41. *Association of Official Analytical Chemists (AOAC) Official Method 950.46 (39.1.02) Moisture (M)*; AOAC International: Arlington, MA, USA, 2006.

42. Zeng, X.; Cheng, K.-W.; Du, Y.; Kong, R.; Lo, C.; Chu, I.K.; Chen, F.; Wang, M. Activities of hydrocolloids as inhibitors of acrylamide formation in model systems and fried potato strips. *Food Chem.* **2010**, *121*, 424–428. [CrossRef]

43. Wang, H.; Feng, F.; Guo, Y.; Shuang, S.; Choi, M.M.F. HPLC-UV quantitative analysis of acrylamide in baked and deep-fried Chinese foods. *J. Food Compos. Anal.* **2013**, *31*, 7–11. [CrossRef]

44. Michalak, J.; Gujska, E.; Kuncewicz, A. RP-HPLC-DAD studies on acrylamide in cereal-based baby foods. *J. Food Compos. Anal.* **2013**, *32*, 68–73. [CrossRef]

45. Pathare, P.B.; Opara, U.L.; Al-Said, F.A.-J. Colour measurement and analysis in fresh and processed foods: A Review. *Food Bioprocess Technol.* **2012**, *6*, 36–60. [CrossRef]

46. Papadakis, S.E.; Abdul-Malek, S.; Kamdem, R.E.; Yam, K.L. A versatile and inexpensive technique for measuring colour of foods. *Food Technol.* **2000**, *54*, 48–51.

47. Palou, E.; Lopez-Malo, A.; Barbosa-Canovas, G.V.; Welti-Chanes, J.; Swanson, B.G. Polyphenoloxidase activity and colour of blanched and high hydrostatic pressure treated banana puree. *J. Food Sci.* **1999**, *64*, 42–45. [CrossRef]

48. Laemmli, U.K. Cleavage of structural proteins during the assembly of the head of Bacteriophage T4. *Nature* **1970**, *227*, 680–985. [CrossRef] [PubMed]

49. Giosafatto, C.V.L.; Rigby, N.M.; Wellner, N.; Ridout, M.; Husband, F.; Mackie, A.R. Microbial transglutaminase-mediated modification of ovalbumin. *Food Hydrocoll.* **2012**, *26*, 261–267. [CrossRef]

50. Bourlieu, C.; Ménard, O.; Bouzerzour, K.; Mandalari, G.; Macierzanka, A.; Mackie, A.R.; Dupont, D. Specificity of Infant Digestive Conditions: Some Clues for Developing Relevant In Vitro Models. *Crit. Rev. Food Sci. Nutr.* **2014**, *54*, 1427–1457. [CrossRef]

51. Minekus, M.; Alminger, M.; Alvito, P.; Balance, S.; Bohn, T.; Bourlieu, C.; Carrière, F.; Boutrou, R.; Corredig, M.; Dupont, D.; et al. A standardised static in vitro digestion method suitable for food—An international consensus. *Food Funct.* **2014**, *5*, 1113–1124. [CrossRef]

52. Ali, S.W.; Joshi, M.; Rajendran, S. Synthesis and characterization of chitosan nanoparticles with enhanced antimicrobial activity. *Int. J. Nanosci.* **2011**, *10*, 979–984. [CrossRef]

53. Antoniou, J.; Liu, F.; Majeed, H.; Zhong, F. Characterization of tara gum edible films incorporated with bulk chitosan and chitosan nanoparticles: A comparative study. *Food Hydrocoll.* **2015**, *44*, 309–319. [CrossRef]

54. Ali, S.W.; Joshi, M.; Rajendran, S. Modulation of size, shape and surface charge of chitosan nanoparticles with reference to antimicrobial activity. American Scientific Publishers. *Adv. Sci. Lett.* **2010**, *3*, 452–460. [CrossRef]

55. Osheba, S.A.; Sorour, A.M.; Abdou, E.S. Effect of chitosan nanoparticles as active coating on chemical quality and oil uptake of fish fingers. *J. Agri. Environ. Sci.* **2013**, *2*, 1–14.

56. Melo, N.F.C.B.; de MendonçaSoares, B.L.; Diniz, K.M.; Leal, C.F.; Canto, D.; Flores, M.A.; da Costa Tavares-Filho, J.H.; Galembeck, A.; Stamford, T.L.M.; Stamford-Arnaud, T.M.; et al. Effects of fungal chitosan nanoparticles as eco-friendly edible coatings on the quality of postharvest table grapes. *Postharvest Biol. Technol.* **2018**, *139*, 56–66. [CrossRef]

57. Angor, M.K.M.; Ajo, R.; Al-Rousan, W.; Al-Abdullah, B. Effect of starchy coating films on the reduction of fat uptake in deep-fat fried potato pellet chips. *Ital. J. Food Sci.* **2013**, *25*, 45–50.

58. Jackson, L.S.; Al-Taher, F. Effects of Consumer Food Preparation on Acrylamide Formation. In *Chemistry and Safety of Acrylamide in Food. Advances in Experimental Medicine and Biology*; Friedman, M., Mottram, D., Eds.; Springer: Boston, MA, USA, 2005; Volume 561, pp. 447–465.

59. FA-1005—Improving Health Properties of Food by Sharing Our Knowledge on the Digestive Process (INFOGEST). Available online: https://www.cost.eu/actions/FA1005/#tabs\T1\textbar{}Name:overview (accessed on 17 February 2020).

Article

Quantitative Analysis of Spectinomycin and Lincomycin in Poultry Eggs by Accelerated Solvent Extraction Coupled with Gas Chromatography Tandem Mass Spectrometry

Bo Wang [1,2], Yajuan Wang [2,3], Xing Xie [4], Zhixiang Diao [2,3], Kaizhou Xie [2,3,*], Genxi Zhang [2,3], Tao Zhang [2,3] and Guojun Dai [2,3]

1 College of Veterinary Medicine, Yangzhou University, Yangzhou 225009, China; dz120180009@yzu.edu.cn
2 Joint International Research Laboratory of Agriculture & Agri-Product Safety, Yangzhou University, Yangzhou 225009, China; yzwyj163@gmail.com (Y.W.); yzdzx163@gmail.com (Z.D.); gxzhang@yzu.edu.cn (G.Z.); zhangt@yzu.edu.cn (T.Z.); daigj@yzu.edu.cn (G.D.)
3 College of Animal Science and Technology, Yangzhou University, Yangzhou 225009, China
4 Institute of Veterinary Medicine, Jiangsu Academy of Agricultural Sciences, Key Laboratory of Veterinary Biological Engineering and Technology, Ministry of Agriculture, Nanjing 210014, China; xiexing@jaas.ac.cn
* Correspondence: kzxie@yzu.edu.cn; Tel.: +86-139-5275-0925

Received: 30 March 2020; Accepted: 13 May 2020; Published: 18 May 2020

Abstract: A method based on accelerated solvent extraction (ASE) coupled with gas chromatography tandem mass spectrometry (GC-MS/MS) was developed for the quantitative analysis of spectinomycin and lincomycin in poultry egg (whole egg, albumen and yolk) samples. In this work, the samples were extracted and purified using an ASE350 instrument and solid-phase extraction (SPE) cartridges, and the parameters of the ASE method were experimentally optimized. The appropriate SPE cartridges were selected, and the conditions for the derivatization reaction were optimized. After derivatization, the poultry egg (whole egg, albumen and yolk) samples were analyzed by GC-MS/MS. This study used blank poultry egg (whole egg, albumen and yolk) samples to evaluate the specificity, sensitivity, linearity, recovery and precision of the method. The linearity (5.6–2000 µg/kg for spectinomycin and 5.9–200 µg/kg for lincomycin), correlation coefficient (≥0.9991), recovery (80.0%–95.7%), precision (relative standard deviations, 1.0%–3.4%), limit of detection (2.3–4.3 µg/kg) and limit of quantification (5.6–9.5 µg/kg) of the method met the requirements for EU parameter verification. Compared with traditional liquid–liquid extraction methods, the proposed method is fast and consumes less reagents, and 24 samples can be processed at a time. Finally, the feasibility of the method was evaluated by testing real samples, and spectinomycin and lincomycin residues in poultry eggs were successfully detected.

Keywords: poultry eggs; spectinomycin; lincomycin; ASE; GC-EI/MS/MS

---

1. Introduction

Spectinomycin and lincomycin are aminoglycoside and lincosamide antibiotics, respectively, and they have synergistic and complementary effects on each other's antibacterial spectra and antibacterial mechanisms. Spectinomycin is an inhibitor of bacterial protein synthesis and acts on the 30S subunit of ribosomes, and its antibacterial mechanism mainly involves preventing the binding of messenger ribonucleic acid and ribosomes, thereby hindering the synthesis of proteins and resulting in bactericidal effects [1]. The antibacterial mechanism of lincomycin mainly consists of binding to the bacterial ribosomal 50S subunit, which inhibits peptide acyltransferase, hinders the synthesis of bacterial proteins and results in bactericidal effects [2]. Spectinomycin has strong antibacterial activity against Gram-negative bacteria and weak activity against Gram-positive bacteria, whereas lincomycin has no effect on Gram-negative bacteria but a strong antibacterial effect on Gram-positive bacteria. Therefore, spectinomycin and lincomycin are usually used in combination to treat infections with Gram-positive bacteria and Gram-negative bacteria and are widely used to treat piglet diarrhea and infection by *Mycoplasma hyopneumoniae* and *Mycoplasma pneumoniae*, which cause chronic respiratory diseases in chickens [3–6]. However, spectinomycin can damage the eighth cranial nerve, exert kidney toxicity and block neuromuscular transmission; lincomycin has strong side effects, damages the gastrointestinal tract and liver, and even causes anaphylactic shock and death. Thus, China and the European Union (EU) have listed spectinomycin as a banned drug for poultry eggs and set maximum residue limits (MRLs) of 300–5000 µg/kg for spectinomycin and 50–1500 µg/kg for lincomycin in animal-derived foods [7,8]. The MRLs in the United States are 100–4000 µg/kg for spectinomycin in chicken and cattle muscle and liver and 100–600 µg/kg for lincomycin in pig muscle and liver. In addition, the presence of the latter two drugs in other animal-derived foods has been banned [9]. Japan has set MRLs for spectinomycin of 500–5000 µg/kg and for lincomycin of 200–1500 µg/kg in animal-derived foods [10]. Thus, it is important to develop fast and efficient analytical methods to detect spectinomycin and lincomycin in poultry eggs.

To date, many methods have been used to measure spectinomycin and lincomycin in animal-derived foods and animal feedstuffs, including fluorescent latex immunoassay (FLI) [11], micellar electrokinetic capillary chromatography combined with ultraviolet detection (MEKC-UVD) [12], enzyme-linked immunosorbent assay (ELISA) [13], high-performance liquid chromatography with electrochemical detection (HPLC-ECD) [14,15], HPLC with fluorescence detection (FLD) [16], HPLC-UVD [17,18], HPLC with evaporative light-scattering detection (ELSD) [19,20], hydrophilic interaction chromatography with mass spectrometry (HILIC-MS) [21], HILIC tandem MS (MS/MS) [22], HPLC-MS [23], HPLC-MS/MS [24–28], gas chromatography–nitrogen phosphorus detection (GC-NPD) [29,30] and GC-MS [30]. The FLI, ELISA, ECD, FLD, UVD and ELSD methods have low sensitivity, specificity, recovery and precision and have many limitations. Molognoni et al. [22] developed a liquid–liquid extraction (LLE) method combined with HILIC-MS/MS for the determination of spectinomycin, halquinol, zilpaterol and melamine residues in animal feedstuffs with good recovery and precision. Juan et al. [27] reported an accelerated solvent extraction (ASE) method for the trace analysis of macrolide and lincosamide antibiotics in meat and milk using HPLC-MS/MS, and the method was fast, sensitive and automatic, making it suitable for the determination of macrolide and lincosamide residues in meat and milk. Tao et al. [30] established an ASE approach for extracting lincomycin and spectinomycin residues from swine and bovine tissues using GC-NPD and GC-MS. ASE is an automated extraction technology that is widely used for veterinary drug residue detection in animal food because of its advantages, such as rapid analysis, low organic solvent use and batch sample processing. Compared to liquid chromatography, gas chromatography has been reported less frequently for the detection of spectinomycin and lincomycin in animal foods. Moreover, the use of single-stage GC-MS has several difficult limitations, such as the inability to effectively exclude sample matrix-derived interferences, to confirm false positives and quasi-deterministic parameters and to quantify target compounds. However, a gas chromatography–tandem mass spectrometry (GC-MS/MS) method can effectively address these issues and accurately quantify target compounds. Thus, an ASE-GC-MS/MS method was developed to determine spectinomycin and lincomycin residues in poultry eggs. The method parameters were validated according to the EU [31] and the Food and Drug Administration (FDA) [32] validation requirements.

2. Materials and Methods

2.1. Chemicals and Reagents

Spectinomycin (97.9% standard) and lincomycin (98.9% standard) were purchased from the Food and Drug Control Agency (Beijing, China). *N,O*-bis(Trimethylsilyl)trifluoroacetamide (BSTFA, >99.0%

standard) was obtained from Sigma-Aldrich (St. Louis, MO, USA). Sodium dodecyl sulfonate (SDS, ≥99.0% standard) was purchased from Sangon Biotech (Shanghai, China). Acetonitrile and methanol (HPLC grade) were acquired from Merck (Fairfield, OH, USA). Analytical-grade phosphoric acid (H_3PO_4), sodium hydroxide, acetic acid, n-hexane, potassium dihydrogen phosphate (KH_2PO_4) and trichloroacetic acid (TCA) were obtained from Sinopharm Chemical Reagent Co. (Shanghai, China). Ultrapure water was obtained from a PURELAB Option-Q synthesis system (ELGA Lab Waters, High Wycombe, Bucks, UK).

Standard stock solutions of spectinomycin and lincomycin at 1 mg/mL were prepared in pure methanol. The standard working solutions were obtained by diluting the standard stock solutions with pure methanol according to the test needs.

2.2. GC-MS/MS Analysis

The GC-MS/MS system consisted of a Trace 1300 gas chromatograph, a TSQ 8000 triple quadrupole tandem mass spectrometer and a Triplus RSH automatic sample injector, and the TraceFinder 3.0 software was used for the analysis (Thermo Fisher Corp., Waltham, MA, USA). GC separation was performed using the following temperature program: 160 °C for 1 min; a ramp at 25 °C/min to 250 °C, followed by a 1 min hold; and a ramp at 15 °C/min to 300 °C, followed by a 5 min hold. A Thermo Fisher TG-5MS amine column (30 m × 0.25 mm; inside diameter (i.d.), 0.25 μm) was used. The GC was operated in splitless mode with a carrier gas (helium, 99.999% standard, 60 psi) flow rate of 1.0 mL/min. The injector temperature was held at 280 °C, and the injection volume was 1.0 μL.

The MS/MS system was equipped with an electron impact (EI) source and used in full scan mode and selected reaction monitoring (SRM) mode. The typical MS parameters were as follows: ionization voltage, 70 eV; ion source temperature, 280 °C; and transfer line temperature, 280 °C. The retention times and relevant MS parameters are presented in Table 1.

Table 1. Retention times and relevant mass spectrometry (MS) parameters for the analytes.

Analyte	Retention Time (min)	Molecular Weight (*m/z*)	Mass Transitions (*m/z*)	Collision Energy (eV)
Spectinomycin	6.93	332.15	201.1 > 75.0 *	16
			201.1 > 185.1	8
Lincomycin	10.53	406. 21	126.1 > 42.0 *	22
			126.1 > 82.0	22

Note: * Ion pair used for quantification.

2.3. Preparation of the Samples

Considering that consumers have separate uses for whole eggs, albumens and yolks in hen, duck and goose eggs, we studied the elimination of spectinomycin and lincomycin residues in whole eggs, albumens and yolks. Because pigeon and quail eggs are relatively small, consumers generally use these as whole eggs. Thus, blank hen, duck and goose eggs were collected as whole eggs, albumen and yolk samples, and blank pigeon and quail eggs were collected as whole egg samples. Blank hen, duck and goose eggs (whole eggs, albumens and yolks) as well as pigeon and quail eggs (whole eggs) were separately homogenized, divided and frozen. In this work, LLE and ASE were used to extract the poultry egg samples, which were then cleaned up by SPE and finally derivatized.

2.3.1. Liquid–Liquid Extraction

Homogenized poultry eggs (2.0 ± 0.02 g) were precisely weighed and then added to 10 mL of 0.01 M KH_2PO_4 solution (pH 4.0). The sample was vortexed for 5 min at 2000× *g*, homogenized ultrasonically for 10 min and then centrifuged for 10 min at 8000× *g*. The extraction solution was collected, and the sample was extracted again. The two extracts were combined, and 5 mL of n-hexane was added. The mixture was vortexed for 5 min at 2000× *g* and then centrifuged for 10 min at 8000× *g*.

After degreasing twice with n-hexane, the extract was added to 5 mL of 3% TCA solution, vortexed for 5 min at 2000× g and then centrifuged for 10 min at 8000× g. The liquid–liquid extraction procedure was performed according to the National Food Safety Standard (GB 29685-2013) [33].

2.3.2. Accelerated Solvent Extraction

Homogenized poultry eggs (2.0 ± 0.02 g) and 4.0 g of diatomaceous earth were fully ground, and then sample preparation was performed. The fat-removal and extraction parameters for the ASE350 instrument (Thermo Fisher Scientific Co. Ltd., Waltham, MA, USA) were as follows: 60 °C, 1500 psi, a static extraction time of 5 min and a nitrogen purge time of 60 s. One extraction was performed with a total solvent rinse of 40% and n-hexane to remove the fat, and two extractions were then performed with a total solvent rinse of 50% and 0.01 M KH_2PO_4 solution (pH 4.0) to extract the analytes, after which the sample extract was collected.

2.3.3. Solid-Phase Extraction

After the sample was processed by LLE or ASE, 10% NaOH solution was added to the extract to adjust the pH to 5.8 ± 0.2, 2 mL of 0.2 M SDS solution was added, and the sample was then vortexed for 1 min. After standing for 15 min, the extract was cleaned up by SPE with an Oasis PRiME HLB cartridge (3 mL/60 mg, Waters Corp., Milford, MA, USA) that had been activated and equilibrated by the addition of 3 mL of methanol, 3 mL of ultra-pure water and 3 mL of 0.02 M SDS solution. After 20 mL of the extracts was added to the Oasis PRiME HLB cartridge at a constant rate (2.0 mL/min) and allowed to completely pass through the cartridge, 9 mL of ultrapure water was added in three portions for rinsing. Finally, 6 mL of methanol was used to elute the two target compounds.

2.4. Derivatization Reaction

After the extract was dried under a stream of nitrogen at 40 °C, 200 μL of BSTFA and 100 μL of acetonitrile were sequentially added to the sample, which was then vortexed for 1 min. Then, the mixture was placed in a 75 °C oven for 60 min. After the derivatization reaction was complete, the mixture was cooled to room temperature and dried under a stream of nitrogen at 40 °C. Finally, 2 mL of n-hexane was added to the sample to dissolve the residue, and the resulting solution was vortexed for 1 min and passed through a 0.22 μm organic phase needle filter into the GC-MS/MS system.

2.5. Quality Parameters

Seven spiked concentration levels for the two analytes were used to establish the linear regression equations: the limit of quantification (LOQ) and 50, 100, 500, 1000, 1500 and 2000 μg/kg for spectinomycin and the LOQ and 10, 20, 50, 100, 150 and 200 μg/kg for lincomycin. The peak areas as a function of the analyte concentration were used to establish standard working curves. The correlation coefficients (R^2 values) were determined and should all have been ≥0.9991. The other parameters were tested according to the EU [31] and the FDA requirements [32], and the TraceFinder 3.0 software (Thermo Fisher Corp., Waltham, MA, USA) was used for the analysis.

3. Results and Discussion

3.1. Optimization of the ASE Conditions

Due to the complexity of the matrices of animal-derived foods, the detection of veterinary drug residues in such foods usually requires sample pretreatment involving extraction and clean-up to avoid clogging the chromatography column and contaminating the instrument. Several methods, such as LLE [22,24,26], solid-phase extraction (SPE) [16,21], core-shell molecularly imprinted solid-phase extraction (CSMISPE) [18] and ASE [27,30], have been developed for the extraction of spectinomycin and lincomycin from animal tissues, meat, milk and animal feedstuffs as well as from swine, calf and chicken plasma. Compared with the LLE, SPE and CSMISPE methods, the ASE method has the advantages of a short extraction time, lower consumption of organic reagents and batch sample processing. Therefore, in this study, the ASE method was used to extract spectinomycin and lincomycin from poultry eggs, and the analyte recoveries were compared with those for the LLE method.

Tao et al. [30] used a 0.01 M KH_2PO_4 solution as an extractant to successfully extract spectinomycin and lincomycin from animal tissues. Based on the chemical properties of spectinomycin and lincomycin, a 0.01 M KH_2PO_4 solution was also selected as the extractant in the present study. In this experiment, the pH of the 0.01 M KH_2PO_4 solution was adjusted with H_3PO_4, and the effects of different pH values (3.0–5.5) on the response values of the two compounds were compared. When the 0.01 M KH_2PO_4 solution (pH 4.0) was used as the extractant, the response values of spectinomycin and lincomycin were the highest (Figure 1a). Thus, the 0.01 M KH_2PO_4 solution (pH 4.0) was finally selected as the extractant in this study. At 1500 psi, the effects of the temperature (40 °C, 60 °C, 80 °C, 100 °C and 120 °C), the amount of extractant (40%, 50%, 60%, 70% and 80% of the extraction cell volume), and the number of extractions (1 and 2 static cycles) on the recovery of spectinomycin and lincomycin from poultry eggs were compared. Firstly, using the 0.01 M KH_2PO_4 solution (pH 4.0) as the extractant and the optimal conditions for ASE extraction temperature were tested under 1500 psi, and the optimal extraction temperature was determined to be 60 °C (Figure 1b). Secondly, under the conditions of 1500 psi, 60 °C and using the 0.01 M KH_2PO_4 solution (pH 4.0) as the extractant, we optimized the amount of extractant and the number of extractions, and a 50% extraction cell volume and two static cycles obtained the best response value (Figure 1c). Thus, the optimal extraction conditions for the ASE method were as follows (Figure 1): 60 °C, 1500 psi, a 0.01 M KH_2PO_4 solution (pH 4.0) as the extractant, 50% extraction cell volume, static extraction for 5 min, one degreasing cycle and two static cycles.

3.2. Optimization of the SPE Conditions

An ion-pair reagent can be combined with the analyte to form an ion-pair and become neutral so that the analyte molecules are retained on the chromatographic column. A test revealed that the ion-pair reagent was susceptible to pH-induced changes: slight changes in pH affected the ion-pair reagent and, consequently, the recoveries of the target compounds. To solve this problem, after LLE or ASE, 10% NaOH was added to the sample extract to adjust the pH (to 5.4, 5.6, 5.8, 6.0 and 6.2), and then 2 mL of 0.2 M SDS solution was added to change the polarity. Adjusting the pH of the extract to 5.8 ± 0.2, the response value of the target was slightly improved. SPE cartridges were used to isolate spectinomycin and lincomycin from poultry eggs. The effects of different ion-pair reagents (sodium hexane sulfonate, sodium heptane sulfonate, sodium octane sulfonate and sodium dodecyl sulfonate) on the recoveries of the target compounds were compared. Sodium dodecyl sulfonate yielded the highest responses for the quantitative ion pairs (spectinomycin: m/z 201.1 > 75.0, lincomycin: m/z 126.1 > 42.0), which resulted in higher analyte recovery (Figure 2). Therefore, a 0.02 M sodium dodecyl sulfonate solution was used to equilibrate the SPE cartridge. This study compared C_{18} cartridges (6 mL/500 mg, Agela Technologies, Tianjin, China), PCX cartridges (6 mL/500 mg, Agela Technologies), and Oasis PRiME HLB cartridges (3 mL/60 mg, Waters Corp) in terms of the target compound recoveries. The C_{18} cartridge (6 mL/500 mg) produced interferences and did not effectively clean up the samples. The PCX cartridge (6 mL/500 mg) resulted in poor peak shapes and recoveries of less than 70%. The Oasis PRiME HLB cartridge (3 mL/60 mg) effectively cleaned up the samples and yielded recoveries above 80%. The Oasis PRiME HLB cartridge is a new type of solid-phase extraction cartridge that can remove 99% of the phospholipid matrix interferences in the sample, which minimizes the matrix effect of mass spectrometry, resulting in more stable data, a longer column life cycle, less instrument maintenance and less downtime. Therefore, the Oasis PRiME HLB cartridge (3 mL/60 mg) was used for sample clean-up.

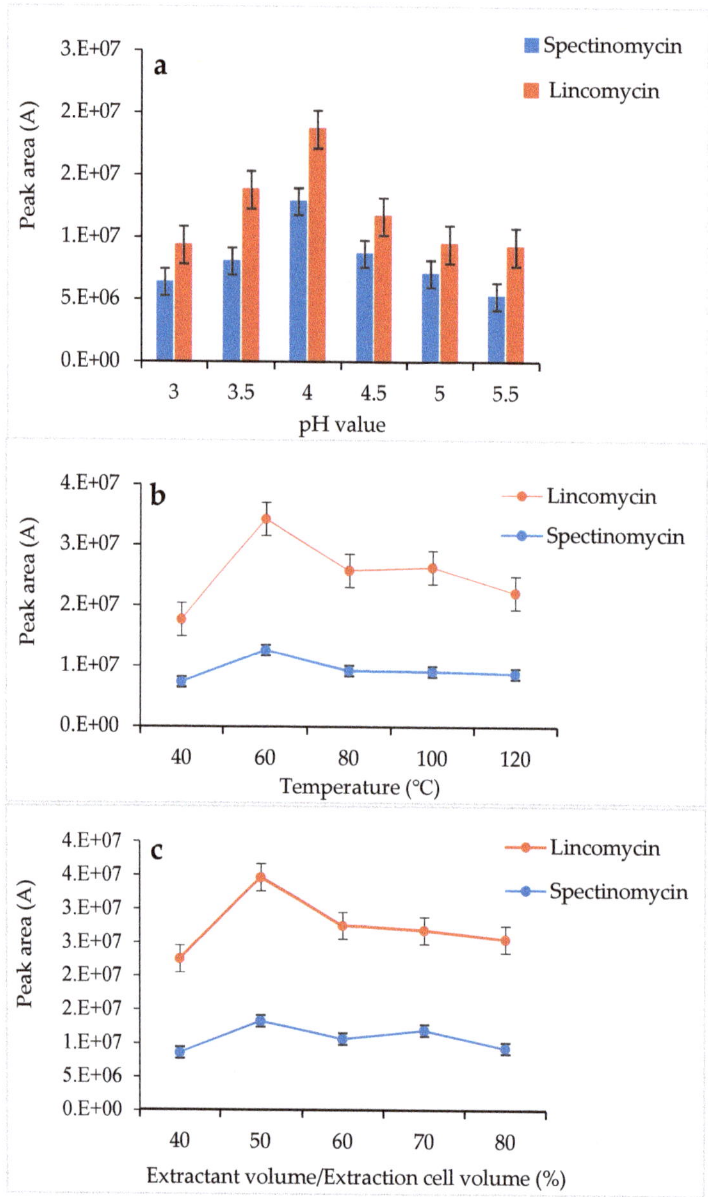

Figure 1. Effects of pH (**a**), temperature (**b**) and extractant volume (**c**) on the extraction efficiency of accelerated solvent extraction (ASE).

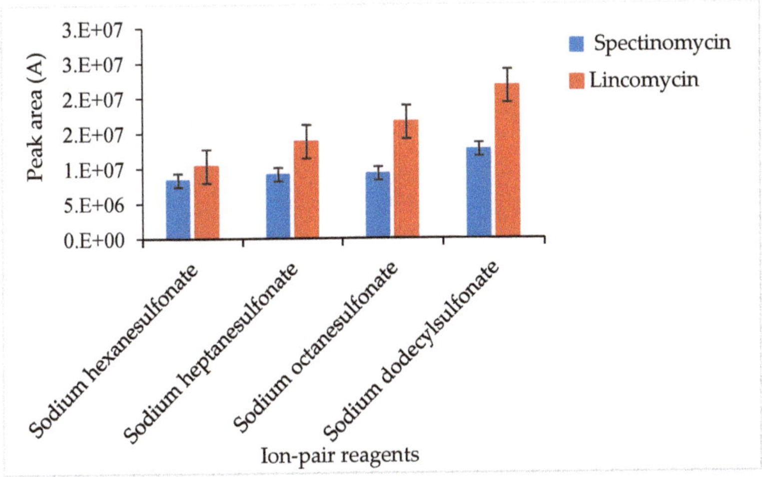

Figure 2. Effects of different ion-pair reagents on the recovery of spectinomycin and lincomycin.

After the optimization of the extraction and clean-up conditions, the effects of the LLE-SPE and ASE-SPE methods on the recoveries of spectinomycin and lincomycin from poultry eggs were compared. The results (Table 2) show that the recoveries for the ASE-SPE method were higher than those for the LLE-SPE method. Therefore, the ASE-SPE method was used to extract and clean up spectinomycin and lincomycin residues in poultry eggs.

Table 2. Comparison of the effects of the extraction method on the recoveries of 50 µg/kg spectinomycin and lincomycin in poultry eggs (%) ($n = 6$). Liquid–liquid extraction, LLE; solid-phase extraction, SPE.

Analyte	Matrix	Sample Preparation Method	
		LLE-SPE	**ASE-SPE**
	Hen whole egg	80.5 ± 1.6	86.0 ± 1.5
	Albumen	81.4 ± 1.4	86.6 ± 2.0
	Yolk	81.2 ± 1.7	86.6 ± 2.1
	Duck whole egg	80.9 ± 3.4	86.1 ± 2.7
	Albumen	74.4 ± 1.4	85.7 ± 2.2
Spectinomycin	Yolk	82.1 ± 3.4	87.1 ± 1.3
	Goose whole egg	79.2 ± 2.2	87.7 ± 1.7
	Albumen	81.1 ± 1.6	83.4 ± 1.4
	Yolk	80.9 ± 3.7	85.9 ± 1.2
	Pigeon whole egg	80.5 ± 2.7	87.7 ± 2.2
	Quail whole egg	79.0 ± 3.2	87.2 ± 1.9
	Hen whole egg	79.4 ± 1.6	84.8 ± 1.6
	Albumen	78.7 ± 2.7	85.8 ± 2.2
	Yolk	81.3 ± 3.3	84.8 ± 2.0
	Duck whole egg	75.4 ± 1.7	85.4 ± 2.6
	Albumen	76.0 ± 2.4	93.4 ± 2.3
Lincomycin	Yolk	83.5 ± 1.6	87.9 ± 2.1
	Goose whole egg	82.3 ± 3.7	87.9 ± 1.1
	Albumen	75.7 ± 2.0	89.1 ± 1.1
	Yolk	76.6 ± 2.2	84.6 ± 2.5
	Pigeon whole egg	83.2 ± 3.6	86.2 ± 2.2
	Quail whole egg	79.3 ± 1.2	86.4 ± 1.5

3.3. Optimization of the GC-MS/MS Analysis

Spectinomycin and lincomycin are highly polar compounds and cannot be detected directly by GC techniques. Usually, derivatization is required to reduce the polarity and boiling point of these compounds before GC detection. Tao et al. [30] reported the successful detection of spectinomycin and lincomycin in animal tissues by a GC method after derivatization by BSTFA. Thus, BSTFA was used as the derivatization reagent in the present work, and the above method of optimizing the ASE parameters was used to optimize the following derivatization conditions: the amount of BSTFA (100–700 μL), amount of acetonitrile (50–300 μL), temperature (35–95 °C) and time (30–90 min). The optimal derivatization conditions (Figure 3) were 75 °C, 60 min, 200 μL of BSTFA and 100 μL of acetonitrile, under which spectinomycin and lincomycin were derivatized to spectinomycin- trimethylsilyl (TMS) and lincomycin-TMS (Figures 4 and 5). After derivatization, BSTFA was removed by drying the sample under a stream of nitrogen. Excess BSTFA crystallizes easily and will plug and damage the column. TMS-derivatized products are easily hydrolyzed and stable for 24 h. Therefore, TMS-derivatized products should be analyzed by GC-MS/MS within 24 h.

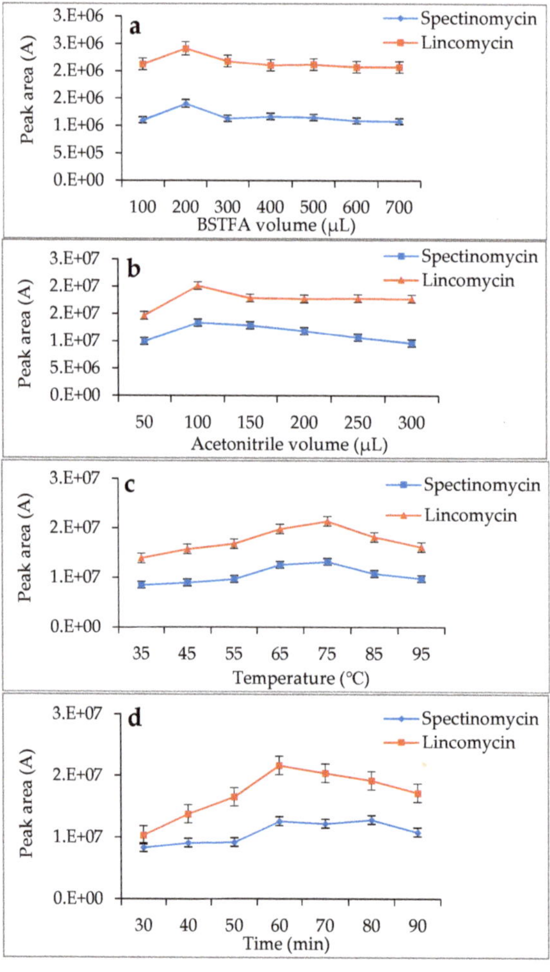

Figure 3. Effects of *N*,*O*-bis(Trimethylsilyl)trifluoroacetamide (BSTFA) volume (**a**), acetonitrile volume (**b**), temperature (**c**) and time (**d**) on the derivatization reaction.

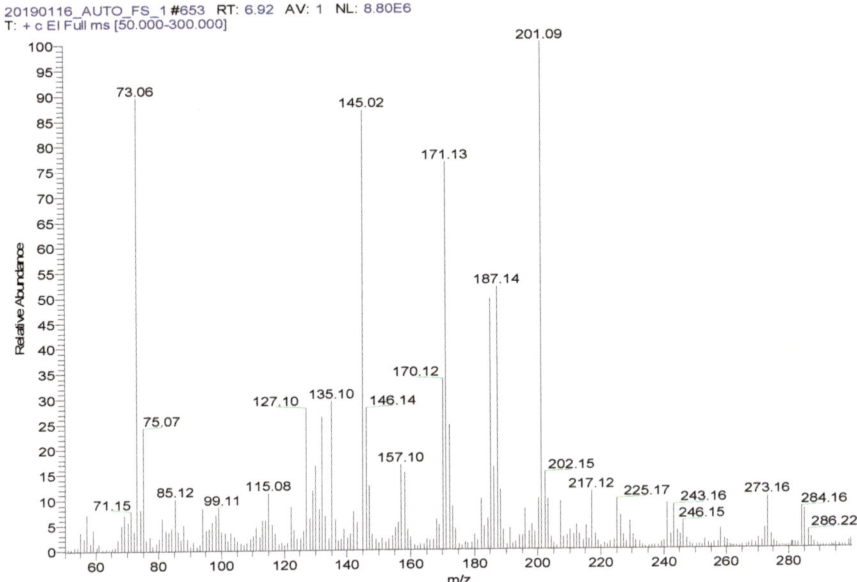

Figure 4. MS spectrum of spectinomycin-TMS.

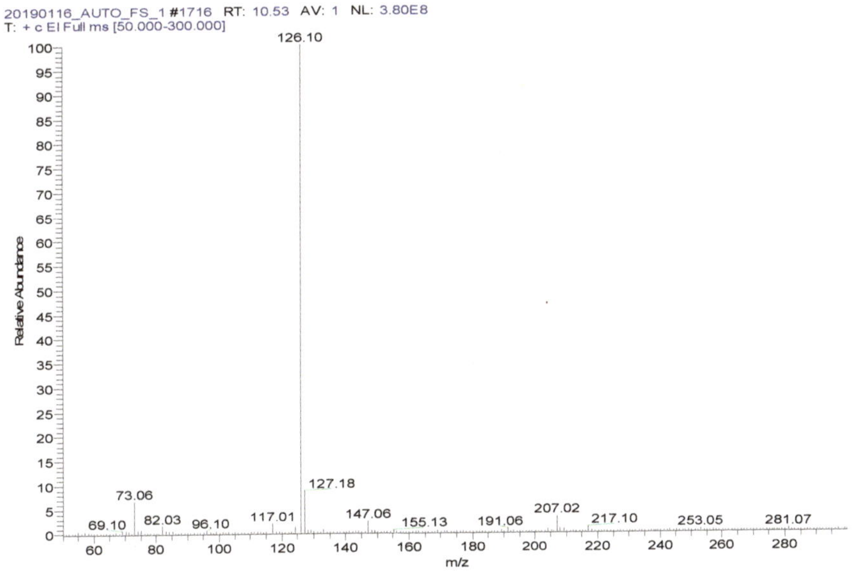

Figure 5. MS spectrum of lincomycin-TMS.

Several capillary columns, including DB-1 (30 m × 0.25 mm i.d., 0.25 μm), HP-5 (30 m × 0.25 mm i.d., 0.25 μm) and Rtx-5 (30 m × 0.25 mm i.d., 0.25 μm), have been reported for the detection of spectinomycin and lincomycin in animal-derived foods and were tested herein. According to previous reports [29,30], lincomycin and spectinomycin derivatives have moderate polarities and low boiling points, so nonpolar and moderately polar capillary columns are usually used to detect these two compounds. The inner surface of the moderately polar TG-5MS (30 m × 0.25 mm i.d., 0.25 μm) capillary column has been chemically treated to reduce the tailing of active basic compounds and increase the detection of amines. Therefore, a TG-5MS (30 m × 0.25 mm i.d., 0.25 μm) capillary column was selected to analyze spectinomycin and lincomycin residues in poultry eggs. Next, the oven temperature program was optimized to decrease the retention time (RT) of the target compounds (spectinomycin and lincomycin, 6.93 and 10.53 min) and shorten the total run time. Analysis was performed in full scan mode and SRM mode to identify precursor and product ions. In this study, two monitored ion pairs were selected for the qualitative and quantitative analysis of the target compounds. The derivatized products were analyzed under the optimized GC-MS/MS conditions. The total ion chromatogram (TIC) and extracted ion chromatograms (XICs) of a blank hen whole egg sample are shown in Figure 6. The TIC and XICs of the quantitative ions from the blank hen whole egg spiked with 50.0 μg/kg spectinomycin and 50.0 μg/kg lincomycin (Figure 7) showed that spectinomycin and lincomycin in hen whole eggs could be effectively separated with sharp peaks and no tailing.

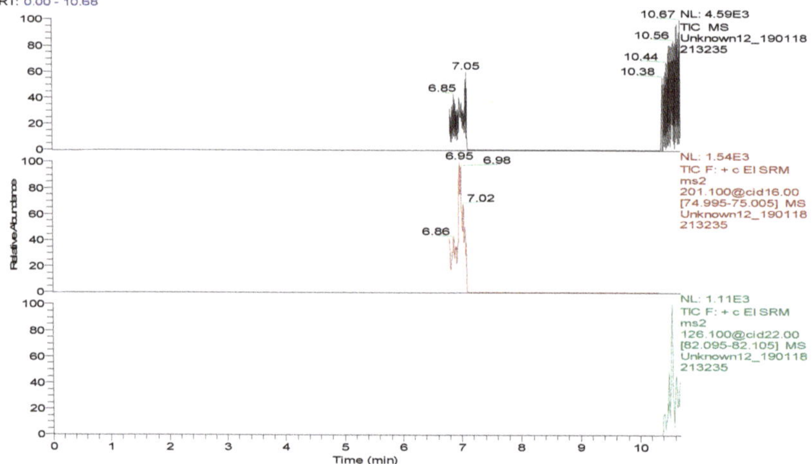

Figure 6. Total ion chromatogram and extracted ion chromatograms of a blank hen whole egg sample.

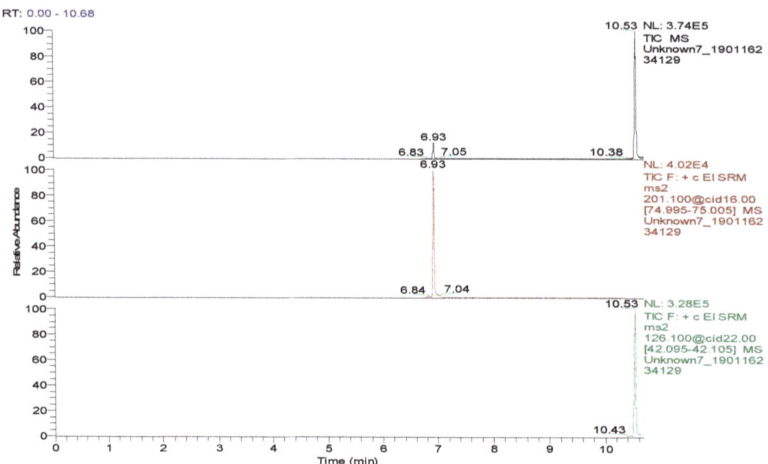

Figure 7. Total ion chromatogram and extracted ion chromatograms of a blank hen whole egg sample spiked with 50.0 μg/kg spectinomycin (retention time (RT), 6.93 min) and 50.0 μg/kg lincomycin (RT, 10.53 min).

3.4. Bioanalytical Method Validation

The specificity of the method for analyzing blank poultry eggs was determined by comparing Figures 6 and 7. Figure 6 shows that the blank hen whole egg sample did not contain spectinomycin and lincomycin. The blank poultry egg samples were extracted and cleaned up by the ASE-SPE method to obtain a blank matrix extract. The standard working solutions of spectinomycin and lincomycin and the reagents required for the abovementioned derivatization reaction were sequentially added to the blank matrix extract for derivatization. The standard curve was constructed from the GC-MS/MS analysis of the samples at the seven concentration levels. The linear ranges of spectinomycin and lincomycin were LOQ–2000 µg/kg and LOQ–200 µg/kg, respectively. The regression equation and determination coefficient data are listed in Table 3. According to the EU guidelines [31], the recovery and precision (intraday precision and interday precision) of the developed GC-MS/MS method were validated at the LOQ and at 0.5, 1.0 and 2 MRL (n = 6 at each level) for each drug in the poultry egg samples. In particular, 4000 µg/kg spectinomycin was added to the blank poultry egg sample; after extraction and purification by the ASE-SPE method, the sample was diluted with blank matrix extract 2-fold before the derivatization reaction was performed to ensure that the detected concentration of the sample was in the linear range. The measured concentration was multiplied by 2 to obtain the actual concentration of the original sample. By this method, the recovery and precision of measuring 4000 µg/kg spectinomycin in poultry eggs were evaluated. As shown in Tables 4 and 5, the recoveries of spectinomycin and lincomycin in the blank hen, duck and goose egg (whole egg, albumen and yolk) samples as well as in the pigeon and quail egg (whole egg) samples were 80.0%–95.7%, and the relative standard deviations (RSDs) were 1.0%–3.4%. In addition, the intraday RSDs were 1.9%–6.0%, and the interday RSDs were 2.2%–6.7%. These data indicate that the recovery and precision of the method meet the EU [31] and FDA [32] requirements for methodological parameters.

Table 3. Linearity, determination coefficient, limit of detection (LOD) and limit of quantification (LOQ) of spectinomycin and lincomycin in poultry eggs.

Analyte	Matrix	Regression Equation	Determination Coefficient (R^2)	Linear Range (µg/kg)	LOD (µg/kg)	LOQ (µg/kg)
Spectinomycin	Hen whole egg	$y = 1251x + 16346$	0.9993	6.0–2000	3.1	6.0
	Albumen	$y = 1298.4x + 34049$	0.9992	6.4–2000	3.0	6.4
	Yolk	$y = 1199.1x - 4540.7$	0.9997	5.6–2000	2.3	5.6
	Duck whole egg	$y = 1059.5x - 3709.3$	0.9999	6.3–2000	3.5	6.3
	Albumen	$y = 1113.1x - 4463.4$	0.9994	7.9–2000	3.8	7.9
	Yolk	$y = 1051.5x - 2573.8$	0.9993	8.0–2000	2.7	8.0
	Goose whole egg	$y = 1113.5x - 3245.8$	0.9996	7.1–2000	3.5	7.1
	Albumen	$y = 1179.6x - 3611.7$	0.9995	7.8–2000	3.2	7.8
	Yolk	$y = 1074.7x - 2777.5$	0.9996	6.7–2000	3.0	6.7
	Pigeon whole egg	$y = 1022.9x - 6310.6$	0.9991	8.0–2000	4.0	8.0
	Quail whole egg	$y = 1073.1x - 4316.6$	0.9998	7.6–2000	3.8	7.6
Lincomycin	Hen whole egg	$y = 10074x - 29019$	0.9994	8.4–200	3.1	8.4
	Albumen	$y = 7829.4x + 98669$	0.9992	6.7–200	2.6	6.7
	Yolk	$y = 13453x - 27487$	0.9994	5.9–200	2.5	5.9
	Duck whole egg	$y = 8127.3x - 16359$	0.9996	6.5–200	2.8	6.5
	Albumen	$y = 7772.3x - 312.45$	0.9992	7.3–200	3.0	7.3
	Yolk	$y = 7759.2x + 32016$	0.9992	8.0–200	4.3	8.0
	Goose whole egg	$y = 8825.4x + 2625.7$	0.9996	8.5–200	3.5	8.5
	Albumen	$y = 9240.1x + 6253.1$	0.9994	9.0–200	3.8	9.0
	Yolk	$y = 8381.4x + 27696$	0.9994	9.2–200	4.0	9.2
	Pigeon whole egg	$y = 8578.4x - 37597$	0.9993	9.5–200	3.9	9.5
	Quail whole egg	$y = 8736.2x - 61525$	0.9994	8.2–200	4.3	8.2

Table 4. Recovery and precision of spectinomycin and lincomycin spiked in blank poultry eggs (whole egg, $n = 6$).

Analyte	Matrix	Spike Level (μg/kg)	Recovery (%)	RSD (%)	Intraday RSD (%)	Interday RSD (%)
Spectinomycin	Hen whole egg	6.0	82.9 ± 1.5	1.8	3.0	4.2
		1000.0	84.5 ± 1.0	1.2	4.1	3.6
		2000.0 [a]	86.0 ± 1.5	1.7	4.0	4.1
		4000.0	95.7 ± 1.2	1.3	3.7	5.5
	Duck whole egg	6.3	80.9 ± 1.3	1.6	4.6	5.0
		1000.0	85.5 ± 1.5	1.8	3.6	4.5
		2000.0 [a]	86.1 ± 2.7	3.1	2.9	3.2
		4000.0	89.4 ± 1.8	2.0	2.9	3.4
	Goose whole egg	7.1	82.1 ± 1.7	2.1	3.2	5.0
		1000.0	86.8 ± 1.3	1.5	2.7	4.9
		2000.0 [a]	87.7 ± 1.7	1.9	2.5	3.0
		4000.0	90.6 ± 1.5	1.7	4.0	4.1
	Pigeon whole egg	8.0	82.3 ± 1.3	1.6	3.3	6.7
		1000.0	83.5 ± 1.1	1.3	3.1	4.6
		2000.0 [a]	87.7 ± 2.2	2.5	3.7	4.8
		4000.0	88.7 ± 1.3	1.5	2.5	3.7
	Quail whole egg	7.6	85.1 ± 1.4	1.6	3.1	6.0
		1000.0	85.3 ± 1.6	1.9	3.2	4.4
		2000.0 [a]	87.1 ± 1.9	2.2	3.0	3.9
		4000.0	94.8 ± 1.1	1.2	2.5	3.5
Lincomycin	Hen whole egg	8.4	82.7 ± 1.3	1.6	4.6	5.7
		25.0	85.4 ± 2.4	2.8	5.0	5.6
		50.0 [a]	84.8 ± 1.6	1.9	3.1	4.5
		100.0	87.1 ± 1.9	2.2	3.2	4.9
	Duck whole egg	6.5	81.8 ± 1.1	1.3	4.3	4.7
		25.0	86.7 ± 2.5	2.9	3.7	5.4
		50.0 [a]	85.4 ± 2.6	3.0	2.7	3.7
		100.0	86.1 ± 2.7	3.1	4.6	5.0
	Goose whole egg	8.5	80.1 ± 1.5	1.9	2.2	4.1
		25.0	84.0 ± 2.6	3.1	3.4	4.1
		50.0 [a]	87.9 ± 1.1	1.3	2.3	3.7
		100.0	93.5 ± 1.8	1.9	3.5	3.9
	Pigeon whole egg	9.5	81.4 ± 1.6	2.0	3.9	6.5
		25.0	82.2 ± 2.1	2.6	2.3	3.1
		50.0 [a]	86.2 ± 2.2	2.6	5.6	6.7
		100.0	91.3 ± 1.6	1.8	3.0	3.9
	Quail whole egg	8.2	80.0 ± 1.3	1.6	2.6	3.3
		25.0	86.7 ± 2.1	2.4	3.9	3.7
		50.0 [a]	86.4 ± 1.5	1.7	2.6	4.4
		100.0	87.3 ± 1.1	1.3	3.5	5.5

Note: [a] Maximum residue limit (MRL). RSD, relative standard deviation.

Table 5. Recovery and precision for spectinomycin and lincomycin spiked in blank poultry eggs (egg albumen and yolk, $n = 6$).

Analyte	Matrix	Spike Level (µg/kg)	Recovery (%)	RSD (%)	Intraday RSD (%)	Interday RSD (%)
Spectinomycin	Hen egg albumen	6.4	83.8 ± 2.4	2.9	3.0	4.6
		1000.0	84.2 ± 2.3	2.7	2.9	3.7
		2000.0 [a]	86.6 ± 2.0	2.3	2.8	3.6
		4000.0	93.0 ± 1.9	2.0	2.2	4.3
	Yolk	5.6	83.9 ± 2.5	3.0	3.5	3.7
		1000.0	84.1 ± 1.7	2.0	3.0	3.4
		2000.0 [a]	86.6 ± 2.1	2.4	3.7	5.5
		4000.0	94.5 ± 1.4	1.5	2.6	3.8
	Duck egg albumen	7.9	83.6 ± 1.3	1.6	2.7	4.7
		1000.0	84.7 ± 1.5	1.8	2.0	2.2
		2000.0 [a]	85.7 ± 2.2	2.6	3.0	3.9
		4000.0	87.3 ± 3.0	3.4	5.2	6.5
	Yolk	8.0	82.9 ± 1.9	2.3	2.8	3.5
		1000.0	85.3 ± 2.1	2.5	3.9	4.2
		2000.0 [a]	87.1 ± 1.3	1.5	2.8	3.3
		4000.0	90.3 ± 2.4	2.7	4.3	6.2
	Goose egg albumen	7.8	80.4 ± 1.5	1.9	2.0	3.7
		1000.0	84.0 ± 2.5	3.0	3.3	6.3
		2000.0 [a]	83.4 ± 1.4	1.7	2.5	3.1
		4000.0	88.1 ± 1.8	2.0	3.9	4.1
	Yolk	6.7	81.6 ± 2.4	2.9	3.1	3.4
		1000.0	82.4 ± 2.0	2.4	2.4	3.2
		2000.0 [a]	85.9 ± 1.2	1.4	2.0	3.5
		4000.0	89.4 ± 1.9	2.1	3.1	4.9
Lincomycin	Hen egg albumen	6.7	82.2 ± 1.2	1.5	2.8	3.4
		25.0	85.2 ± 1.6	1.9	3.2	4.1
		50.0 [a]	85.8 ± 2.2	2.6	3.4	4.5
		100.0	86.5 ± 2.5	2.9	3.5	5.7
	Yolk	5.9	82.7 ± 1.4	1.7	3.9	4.8
		25.0	85.4 ± 2.0	2.3	3.0	3.6
		50.0 [a]	84.8 ± 2.0	2.4	2.6	3.8
		100.0	91.2 ± 2.1	2.3	4.1	5.5
	Duck egg albumen	7.3	84.9 ± 1.5	1.8	1.9	3.5
		25.0	91.2 ± 2.4	2.6	6.0	5.5
		50.0 [a]	93.4 ± 2.3	2.5	3.7	4.3
		100.0	95.1 ± 2.2	2.3	3.2	4.9
	Yolk	8.0	85.2 ± 1.3	1.5	3.2	5.3
		25.0	86.5 ± 0.9	1.0	2.5	3.6
		50.0 [a]	87.9 ± 2.1	2.4	3.8	4.1
		100.0	89.1 ± 2.3	2.6	3.5	3.7
	Goose egg albumen	9.0	80.8 ± 1.6	2.0	2.8	5.3
		25.0	85.4 ± 1.2	1.4	2.6	4.2
		50.0 [a]	89.1 ± 1.1	1.2	2.7	3.5
		100.0	92.1 ± 1.2	1.3	3.9	3.6
	Yolk	9.2	80.2 ± 1.9	2.4	4.8	4.4
		25.0	85.6 ± 2.1	2.5	4.3	3.9
		50.0 [a]	84.6 ± 2.5	3.0	4.0	4.1
		100.0	87.9 ± 1.6	1.8	2.7	4.1

Note: [a] MRL.

Blank matrix extracts of the hen, duck and goose egg (whole egg, albumen and yolk) samples as well as of the pigeon and quail egg (whole egg) samples were prepared, and the spectinomycin and lincomycin standard working solutions were added to the blank matrix extract, derivatized and detected by GC-MS/MS. The concentrations corresponding to signal-to-noise (S/N) ratios of 3 and 10 for the target compounds were set as the limit of detection (LOD) and LOQ, respectively, of the target compounds in the hen, duck and goose egg (whole egg, albumen and yolk) samples as well as in the pigeon and quail egg (whole egg) samples. As shown in Table 3, the LODs of spectinomycin and lincomycin in the hen, duck, goose, pigeon and quail egg (whole egg) samples were 3.1, 3.5, 3.5, 4.0 and 3.8 µg/kg and 3.1, 2.8, 3.5, 3.9 and 4.3 µg/kg, respectively, and the LOQs of spectinomycin and lincomycin in the same poultry egg samples were 6.0, 6.3, 7.1, 8.0 and 7.6 µg/kg and 8.4, 6.5, 8.5, 9.5 and 8.2 µg/kg, respectively. The results for the LOD and LOQ of spectinomycin and lincomycin in the hen, duck and goose egg (albumen and yolk) samples are shown in Table 3. These LOQs and LOQs are relatively low, and the method is therefore highly sensitive and accurate.

3.5. Comparison of Different Detection Methods

Various analytical methods, including HPLC-FLD [16], HPLC-UVD [18], HILIC-MS/MS [22], HPLC-MS [23], HPLC-MS/MS [24,27], GC-NPD [30] and GC-MS [30], have been used to detect spectinomycin and lincomycin in meat, milk, feedstuffs, honey and animal tissues as well as in swine, calf and chicken plasma. Negarian et al. [18] established an HPLC-UVD method that showed better recovery (80.0%–89.0%) and precision (3.0%–3.9%) for the detection of lincomycin in milk and used CSMISPE to extract and clean up milk samples. Sin et al. [24] developed an LLE method to extract lincomycin from animal tissues and bovine milk. The average recoveries of lincomycin from animal tissues and bovine milk samples were 93.9%–107%, with a precision of 1.3%–7.8%. The LODs and LOQs of this method were 1.5–8.8 µg/kg and 25.0–50.0 µg/kg, respectively. Juan et al. [27] reported an ASE-HPLC-MS/MS method for the simultaneous determination of macrolide and lincosamide antibiotics in meat and milk. ASE is an automated technology that uses solvents at a relatively high pressure and a temperature below the critical points. Compared with LLE and SPE, ASE improves the work efficiency and reduces the amount of extractant required for analysis. Tao et al. [30] established GC-NPD and GC-MS methods for the determination of spectinomycin and lincomycin residues in animal tissues. Animal tissue samples were extracted by ASE, cleaned up with SPE cartridges and detected by GC-NPD and GC-MS. The average recoveries with the GC-NPD and GC-MS methods were 73.0%–97.0% and 70.0%–93.0%, and the RSDs were less than 17% and 21%, respectively. We compared the analysis time, sensitivity and recovery for spectinomycin and lincomycin analysis using different extraction and detection methods. As shown in Table 6, HPLC or GC with MS or MS/MS detection yielded higher sensitivity and precision than FLD, UVD and NPD.

ASE is an automated extraction technology that effectively improves the work efficiency, and 24 samples can be processed simultaneously in the same batch. In this study, LLE and ASE were used to effectively extract poultry egg samples. However, the LLE method is complicated and time- and reagent-consuming. After comparing sample pretreatment methods, we selected ASE for the extraction of spectinomycin and lincomycin residues from poultry eggs. Moreover, GC-MS/MS has higher sensitivity and precision than GC-MS. In this study, the parameters of ASE and GC-MS/MS were optimized to successfully detect spectinomycin and lincomycin in poultry eggs. The newly developed ASE-GC-MS/MS method provides new techniques and a scientific basis for the detection of spectinomycin and lincomycin residues in poultry eggs.

Table 6. Comparison of the present method with other methods.

Detection Method	Sample Preparation Method	Analyte	Animal-Derived Food	Analysis Time (min)	LOD (µg/kg)	LOQ (µg/kg)	Recovery (%)
HPLC-FLD [16]	SPE	Spectinomycin	Swine, calf and chicken plasma	12.0	-	-	91.0–104
HPLC-UVD [18]	CSMISPE	Lincomycin	Milk	9.0	20.0	80.0	80.0–89.0
HILIC-MS/MS [22]	LLE	Spectinomycin	Feedstuffs	10.0	-	-	80.0–92.0
HPLC-MS [23]	SPE	Lincomycin	Honey	10.0	7.0	10.0	102–105
HPLC-MS/MS [24]	LLE	Lincomycin	Animal tissues and milk	10.0	1.5–8.8	25.0–50.0	93.9–107
HPLC-MS/MS [27]	ASE	Lincomycin	Meat and milk	40.0	5.0–10.0	10.0–15.0	86.0–91.0
GC-NPD [30]	ASE	Spectinomycin Lincomycin	Animal tissues	13.0	8.1–9.4	16.4–21.4	73.0–97.0
GC-MS [30]	ASE	Spectinomycin Lincomycin	Animal tissues	15.0	1.9–3.1	4.7–5.7	70.0–93.0
GC-MS/MS	ASE	Spectinomycin Lincomycin	Poultry eggs	10.8	2.3–4.3	5.6–9.5	80.0–95.7

Note: "-" Not reported. FLD, fluorescence detection; UVD, ultraviolet detection; NPD, nitrogen phosphorus detection; CSMISPE, core-shell molecularly imprinted solid-phase extraction.

3.6. Real Sample Analysis

To evaluate the feasibility and accuracy of the newly developed method, we analyzed real samples using ASE-GC-MS/MS. One hundred and fifty commercial poultry eggs (30 hen eggs, 30 duck eggs, 30 goose eggs, 30 pigeon eggs and 30 quail eggs) were purchased from a local supermarket. Each poultry egg sample was processed in accordance with the sample pretreatment method described above and labeled, and each sample was detected and analyzed by the GC-MS/MS method. The target compounds were not detected in duck, goose, pigeon and quail eggs; only hen eggs were found to contain lincomycin residues (11.5 μg/kg less than the MRL). Therefore, the developed ASE-GC-MS/MS method can be applied to quantify spectinomycin and lincomycin in poultry egg samples.

4. Conclusions

In this study, we successfully developed a rapid, sensitive and specific ASE-GC-MS/MS method for the determination of spectinomycin and lincomycin residues in poultry egg samples. ASE is a promising technique for the preparation of animal-derived food samples. The developed method is accurate, has high recovery and precision, and fulfills the validation requirements of the Ministry of Agriculture of the People's Republic of China, the EU and the FDA. The analysis of real samples showed that this new method is feasible and can detect spectinomycin and lincomycin residues in poultry egg samples.

Author Contributions: Conceptualization, K.X.; Data Curation, B.W. and Y.W.; Formal Analysis, G.Z., T.Z. and G.D.; Funding Acquisition, B.W., X.X. and K.X.; Investigation, Y.W.; Methodology, B.W., Y.W., X.X. and K.X.; Resources, X.X. and Z.D.; Software, Z.D. and G.D.; Validation, Z.D., G.Z. and T.Z.; Writing of Original Draft, B.W. and Y.W.; and Writing of Review & Editing, B.W. All authors have read and agreed to the published version of the manuscript.

Funding: This research was financially supported by the China Agriculture Research System (CARS-41-G23), the Priority Academic Program Development of Jiangsu Higher Education Institutions (PAPD), the National Natural and Science Foundation of China (31800161), the Natural Sciences Foundation of Jiangsu Province (BK20180297), the Yangzhou University High-End Talent Support Program and the Yangzhou University International Academic Exchange Foundation.

Conflicts of Interest: The authors declare no conflict of interest.

References

1. Butler, M.M.; Waidyarachchi, S.L.; Connolly, K.L.; Jerse, A.E.; Chai, W.; Lee, R.E.; Kohlhoff, S.A.; Shinabarger, D.L.; Bowlin, T.L. Aminomethyl spectinomycins as therapeutics for drug-resistant gonorrhea and chlamydia coinfections. *Antimicrob. Agents Chemother.* **2018**, *62*, e00325-18. [CrossRef] [PubMed]

2. Lin, A.H.; Murray, R.W.; Vidmar, T.J.; Marotti, K.R. The oxazolidinone eperezolid binds to the 50S ribosomal subunit and competes with binding of chloramphenicol and lincomycin. *Antimicrob. Agents Chemother.* **1997**, *41*, 2127–2131. [CrossRef] [PubMed]

3. Gao, P.; Hou, Q.; Kwok, L.-Y.; Huo, D.; Feng, S.; Zhang, H. Effect of feeding Lactobacillus plantarum P-8 on the faecal microbiota of broiler chickens exposed to lincomycin. *Sci. Bull.* **2017**, *62*, 105–113. [CrossRef]

4. Yu, Y.; Fang, J.-T.; Zheng, M.; Zhang, Q.; Walsh, T.R.; Liao, X.-P.; Sun, J.; Liu, Y.-H. Combination therapy strategies against multiple-resistant Streptococcus suis. *Front. Pharmacol.* **2018**, *9*, 489. [CrossRef]

5. Pokrant, E.; Maddaleno, A.; Lobos, R.; Trincado, L.; Lapierre, L.; San Martín, B.; Cornejo, J. Assessing the depletion of lincomycin in feathers from treated broiler chickens: A comparison with the concentration of its residues in edible tissues. *Food Addit. Contam.* **2019**, *36*, 1647–1653. [CrossRef]

6. Verrette, L.; Fairbrother, J.M.; Boulianne, M. Effect of cessation of ceftiofur and substitution with lincomycin-spectinomycin on extended-spectrum-β-lactamase/AmpC genes and multidrug resistance in escherichia coli from a canadian broiler production pyramid. *Appl. Environ. Microbiol.* **2019**, *85*, e00037-19. [CrossRef]

7. Ministry of Agriculture of the People's Republic of China. *Maxium Residue Level of Veterinary Drugs in Food of Animal Origin*; Notice no. 235 (Appendix 4); Ministry of Agriculture of the People's Republic of China: Beijing, China, 2002.

8. The European Medicines Agency. Pharmacologically Active Substances and their Classification Regarding Maximum Residue Limits in Foodstuffs of Animal Origin. In *Commission Regulation (EU)*; No. 37/2010 of 22 December 2009; The European Medicines Agency: Amsterdam, The Netherlands, 2010.

9. U.S. Food and Drug Administration. *CFR-Code of Federal Regulations Title 21 Part 556 Tolerances for Residue of New Animal Drugs in Food*; U.S. Food and Drug Administration: Rockville, MD, USA, 2014.

10. The Japan Food Chemical Research Foundation. *Maximum Residue Limits (MRLs) List of Agricultural Chemicals in Foods*; The Japan Food Chemical Research Foundation: Tokyo, Japan, 2015.

11. Medina, M.B. Development of a fluorescent latex immunoassay for detection of a spectinomycin antibiotic. *J. Agric. Food Chem.* **2004**, *52*, 3231–3236. [CrossRef]

12. Kowalski, P.; Konieczna, L.; Olędzka, I.; Plenis, A.; Bączek, T. Development and validation of electromigration technique for the determination of lincomycin and clindamycin residues in poultry tissues. *Food Anal. Methods* **2014**, *7*, 276–282. [CrossRef]

13. Cao, S.; Song, S.; Liu, L.; Kong, N.; Kuang, H.; Xu, C. Comparison of an enzyme-linked immunosorbent assay with an immunochromatographic assay for detection of lincomycin in milk and honey. *Immunol. Investig.* **2015**, *44*, 438–450. [CrossRef]

14. Schermerhorn, P.G.; Chu, P.-S.; Kijak, P.J. Determination of spectinomycin residues in bovine milk using liquid chromatography with electrochemical detection. *J. Agric. Food Chem.* **1995**, *43*, 2122–2125. [CrossRef]

15. Szúnyog, J.; Adams, E.; Liekens, K.; Roets, E.; Hoogmartens, J. Analysis of a formulation containing lincomycin and spectinomycin by liquid chromatography with pulsed electrochemical detection. *J. Pharm. Biomed. Anal.* **2002**, *29*, 213–220. [CrossRef]

16. Haagsna, N.; Scherpenisse, P.; Simmonds, R.J.; Wood, S.A.; Rees, S.A. High-performance liquid chromatographic determination of spectinomycin in swine, calf and chicken plasma using post-column derivatization. *J. Chromatogr. B Biomed. Sci. Appl.* **1995**, *672*, 165–171. [CrossRef]

17. Hamamoto, K.; Mizuno, Y.; Koike, R.; Yamaoka, R.; Takahashi, T.; Takahashi, Y. Residue analysis of spectinomycin in tissues of chicken and swine by HPLC. *J. Food Hyg. Soc. Jpn.* **2003**, *44*, 114–118. [CrossRef]

18. Negarian, M.; Mohammadinejad, A.; Mohajeri, S.A. Preparation, evaluation and application of core–shell molecularly imprinted particles as the sorbent in solid-phase extraction and analysis of lincomycin residue in pasteurized milk. *Food Chem.* **2019**, *288*, 29–38. [CrossRef]

19. Wang, M.J.; Hu, C.Q. Analysis of spectinomycin by HPLC with evaporative light-scattering detection. *Chromatographia* **2006**, *63*, 255–260. [CrossRef]

20. Zhou, J.; Zhang, L.; Wang, Y.; Yan, C. HPLC-ELSD analysis of spectinomycin dihydrochloride and its impurities. *J. Sep. Sci.* **2011**, *34*, 1811–1819. [CrossRef]

21. Peru, K.M.; Kuchta, S.L.; Headley, J.V.; Cessna, A.J. Development of a hydrophilic interaction chromatography–mass spectrometry assay for spectinomycin and lincomycin in liquid hog manure supernatant and run-off from cropland. *J. Chromatogr. A* **2006**, *1107*, 152–158. [CrossRef]

22. Molognoni, L.; de Souza, N.C.; de Sá Ploêncio, L.A.; Micke, G.A.; Daguer, H. Simultaneous analysis of spectinomycin, halquinol, zilpaterol, and melamine in feedingstuffs by ion-pair liquid chromatography–tandem mass spectrometry. *J. Chromatogr. A* **2018**, *1569*, 110–117. [CrossRef]

23. Thompson, T.S.; Noot, D.K.; Calvert, J.; Pernal, S.F. Determination of lincomycin and tylosin residues in honey using solid-phase extraction and liquid chromatography-atmospheric pressure chemical ionization mass spectrometry. *J. Chromatogr. A* **2003**, *1020*, 241–250. [CrossRef]

24. Sin, D.W.-M.; Wong, Y.-C.; Ip, A.C.-B. Quantitative analysis of lincomycin in animal tissues and bovine milk by liquid chromatography electrospray ionization tandem mass spectrometry. *J. Pharm. Biomed. Anal.* **2004**, *34*, 651–659. [CrossRef]

25. Sin, D.W.M.; Ho, C.; Wong, Y.C.; Ho, S.K.; Ip, A.C.B. Simultaneous determination of lincomycin and virginiamycin M1 in swine muscle, liver and kidney by liquid chromatography–electrospray ionization tandem mass spectrometry. *Anal. Chim. Acta* **2004**, *517*, 39–45. [CrossRef]

26. van Holthoon, F.L.; Essers, M.L.; Mulder, P.J.; Stead, S.L.; Caldow, M.; Ashwin, H.M.; Sharman, M. A generic method for the quantitative analysis of aminoglycosides (and spectinomycin) in animal tissue using methylated internal standards and liquid chromatography tandem mass spectrometry. *Anal. Chim. Acta* **2009**, *637*, 135–143. [CrossRef] [PubMed]

27. Juan, C.; Moltó, J.C.; Mañes, J.; Font, G. Determination of macrolide and lincosamide antibiotics by pressurised liquid extraction and liquid chromatography-tandem mass spectrometry in meat and milk. *Food Control* **2010**, *21*, 1703–1709. [CrossRef]

28. Maddaleno, A.; Pokrant, E.; Yanten, F.; San Martin, B.; Cornejo, J. Implementation and validation of an analytical method for lincomycin determination in feathers and edible tissues of broiler chickens by liquid chromatography tandem mass spectrometry. *J. Anal. Methods Chem.* **2019**, *2019*, 4569707. [CrossRef]

29. Luo, W.; Yin, B.; Ang, C.Y.W.; Rushing, L.; Thompson, H.C. Determination of lincomycin residues in salmon tissues by gas chromatography with nitrogen-phosphorus detection. *J. Chromatogr. B Biomed. Sci. Appl.* **1996**, *687*, 405–411. [CrossRef]

30. Tao, Y.; Chen, D.; Yu, G.; Yu, H.; Pan, Y.; Wang, Y.; Huang, L.; Yuan, Z. Simultaneous determination of lincomycin and spectinomycin residues in animal tissues by gas chromatography-nitrogen phosphorus detection and gas chromatography-mass spectrometry with accelerated solvent extraction. *Food Addit. Contam. A* **2011**, *28*, 145–154. [CrossRef]

31. The European Communities. Commission decision 2002/657/EC of 12 August 2002 implementing council directive 96/23/EC concerning the performance of analytical methods and the interpretation of results. *Off. J. Eur. Communities* **2002**, *221*, 8–36.

32. U.S. Department of Health and Human Services; Food and Drug Administration; Center for Drug Evaluation and Research; Center for Veterinary Medicine. *Guidance for Industry: Bioanalytical Method Validation*; US Department of Health and Human Services: Washington, DC, USA, 2001.

33. National Food Safety Standard. *GB 29685-2013: Determination of Lincomycin, Clindamycin and Spectinomycin in Animal Derived Food by Gas Chromatography-Mass Spectrometry Method*; Ministry of Agriculture of the People's Republic of China: Beijing, China, 2013.

 *foods*

Article

Comparison between Pressurized Liquid Extraction and Conventional Soxhlet Extraction for Rosemary Antioxidants, Yield, Composition, and Environmental Footprint

Mathilde Hirondart [1,2], Natacha Rombaut [1,2], Anne Sylvie Fabiano-Tixier [1,2], Antoine Bily [2,3] and Farid Chemat [1,2,*]

[1] Avignon University, INRAE, UMR408, GREEN Team Extraction, F-84000 Avignon, France; mathilde.hirondart@sigma-clermont.fr (M.H.); rombaut.natacha@gmail.com (N.R.); anne-sylvie.fabiano@univ-avignon.fr (A.S.F.-T.)

[2] ORTESA, LabCom Naturex-Avignon University, F-84000 Avignon, France; antoine.bily@givaudan.com

[3] Naturex-Givaudan, 250 rue Pierre Bayle, BP 81218, CEDEX 9, F-84911 Avignon, France

[*] Correspondence: farid.chemat@univ-avignon.fr; Tel.: +33-(0)4-90-14-44-65; Fax: +33-(0)4-90-14-44-41

Received: 30 March 2020; Accepted: 1 May 2020; Published: 5 May 2020

Abstract: Nowadays, "green analytical chemistry" challenges are to develop techniques which reduce the environmental impact not only in term of analysis but also in the sample preparation step. Within this objective, pressurized liquid extraction (PLE) was investigated to determine the initial composition of key antioxidants contained in rosemary leaves: Rosmarinic acid (RA), carnosic acid (CA), and carnosol (CO). An experimental design was applied to identify an optimized PLE set of extraction parameters: A temperature of 183 °C, a pressure of 130 bar, and an extraction duration of 3 min enabled recovering rosemary antioxidants. PLE was further compared to conventional Soxhlet extraction (CSE) in term of global processing time, energy used, solvent recovery, raw material used, accuracy, reproducibility, and robustness to extract quantitatively RA, CA, and CO from rosemary leaves. A statistical comparison of the two extraction procedure (PLE and CSE) was achieved and showed no significant difference between the two procedures in terms of RA, CA, and CO extraction. To complete the study showing that the use of PLE is an advantageous alternative to CSE, the eco-footprint of the PLE process was evaluated. Results demonstrate that it is a rapid, clean, and environmentally friendly extraction technique.

Keywords: Pressurized liquid extraction; soxhlet; solvent extraction; green analytical chemistry; Rosemary

---

1. Introduction

In the field of raw material extraction, the first challenge consists of determining the potential of the plant matrix that means what can be extracted and valorized. The chemical composition of the plant material may highly vary depending on the local environmental conditions, development stages, plant part, harvesting season, the technique used for drying, and the storage condition. Therefore, for each batch of plant material used for industrial extraction, an analysis has to be performed to determine the amount of available extractives.

In general, an analytical procedure for antioxidants from plants or spices comprises two steps: Extraction (Soxhlet, maceration, percolation) followed by analysis (spectrophotometry, high performance liquid chromatography coupled or not to mass spectrometry (HPLC-MS), gas chromatography coupled or not to mass spectrometry (GC–MS)). Whereas the last step is finished after only 15 to 30 min, extraction takes at least several hours. Conventional Soxhlet extraction (CSE) is the most used method for solid-liquid extraction in natural product chemistry and is a reference procedure for the extraction of fat and oil according to International Organization for Standardization (ISO standards) [1–3]. It has several disadvantages such as long operation time requiring a minimum of hours or days, large solvent volumes involved, time and energy consuming for the concentration step by evaporation to recover the final extract, and inadequacy for thermolabile analytes.

Pressurized liquid extraction (PLE) has been intensively studied as an efficient extraction technique to substitute CSE [2,4]. It is based on the ability to perform rapid (less than 30 min) and clean extraction at high pressure and temperature. Various parameters of extraction can be modified to improve extraction performance (solvent, pressure, temperature, time of extraction, etc.) [5–9]. High temperature and pressure increase analytes' solubility and solvent diffusion rate, while solvent viscosity and surface tension decrease, resulting in a drained matrix after extraction [10]. With PLE, extractions can be programmed and automatically run, which is convenient for quality control.

In this study we focused on rosemary (*Rosmarinus Officinalis* L.), which is mostly studied and used in the food industry due to its richness in antioxidants' compounds [11,12], particularly rosmarinic acid (RA), carnosic acid (CA), and carnosol (CO) (Figure 1).

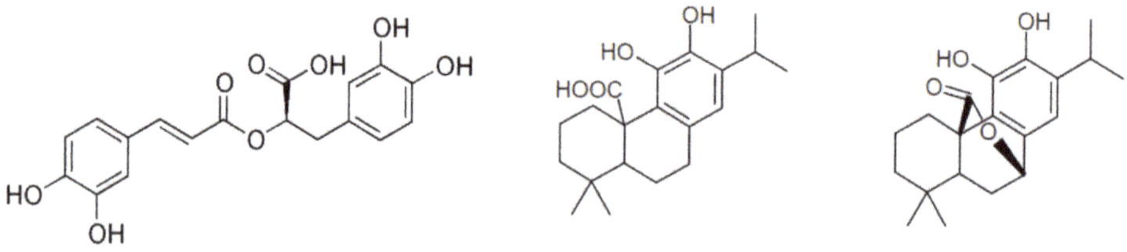

Figure 1. Structures of rosmarinic acid (**a**), carnosic acid, (**b**) and carnosol (**c**).

These compounds are extracted at industrial scale and are dedicated to food applications since the antioxidant extract of rosemary has been authorized in 2010 by the European Union as food additive E392 (directive No. 2010/69/EU). Throughout literature, extraction of chemical compounds from rosemary leaves has been investigated using PLE [13–16]. These studies were mainly focused on maximization of antioxidant activity of rosemary extracts and no complete parametric study of extraction of monitored compounds by PLE has been performed. Additionally, evaluation of the green aspects of PLE is not found in literature.

A major problem in the field of extraction remains the characterization of the raw material studied. The objective of our study was to propose a new method of raw material characterization by optimizing the extraction process in order to be sure to have exhausted the studied raw material. Numerous studies have already been carried out on the extraction of rosemary with innovative technologies such as supercritical fluid extraction (SFE) and pressurized liquid extraction (PLE) coupled with a new quantitative Ultra Performance Liquid Chromatography coupled to Tandem Mass Spectrometry (UPLC-MS/MS) method [13,14]. The difference with the work cited above is that we wanted to propose a green method that could replace Soxhlet in order to optimize the characterization of the raw material studied in the analytical laboratory. The procedure used minimizes the use of organic solvents, which makes it attractive in the analytical field.

In the present work, PLE was studied as a green alternative to Soxhlet extraction of antioxidants from rosemary leaves to extract qualitatively and quantitatively RA, CA, and CO. PLE was optimized via a response surface methodology and a desirability function, which simultaneously maximized extraction, was used. We ran statistical tests in order to check the reliability and the reproducibility of this new procedure. Finally, the eco-footprint of the PLE process was evaluated to demonstrate that it is a rapid, environmentally friendly, and clean extraction technique.

2. Materials and Methods

2.1. Plant Material and Chemicals

Rosemary leaves (*Rosmarinus officinalis* L.) were provided by the company Naturex (Avignon France), and rosemary leaves were collected in Morocco in 2015. Initial moisture was 8.2 ± 0.2%. Leaves were ground before extraction using a grinder (MF 10 basic, IKA, Staufen, Germany) with a 0.5-mm sieve. Granulometry of the rosemary powder was 610 ± 22 µm.

For the extraction solvent, food grade ethanol 96° *v/v* (Cristalco, FranceAlcools, Paris, France) and demineralized water were used. For HPLC analysis, solvent used were all HPLC grade: Methanol, water, acetonitrile, and tetrahydrofuran. Phosphoric acid 85% ACS grade (according to American Chemical Society specifications) and trifluoroacetic acid 99% were purchased from Sigma-Aldrich, USA. Standards used were rosmarinic acid (Extrasynthese, Genay, France) and carnosic acid (Sigma-Aldrich, St. Louis, MO, USA). Nitrogen used had a purity of 99.999% (Alphagaz 1 Smartop, Air Liquid, Paris, France).

2.2. Extraction Procedures

In this study, a procedure of PLE was developed and optimized for analytical determination of RA and CA contents in rosemary leaves. PLE performance was compared to the reference method of Soxhlet extraction. Those processes are illustrated in Figure 2.

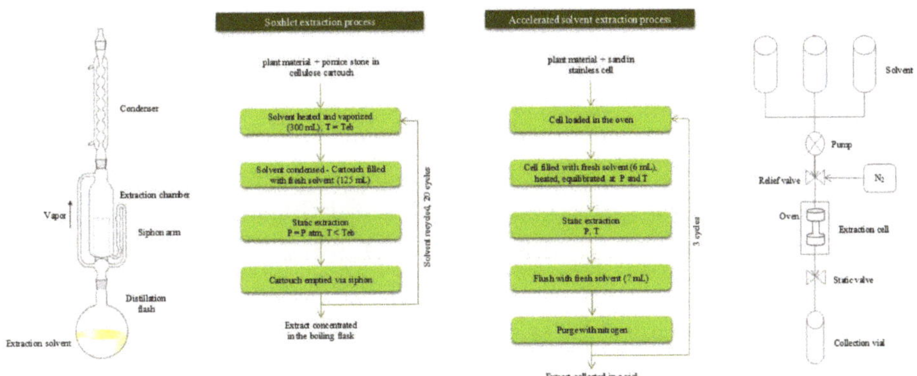

Figure 2. Comparison of Soxhlet and Accelerated Solvent Extraction (ASE) processes.

2.2.1. Reference Procedure: Conventional Soxhlet Extraction (CSE)

For CSE, 10 g of ground rosemary leaves and 5 g of pumice stone were mixed in a 34 × 130 mm cellulose thimble (plugged with cotton in order to avoid transfer of sample particles in the distillation flask) and placed in Soxhlet apparatus with flask containing 300 mL of solvent. Extractions were performed using a solid to liquid ratio of 1 to 12 (g/mL). Extraction was performed during 8 h. After extraction, the extract was concentrated under vacuum (Laborota 4001, Heidolph, Germany) and conserved at 4 °C before analysis. All extractions were done at least in duplicate and the mean values were reported.

2.2.2. Pressurized Liquid Extraction (PLE)

An accelerated solvent extractor ASE200 model was used (Dionex, Thermo Fisher Scientific, Waltham, MA, USA). This apparatus allows extraction of plant material at high pressure (up to 130 bar) and high temperature (up to 200 °C). Preliminary trials were made in order to determine the optimal parameters (loading of the cell, flushing volume, and percentage of dispersant), and will be discussed in the result section. Optimal loading was determined to be 3.1 g of ground rosemary leaves, and 7.3 g of

Fontainebleau sand (VWR Chemicals, Radnor, PA, USA) were homogenized in an 11-mL stainless-steel cell. The cells were equipped with stainless steel frits on both sides, and a cellulose filter at the bottom to obtain a filtered extract. The extraction procedure cycle was done as follows: First, the cell was filled with extraction solvent via an HPLC pump, pressurized, and placed into the preheated oven. Depending on the set extraction temperature, the cell preheating duration was between 5 and 9 min, followed by a static period of extraction. Then, the cell was flushed with fresh solvent (60% of the extraction cell volume) and purged with a flow of nitrogen during 1 min. Several cycles of extraction can be performed to drain active compounds from the plant matrix. Extracts were collected into a glass vial and analyzed without a concentration step. The dry matter content of each extract was determined by drying 5 mL of extract at 130 °C during 3 h, to calculate the mass extraction yield.

Preliminary trials were performed to evaluate the impact of some PLE parameters on extraction performance: Solvent, percentage of dispersant, and flushing volume. For these trials, the other extraction parameters were fixed according to literature [17]: Temperature (T) = 100 °C, Pressure (P) = 80 bar, static time of extraction = 5 min, and 3 cycles of extraction.

2.3. Statistical Analysis

2.3.1. Experimental Design

To investigate the influence of PLE extraction parameters on the extraction of rosemary antioxidants, a response surface methodology was used. Three independent factors, namely the temperature (A), the pressure (B), and the extraction time (C), were studied to evaluate their impact on several responses: The mass yield (%) and the contents in RA, CA, and CO (mg/g). The independent variables, given in Table 1, were coded according to Equation (1):

$$X_i = \frac{x_i - x_{i0}}{\Delta x_i} \tag{1}$$

where X_i and x_i are, respectively, the dimensionless and the actual values of the independent variable i, x_{i0} is the actual value of the independent variable i at the central point, and Δx_i is the step change of x_i corresponding to a unit variation of the dimensionless value. For the three variables, the design yielded randomized experiments with eight (2^3) factorial points, six axial points ($-\alpha$ and $+\alpha$ (in our case 1.68)) to form a central composite design, and six center points for replications and estimation of the experimental error and to prove the suitability of the model. Coded values of the independent variables are listed in Table 1.

$$Y = \beta_0 + \sum_{i=1}^{2} \beta_i X_i + \sum_{i=1}^{2} \beta_{ii} X_i^2 + \sum_{i} \sum_{j=i+1} \beta_{ij} X_i X_j \tag{2}$$

The responses are related to the coded independent variables X_i and X_j according to the second order polynomial expressed in Equation (2) with β_0 the interception coefficient, β_i the linear terms, β_{ii} the quadratic terms, and β_{ij} the interaction terms. Fisher's test for analysis of variance (ANOVA) performed on experimental data was used to assess the statistical significance of the proposed model. The experimental design was analyzed using the software Statgraphics (StatPoint Technologies, Inc., Warrenton, VA, USA) for Windows.

Table 1. Central composite design (CCD) matrix with experimental responses obtained (mass extraction yield and leaf content in rosmarinic acid, carnosic acid, and carnosol).

Variables						Responses			
Temperature		Pressure		Extraction Time		Mass Yield	RA Content	CA Content	CO Content
Actual Value (°C)	Coded Value	Actual Value (bar)	Coded Value	Actual Value (min)	Coded Value	%	mg/g	mg/g	mg/g
160	+1	111.8	+1	27	+1	44.0	11.56	21.76	2.01
160	+1	111.8	+1	9	−1	41.9	11.57	22.54	2.10
160	+1	58.2	-1	27	+1	44.9	11.18	22.32	2.07
160	+1	58.2	−1	9	−1	44.0	11.49	22.62	2.17
70	−1	111.8	+1	27	+1	31.3	11.13	20.55	2.22
70	−1	111.8	+1	9	−1	30.5	10.83	20.49	2.40
70	−1	58.2	−1	27	+1	31.3	11.84	21.20	2.16
70	−1	58.2	−1	9	−1	31.5	11.67	20.68	2.12
115	0	85	0	18	0	34.3	11.29	20.98	2.17
115	0	85	0	18	0	34.5	11.70	20.12	2.17
115	0	85	0	18	0	34.3	11.72	20.63	2.05
115	0	85	0	18	0	34.8	11.71	20.41	2.11
115	0	85	0	18	0	34.3	11.68	20.39	2.18
115	0	85	0	18	0	34.1	12.01	20.34	2.09
39.3	−α	85	0	18	0	26.3	10.56	19.90	2.29
190.7	+α	85	0	18	0	49.6	9.93	22.12	2.22
115	0	40	−α	18	0	34.0	11.32	21.28	2.30
115	0	130	+α	18	0	35.1	12.24	21.35	1.98
115	0	85	0	3	−α	34.5	12.19	21.88	2.07
115	0	85	0	33	+α	34.7	11.99	21.63	2.10

2.3.2. Reproducibility and Statistical Comparison

The optimized PLE method compared to the CSE method was performed for the extraction of antioxidants from rosemary. It consisted of a series of eight successive experiments performed for each extraction procedure. Then the statistical study was performed in two steps: First, the Fisher–Snedecor's test to compare the variability of the results and then the student test in order to compare the mean values obtained by the two different extraction procedures. Those two tests were performed with $\alpha = 0.05$.

2.4. HPLC Analysis

Analyses of RA, CA, and CO were performed by HPLC (Agilent 1100, Agilent Technologies, Santa Clara, CA, USA) equipped with a Diode Array Detector (DAD) detector. HPLC analyses were made according to previously reported procedures without further optimization and specific procedures for each compound are described below [18].

2.4.1. Rosmarinic Acid Analysis

The column used was a C_{18} column (5 μm, 4.6 mm × 250 mm, Zorbax SB, Agilent Technologies, Santa Clara, CA, USA). The mobile phase was composed of 32% acetonitrile and 68% water with 0.1% trifluoroacetic acid (mL/mL) and the flow rate was set at 1 mL/min. The column oven temperature was 20 °C and the run time was 10 min. Five μL were injected. Rosmarinic acid was detected at a wavelength of 328 nm. For quantification of rosmarinic acid in the extract, a calibration curve was calculated by linear regression analysis for rosmarinic acid standard.

2.4.2. Carnosic Acid and Carnosol Analysis

The column used was a C_{18} column (1.8 μm, 4.6 mm × 50 mm, Zorbax Eclipse XBD-C18, Agilent Technologies, France). The mobile phase was isocratic and composed of 0.5% H_3PO_4

(in water)/acetonitrile (35/65, mL/mL), and the flow rate was set at 1.5 mL/min. The column oven temperature was 25 °C and the run time was 15 min. Five µL were injected. Carnosic acid and carnosol were detected at a wavelength of 230 nm. For quantification of carnosic acid in the extract, a calibration curve was calculated by linear regression analysis for carnosic acid standard. Carnosol was expressed as carnosic acid.

2.5. Calculations

In order to assess the extraction performances of the evaluated processes, mass extraction yield, purity, and content in each compound of interest were calculated. Each mass included in equations below was expressed in dry weight.

$$mass\ extraction\ yield \left(\%, \frac{g}{100g}\right) = \frac{weight\ of\ extract}{weight\ of\ rosemary\ leaves} \times 100 \tag{3}$$

$$purity \left(\%, \frac{g}{100g}\right) = \frac{weight\ of\ RA,\ CA\ or\ CO}{weight\ of\ extract} \times 100 \tag{4}$$

$$content\ in\ RA,\ CA\ and\ CO\ (mg/g\ rosemary) = \frac{purity \times weight\ of\ extract}{weight\ of\ rosemary\ leaves} \tag{5}$$

3. Results and Discussion

3.1. Pressurized Liquid Extraction (PLE): Preliminary Study

3.1.1. Solvent Evaluation of Ethanol/Water Ratio on Extraction Efficiency

To determine the solvent that maximized the extraction of both RA and CA, PLE was performed with various percentages of ethanol in water: 0, 20, 40, 60, 80, and 100% (g/g). Hydro-alcoholic solutions as solvent offer many advantages. Indeed, they can solubilize both hydrophilic and lipophilic active compounds. To test different ethanolic solvent proportions, the flushing volume was fixed at 60% mL/mL of the extraction cell. The extraction cell was filled with 30% g/g of ground rosemary leaves and 70% g/g of Fontainebleau sand.

The influence of ethanol proportion in the extraction solvent on extract composition is reported in Figure 3.

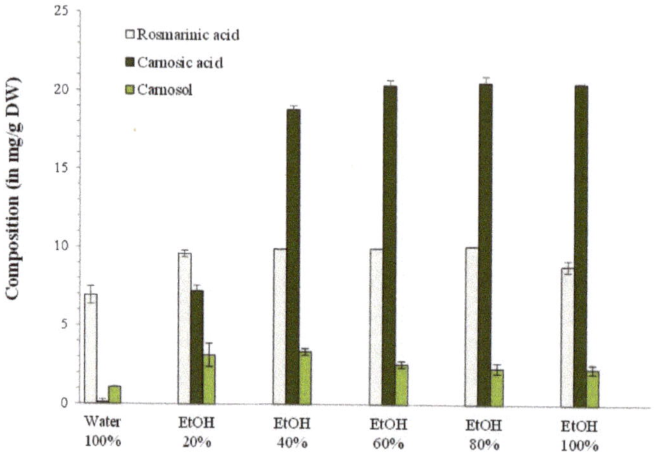

Figure 3. Influence of different ratios of ethanol/water as extraction solvent on the antioxidants composition of Pressurized Liquid Extraction (PLE) extracts of rosemary.

At low ethanol percentage (0 and 20%), RA was extracted but no CA, while from 40% ethanol, CA was extracted as well. These results showed that the solvent maximizing both RA and CA extraction was 80% ethanol, with 10.13 ± 0.02 mg RA/g rosemary and 20.6 ± 0.4 mg CA/g rosemary extracted. Maximal extraction and solubilization of both compounds was possible with 80% ethanol thanks to its intermediate polarity, despite the different chemical structures of RA and CA. Indeed, RA is a caffeic acid ester [19], rather hydrophilic, so preferentially extracted and solubilized in solvents that are relatively polar, as ethanol [20]. CA is a phenolic diterpene [21] and is relatively lipophilic, but still soluble in intermediate polarity solvents such as acetone or ethanol [22,23].

3.1.2. Dispersant

Fontainebleau sand was used as a dispersant in order to favor a uniform distribution of sample and maximize the extraction yield. It was mixed with ground rosemary leaves in different proportions to quantify the impact of dispersant on extraction yield. Trials were carried out with 30, 50, 70, and 90% g/g of dispersant. This parameter is usually fixed or not specified in literature, suggesting that it is not impacting the extraction performances. In this study, we measured its impact only on mass yield to verify this hypothesis. The flushing volume was fixed at 60% mL/mL of the extraction cell, and 80% ethanol was used as extraction solvent. It can be seen in Figure 4 that the proportion of dispersant in the cell had a small impact on extraction mass yield, which varied between 31 ± 2% and 37 ± 2%. The 70% dispersant must be selected to maximize the mass yield (37 ± 2%).

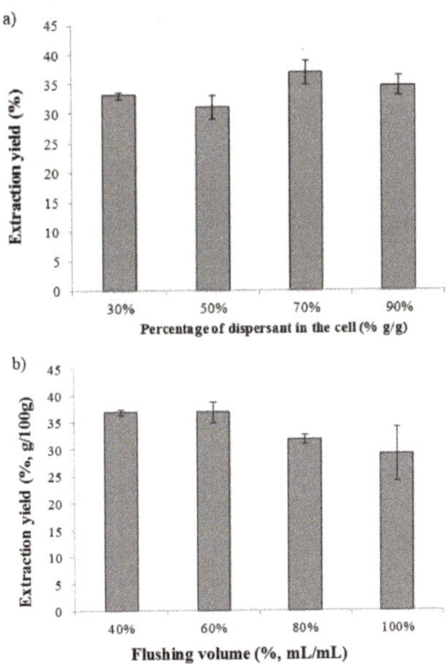

Figure 4. Influence of dispersant proportion in the extraction cell (**a**) and flushing volume (**b**) on the mass extraction yield of PLE of rosemary leaves. The bars with range mean standard deviation between three experiments.

3.1.3. Flushing Volume

The flushing volume is the amount of fresh solvent injected during PLE after static extraction (Figure 2). It is measured as a percentage mL/mL of the extraction cell volume (11 mL). Extractions were performed with 40, 60, 80, and 100% mL/mL (Figure 4).

As for the proportion of dispersant, the flushing volume is usually fixed in literature [14,15]. In this study we measured its impact on extraction mass yield. A flushing volume of 60% mL/mL maximized the extraction mass yield (36.9 ± 2.0%), while higher flushing volume decreased it. A 60% flushing volume was commonly applied throughout literature, which confirms the results obtained [14].

3.2. PLE Extraction: Experimental Design and Statistical Analysis

Three variables that could impact extraction efficiency of antioxidants from rosemary by PLE were studied in a central composite design, namely, temperature (A), pressure (B), and extraction time (C). The choice of the A and B variation domain was selected considering the limits of the ASE equipment. A range from 40 to 190 °C was chosen for the temperature (A), and a range from 40 to 130 bar for the pressure (B). This wide temperature range was chosen to thoroughly evaluate temperature impact on extraction. Within this range, thermal degradation of compounds could also be assessed. The total extraction duration depends on the duration of equilibration of the cell, which varies from 5 to 9 min according to the temperature of extraction. However, as the temperature is a factor impacting the extraction, we chose to consider the duration of static extraction as an independent variable (C). As we performed three cycles of PLE for each extraction cell (Figure 2), the extraction time (C) was the sum of the three static extraction periods. A range from 3 = 3 × 1 min to 33 = 3 × 11 min of static extraction was selected. The controlled variables were studied in a multivariate study with 20 experiments, as shown in Table 1.

Responses varied greatly as a function of the combination of parameter settings. Mass extraction yield ranged from 26.3 to 49.6%, RA content ranged from 9.93 mg/g rosemary to 12.24 mg/g rosemary, and CA content ranged from 19.90 mg/g rosemary to 22.12 mg/g rosemary (Table 1). Experimentally, the formation of Maillard reaction product at temperature of 150 °C or above occurred as evidenced by the brown color of the extract and the burnt smell. The presence of these toxic compounds in extracts must be avoided, so extractions carried out at temperature in the range of 150–200 °C are usually not recommended [15]. However, in this study extraction was investigated for analytical purpose and sensory characteristics of the extracts was not considered. Moreover, the degradation of targeted compounds did not occur because the RA and CA content in extracts were not lower at 190 °C than at 115 °C (Table 1). As shown in Table 1, there was no significant difference in CO content, which indicates that there was no degradation of CA into CO due to oxidation, as it is suggested in published studies [18,22].

By considering a confidence level of 95%, the linear effects of the temperature (A) as well as all quadratic effects (A^2, B^2, and C^2) were significant (Table 2) with *p*-value below 0.05. There were also interactions between variables A and B in significant scale, with a *p*-value lower than 0.05 (Table 2). Empirical relationships allowed linking responses studied and key variables involved in the model. From ANOVA, the coefficient of determination R^2 was determined to be higher than 80% for the three considered responses.

Table 2. Summary of the ANOVA for the central composite design.

Variables	Responses					
	Mass Yield		**RA Content**		**CA Content**	
	F-Ratio	*p-Value*	*F-Ratio*	*p-Value*	*F-Ratio*	*p-Value*
Temperature (A)	445.98	0	0.31	0.5871	106.25	0
Pressure (B)	0.26	0.6225	0.12	0.7389	2.12	0.1756
Extraction time (C)	0.86	0.3743	0.04	0.8487	0.89	0.3685
A^2	28.41	0.0003	32.84	0.0002	9.79	0.0107
A × B	0.38	0.5508	5.17	0.0463	0.07	0.7911
A × C	0.55	0.4758	0.57	0.4661	4.74	0.0546
B^2	1.81	0.208	0.19	0.6738	21.88	0.0009
B × C	0.46	0.5125	0.29	0.6002	1.18	0.3020
C^2	1.97	0.1904	3.50	0.0907	49.02	0
Error R^2 (%)	97.9518		82.1394		94.8633	
R^2 adjusted for d.f (%)	96.1084		66.0649		90.2402	
Optimal conditions predicted	A = 190 °C B = 40 bar C = 33 min		A = 100 °C B = 40 bar C = 3 min		A = 190 °C B = 130 bar C = 4 min	

Three-dimensional surface responses of a multiple nonlinear regression model (Figure 5) illustrate the linear and quadratic effects together with the interaction effects on the responses given in Table 1. Figure 5 highlights the behavior of the three responses as a function of two variables: Temperature (A) and pressure (B). In each plot, the extraction time (C) was fixed at the central value ("0"). The most influential effect was the linear terms of temperature (A) as can be seen in Table 2, with low *p*-values: 0, 0.5871, 0 for mass yield, RA content, and CA content, respectively.

As expected, the model confirmed that mass extraction yield increased with temperature (B). Influence of quadratic terms given in Table 2 is illustrated in Figure 5 by observation of the surface curvatures of the plots. Optimal settings for the maximization of each response are presented in Table 2. An optimization of the desirability was carried out to obtain optimal factors' settings for the multi-responses' maximization. The settings which simultaneously maximized mass yield, RA content, and CA content were: Temperature A = 183 °C, pressure B = 130 bar, extraction time C = 3 × 1 min.

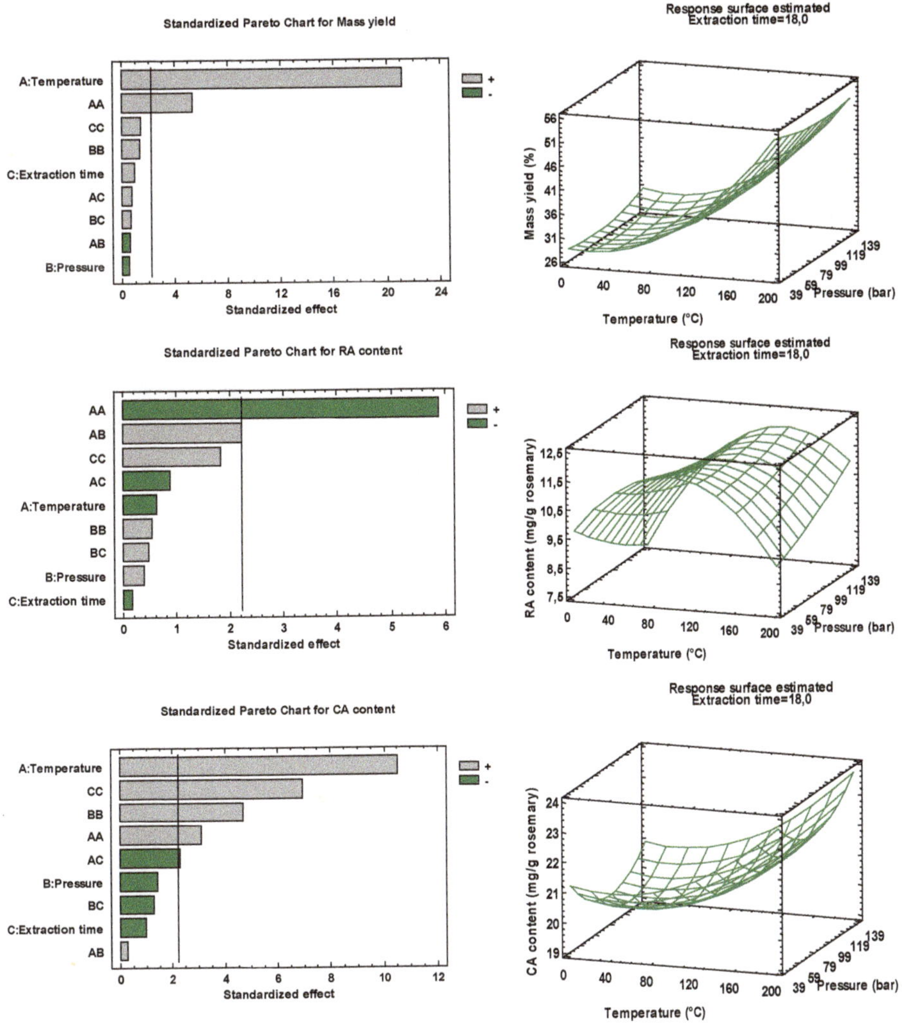

Figure 5. Standardized Pareto charts and response surfaces estimated for the optimization of PLE parameters.

3.3. Statistical Comparison of CSE and PLE

In order to assess the reliability of the PLE extraction technique to replace CSE for raw material active content determination, statistical analysis was performed on repeated trials. For this purpose, 16 experiments were run, 8 with the conventional Soxhlet technique and 8 with the optimized PLE extraction (Table 3). The RA, CA, and CO contents (mg/g rosemary) were analyzed and reported.

The mean value of RA content obtained with PLE was similar to the value obtained using CSE (10 ± 1 and 9.9 ± 0.5 mg/g rosemary, respectively) and the mean value of CA content obtained with PLE was higher than the value obtained using CSE (21 ± 1 and 17.7 ± 0.9 mg/g rosemary, respectively). The relative standard deviations were 10% for PLE and CSE, which means that the results were little dispersed around the mean value. This was confirmed using the Fisher–Snedecor's test, which gave no significant difference in the dispersion of the results for the two different processes ($\alpha = 0.05$) for both RA and CA content. The student test was then applied in order to check if there was any significant difference between the mean values of RA and CA content obtained by the two processes. The tabular value obtained by the student test table for $\alpha = 0.05$ was 2.10 and the calculated value was 0.35 for RA content and 1.83×10^{-6} for CA content, which meant that there was no significant difference between the mean values from a statistical point of view. The statistical tests validated the PLE technique as a good alternative to the conventional one for the determination of active compounds' content in rosemary.

Extraction performance of CSE and PLE are compared in Table 3 in terms of mass extraction yield and active contents. Mass yield of optimized PLE extraction ($47.6 \pm 0.5\%$) was higher than mass yield obtained with CSE ($26 \pm 1\%$). High pressure and high temperature generated during PLE enabled extracting more compounds from the plant material [16]. Extraction of active compounds such as RA and CA was improved with PLE (Table 3). The use of drastic extraction conditions during PLE did not lead to the degradation of compounds, even the most thermosensitive such as CA. This absence of degradation could be explained by the absence of oxygen during PLE due to nitrogen flushing and a short contacting duration between the solvent and the matrix (around 45 min against 8 h with CSE). A higher CO content was obtained with CSE than with PLE, respectively 4.4 ± 0.8 mg/g rosemary and 1.9 ± 0.3 mg/g rosemary, suggesting the degradation of CA into CO during CSE (Table 3). Due to higher extraction yields achieved with PLE, purity of the active compounds in extracts was lower, this extraction technique was less selective regarding these compounds. However, the goal of this analytical technique was not to reach high purities of RA and CA in extracts, but to drain completely the plant material. High purity in RA and CA in the final extract was not considered as a response to maximize. PLE with optimized conditions seems to be a good technique to quantitatively extract RA, CA, and CO from ground rosemary leaves, with better performance than CSE in term of active content yields.

Table 3. Reproducibility of extraction and statistical comparison test between Pressurized Liquid Extraction (PLE) and Conventional Soxhlet Extraction (CSE).

Experiments		1	2	3	4	5	6	7	8	Mean (%)	SD (%)	RSD (%)	Variance S^2	F_{CAL}	F_{TAB}	t_{CAL}	t_{TAB}
Extraction yield (%)	CSE	26.5	26.2	28.9	26.3	28.2	28.3	28.7	30.2	27.91	1.45	5.21	1.85	0.31		2.55×10^{-14}	
	PLE	47.8	44.7	46.7	46.2	46.5	45.5	45.3	46.7	46.20	0.97	2.11	0.83				
RA content (mg/g)	CSE	9.71	9.23	9.98	9.14	9.70	10.39	10.23	10.49	9.86	0.51	5.13	0.22	0.11		0.35	
	PLE	11.74	10.88	9.15	9.95	8.59	9.96	9.87	9.92	10.01	0.97	9.65	0.82		3.79		2.1
CA content (mg/g)	CSE	18.18	17.08	17.38	17.56	17.61	16.16	18.67	18.78	17.68	0.86	4.88	0.65	0.69		1.83×10^{-6}	
	PLE	22.46	20.18	22.38	20.62	21.26	19.99	20.25	21.88	21.13	1.01	4.78	0.89				
CO content (mg/g)	CSE	3.51	4.29	4.34	5.64	4.25	5.36	3.62	4.14	4.39	0.75	17.14	0.50	0.01		1.67×10^{-7}	
	PLE	2.20	1.72	1.69	2.00	1.68	1.96	1.50	2.16	1.86	0.25	13.58	0.06				

SD, standard deviation; RSD, relative standard deviation; F_{CAL}, F value calculated using Fisher-Snedecor's test; F_{TAB}, F value tabulated for $\alpha = 0.05$ and 7 of freedom; $t_{CAL} = t$ value calculated using student's test; $t_{TAB} = t$ value tabulated for $\alpha = 0.05$ and 14 degrees of freedom.

3.4. Eco-Footprint: CSE vs. PLE Processes

The two extraction techniques were evaluated according to the six principles of green extraction developed by Chemat et al. [24]. The six parameters considered were calculated as follows:

- Raw material (Principle 1): Mass of plant material required for an analysis (in g).
- Solvent (Principle 2): Mass of solvent required for an analysis (in g).
- Energy (Principle 3): Energy consumption for the analysis considering extraction and evaporation steps based on the energy transfer equation [25–27] (in kWh).
- By-products (Principle 4): Amount of waste generated by an analysis (solvent and plant material) (in g).
- Process (Principle 5): Time of an analysis including steps of preparation, extraction, evaporation, and cleaning (in h).
- Product recovery (Principle 6): (Mass of final product recovered) / (mass of available product in the plant material) (in %).

In Figure 6, it is important to notice that for each principle, a value close to the center is a positive result whereas a value far from the center corresponds to a negative result. Thus, for "Product recovery", the center corresponds to a maximum of actives extracted.

Compared to CSE, PLE enabled reducing extraction time by a factor 8, from 9 h 40 min to 1 h 10 min. As well, PLE required less solvent, around 50 mL against 300 mL for CSE. Waste of solvent is a real problem in analytical laboratories because usually solvents are not recycled. It is all the more important to minimize the amount of solvent required for an analysis. Thus, during PLE less waste is produced by an analysis in terms of solvent and spent residue. In PLE extraction, less raw material was needed (3 g against 10 g for CSE).

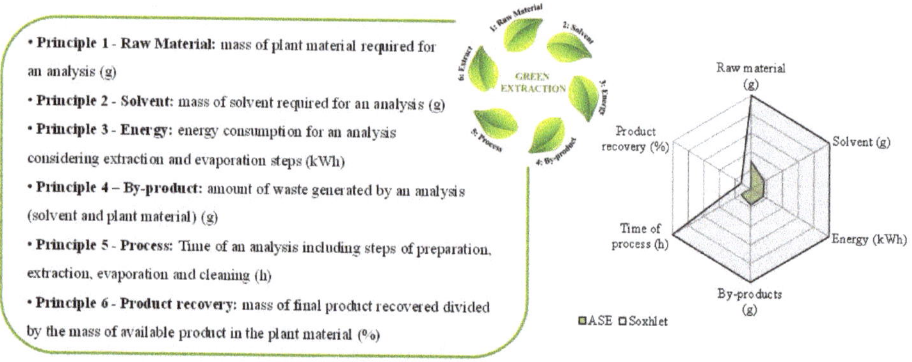

Figure 6. Eco-footprint of PLE vs. CSE processes.

More than the economical aspect, it can be very practical in a sourcing demarche where, regularly, only few quantities of raw material are available. Energy consumption was lower for PLE extraction, because even if the extraction temperature was higher (183 °C instead of 78 °C for CSE), less solvent had to be heated (50 mL against 300 mL), and there were only 3 heating cycles during PLE against 20 cycles during CSE. Another positive aspect of PLE compared to CSE was the percentage of product recovery (100% for PLE and 87.7% for CSE). Higher pressure and temperature during PLE allowed extracting more actives, and their degradation was avoided thanks to the absence of oxygen in the system and the short time of contact between the matrix and the solvent. Finally, the reduced cost of extraction was advantageous for the PLE method in terms of time, amount of raw material and solvent, product recovery, and waste generated. The eco-footprint of PLE was 33 times lower than CSE, with 2.96 area units for PLE against 100 area units for CSE, represented in Figure 6. The implementation of this technique in industrial quality control laboratories could be advantageous compared to CSE in terms of capital expenses and economic savings.

4. Conclusions

Optimization of PLE was carried out using a central composite design methodology. Maximization of extraction was obtained combining three PLE parameters: Temperature (183 °C), pressure (130 bar), and static duration of extraction (3 min). Given the high temperatures tested, carnosol was monitored to follow degradation of carnosic acid and no increase of carnosol was evidenced. To evaluate if PLE could replace CSE, a statistical comparison of extraction performances of the two processes was performed. Ultimately, the eco-footprint of PLE and CSE were determined considering consumption of raw material, solvent, energy, and time. PLE proved to be a rapid, clean, and environmentally friendly technique for determination of active content in plant matrices.

Author Contributions: Conceptualization, M.H., N.R., A.B., A.S.F.-T. and F.C.; methodology, M.H., N.R., A.B., A.S.F.-T. and F.C.; software, M.H., N.R., A.B., A.S.F.-T. and F.C.; validation, M.H., N.R., A.B., A.S.F.-T. and F.C.; formal analysis, M.H., N.R., A.B., A.S.F.-T. and F.C.; investigation M.H., N.R., A.B., A.S.F.-T. and F.C.; resources M.H., N.R., A.B., A.S.F.-T. and F.C.; data curation, M.H., N.R., A.B., A.S.F.-T. and F.C.; writing—original draft preparation, M.H., N.R., A.B., A.S.F.-T. and F.C.; writing—review and editing, M.H., N.R., A.B., A.S.F.-T. and F.C.; visualization, M.H., N.R., A.B., A.S.F.-T. and F.C.; supervision, M.H., N.R., A.B., A.S.F.-T. and F.C.; project administration, M.H., N.R., A.B., A.S.F.-T. and F.C.; funding acquisition, A.B. and F.C. All authors have read and agreed to the published version of the manuscript.

Funding: The authors thank the Agence Nationale de la Recherche (ANR) program for financial contribution to the project ANR Labcom ORTESA (Optimization and Research of Technologies for Extraction and Alternative Solvents).

Acknowledgments: Authors acknowledge Anthony Aldebert and Cindy Gonzalez for their help on the analytical aspects of this work.

Conflicts of Interest: The authors declare no conflicts of interest.

References

1. Luque de Castro, M.; Ayuso, L. Soxhlet extraction. In *Encyclopedia of Separation Science*; Wilson, I., Ed.; Academic Press: San Diego, CA, USA, 2000; pp. 2701–2709.

2. Rodríguez-Solana, R.; Salgado, J.; Domínguez, J.; Cortés-Diéguez, S. Characterization of fennel extracts and quantification of estragole: Optimization and comparison of accelerated solvent extraction and soxhlet techniques. *Ind. Crops Prod.* **2014**, *52*, 528–536. [CrossRef]

3. Wang, L.; Weller, C. Recent advances in extraction of nutraceuticals from plants. *Trends Food Sci. Technol.* **2006**, *17*, 300–312. [CrossRef]

4. Luque de Castro, M.; Priego-Capote, F. 2.05-soxhlet extraction versus accelerated solvent extraction a2. In *Comprehensive Sampling and Sample Preparation*; Pawliszyn, J., Ed.; Academic Press: Oxford, UK, 2012; pp. 83–103.

5. Steyaningsih, W.; Saputro, I.E.; Palma, M.; Barroso, C.G. Pressurized liquid extraction of phenolic compounds from rice (Oryza sativa) grains. *Food Chem.* **2016**, *192*, 452–459. [CrossRef] [PubMed]

6. Feuereisen, M.M.; Barraza, M.G.; Zimmermann, B.F.; Schieber, A.; Schulze-Kaysers, N. Pressurized liquid extraction of anthocyanins and biflavonoids from *Schinus terebinthifolius* Raddi: A multivariate optimization. *Food Chem.* **2017**, *214*, 564–571. [CrossRef] [PubMed]

7. Andreu, V.; Pico, Y. Pressurized liquid extraction of organic contaminants in environmental and food samples. *Trends Anal. Chem.* **2019**, *118*, 709–721. [CrossRef]

8. Saldaña, M.; Valdivieso-Ramírez, C. Pressurized fluid systems: Phytochemical production from biomass. *J. Supercrit. Fluids* **2015**, *96*, 228–244. [CrossRef]

9. Liang, X.; Nielsen, N.J.; Christensen, J.H. Selective pressurized liquid extraction of plant secondary matabolites; *Convallaria majalis* L. as a case. *Anal. Chim. Acta* **2020**, *X4*, 100040.

10. Pereira, D.T.V.; Tarone, A.G.; Cazarin, C.B.B.; Barbero, G.F.; Martinez, J. Pressurized liquid extraction of bioactive compounds from grape marc. *J. Food Eng.* **2019**, *240*, 105–113. [CrossRef]

11. De Oliveira, G.A.; de Oliveira, A.; da Conceição, E.; Leles, M. Multiresponse optimization of an extraction procedure of carnosol and rosmarinic and carnosic acids from rosemary. *Food Chem.* **2016**, *211*, 465–473. [CrossRef]

12. Ribeiro-Santos, R.; Carvalho-Costa, D.; Cavaleiro, C.; Costa, H.; Albuquerque, T.; Castilho, M.; Ramos, F.; Melo, N.; Sanches-Silva, A. A novel insight on an ancient aromatic plant: The rosemary (*rosmarinus officinalis* L.). *Trends Food Sci. Technol.* **2015**, *45*, 355–368. [CrossRef]

13. Borrás, L.; Arráez-Román, D.; Herrero, M.; Ibáñez, E.; Segura-Carretero, A.; Fernández-Gutiérrez, A. Comparison of different extraction procedures for the comprehensive characterization of bioactive phenolic compounds in *rosmarinus officinalis* by reversed-phase high-performance liquid chromatography with diode array detection coupled to electrospray time-of-flight mass spectrometry. *J. Chromatogr. A.* **2011**, *1218*, 7682–7690.

14. Herrero, M.; Plaza, M.; Cifuentes, A.; Ibanez, E. Green processes for the extraction of bioactives from rosemary: Chemical and functional characterization via ultra-performance liquid chromatography-tandem mass spectrometry and in-vitro assays. *J. Chromatogr. A* **2010**, *1217*, 2512–2520. [CrossRef]

15. Hossain, B.; Barry-Ryan, C.; Martin-Diana, A.; Brunton, N. Optimisation of accelerated solvent extraction of antioxidant compounds from rosemary (*rosmarinus officinalis* L.), marjoram (*origanum majorana* L.) and oregano (*origanum vulgare* L.) using response surface methodology. *Food Chem.* **2011**, *126*, 339–346. [CrossRef]

16. Rodríguez-Solana, R.; Salgado, J.; Domínguez, J.; Cortés-Diéguez, S. Comparison of soxhlet, accelerated solvent and supercritical fluid extraction techniques for volatile (gc–ms and gc/fid) and phenolic compounds (hplc–esi/ms/ms) from lamiaceae species. *Phytochem. Anal.* **2015**, *26*, 61–71. [CrossRef] [PubMed]

17. Herrero, M.; Arráez-Román, D.; Segura, A.; Kenndler, E.; Gius, B.; Raggi, M.; Ibáñez, E.; Cifuentes, A. Pressurized liquid extraction–capillary electrophoresis–mass spectrometry for the analysis of polar antioxidants in rosemary extracts. *J. Chromatogr. A* **2005**, *1084*, 54–62. [CrossRef] [PubMed]

18. Jacotet-Navarro, M.; Rombaut, N.; Fabiano-Tixier, A.; Danguien, M.; Bily, A.; Chemat, F. Ultrasound versus microwave as green processes for extraction of rosmarinic, carnosic and ursolic acids from rosemary. *Ultrason. Sonochem.* **2015**, *27*, 102–109. [CrossRef]

19. Petersen, M.; Abdullah, Y.; Benner, J.; Eberle, D.; Gehlen, K.; Hücherig, S.; Janiak, V.; Kim, K.; Sander, M.; Weitzel, C.; et al. Evolution of rosmarinic acid biosynthesis. *Phytochemistry* **2009**, *70*, 1663–1679. [CrossRef]

20. Bernatoniene, J.; Cizauskaite, U.; Ivanauskas, L.; Jakstas, V.; Kalveniene, Z.; Kopustinskiene, D. Novel approaches to optimize extraction processes of ursolic, oleanolic and rosmarinic acids from *rosmarinus officinalis* leaves. *Ind. Crops Prod.* **2016**, *84*, 72–79. [CrossRef]

21. Birtić, S.; Dussort, P.; Pierre, F.; Bily, A.; Roller, M. Carnosic acid. *Phytochemistry* **2015**, *115*, 9–19. [CrossRef]

22. Ho, C.; Osawa, T.; Huang, M.; Rosen, R. *Food Phytochemicals for Cancer Prevention II; Teas, Spices, and Herbs*; American Chemical Society: Washington, DC, USA, 1994; Volume 547, p. 388.

23. Rodríguez-Rojo, S.; Visentin, A.; Maestri, D.; Cocero, M. Assisted extraction of rosemary antioxidants with green solvents. *J. Food Eng.* **2012**, *109*, 98–103. [CrossRef]

24. Chemat, F.; Abert-Vian, M.; Cravotto, G. Green extraction of natural products: Concept and principles. *Int. J. Mol. Sci.* **2012**, *13*, 8615–8627. [CrossRef] [PubMed]

25. Larkin, J. Thermodynamic properties of aqueous non-electrolyte mixtures i. Excess enthalpy for water + ethanol at 298.15 to 383.15 k. *J. Chem. Thermodyn.* **1975**, *7*, 137–148. [CrossRef]

26. Newsham, D.; Mendez-Lecanda, E. Isobaric enthalpies of vaporization of water, methanol, ethanol, propan-2-ol, and their mixtures. *J. Chem. Thermodyn.* **1982**, *14*, 291–301. [CrossRef]

27. Pingret, D.; Fabiano-Tixier, A.; Le Bourvellec, C.; Renard, C.; Chemat, F. Lab and pilot-scale ultrasound-assisted water extraction of polyphenols from apple pomace. *J. Food Eng.* **2012**, *111*, 73–81. [CrossRef]

Article

Phytochemical Profile and Biological Properties of *Colchicum triphyllum* (Meadow Saffron)

Biancamaria Senizza [1], Gabriele Rocchetti [1,*], Murat Ali Okur [2], Gokhan Zengin [2], Evren Yıldıztugay [3], Gunes Ak [2], Domenico Montesano [4,*] and Luigi Lucini [1]

[1] Department for Sustainable Food Process, Università Cattolica del Sacro Cuore, Via Emilia Parmense 84, 29122 Piacenza, Italy; biancamaria.senizza@unicatt.it (B.S.); luigi.lucini@unicatt.it (L.L.)

[2] Department of Biology, Science Faculty, Selcuk University, Campus, 4213 Konya, Turkey; murataliokur@hotmail.com (M.A.O.); gokhanzengin@selcuk.edu.tr (G.Z.); akguneselcuk@gmail.com (G.A.)

[3] Department of Biotechnology Science Faculty, Selcuk University, Campus, 4213 Konya, Turkey; eytugay@gmail.com

[4] Department of Pharmaceutical Sciences, Food Science and Nutrition Section, University of Perugia, Via S. Costanzo 1, 06126 Perugia, Italy

* Correspondence: gabriele.rocchetti@unicatt.it (G.R.); domenico.montesano@unipg.it (D.M.)

Received: 6 March 2020; Accepted: 7 April 2020; Published: 8 April 2020

Abstract: In this work, the phytochemical profile and the biological properties of *Colchicum triphyllum* (an unexplored Turkish cultivar belonging to Colchicaceae) have been comprehensively investigated for the first time. Herein, we focused on the evaluation of the in vitro antioxidant and enzyme inhibitory effects of flower, tuber, and leaf extracts, obtained using different extraction methods, namely maceration (both aqueous and methanolic), infusion, and Soxhlet. Besides, the complete phenolic and alkaloid untargeted metabolomic profiling of the different extracts was investigated. In this regard, ultra-high-performance liquid chromatography coupled with quadrupole time-of-flight mass spectrometry (UHPLC-QTOF-MS) allowed us to putatively annotate 285 compounds when considering the different matrix extracts, including mainly alkaloids, flavonoids, lignans, phenolic acids, and tyrosol equivalents. The most abundant polyphenols were flavonoids (119 compounds), while colchicine, demecolcine, and lumicolchicine isomers were some of the most widespread alkaloids in each extract analyzed. In addition, our findings showed that *C. triphyllum* tuber extracts were a superior source of both total alkaloids and total polyphenols, being on average 2.89 and 10.41 mg/g, respectively. Multivariate statistics following metabolomics allowed for the detection of those compounds most affected by the different extraction methods. Overall, *C. triphyllum* leaf extracts showed a strong in vitro antioxidant capacity, in terms of cupric reducing antioxidant power (CUPRAC; on average 96.45 mg Trolox Equivalents (TE)/g) and ferric reducing antioxidant power (FRAP) reducing power (on average 66.86 mg TE/g). Interestingly, each *C. triphyllum* methanolic extract analyzed (i.e., from tuber, leaf, and flower) was active against the tyrosinase in terms of inhibition, recording the higher values for methanolic macerated leaves (i.e., 125.78 mg kojic acid equivalent (KAE)/g). On the other hand, moderate inhibitory activities were observed against AChE and α-amylase. Strong correlations ($p < 0.01$) were also observed between the phytochemical profiles and the biological activities determined. Therefore, our findings highlighted, for the first time, the potential of *C. triphhyllum* extracts in food and pharmaceutical applications.

Keywords: meadow saffron; metabolomics; UHPLC-QTOF-mass spectrometry; extraction methods; bioactive compounds; antioxidants

1. Introduction

Colchicum triphyllum Kunze is a spring/autumn-flowering species belonging to the Colchicaceae family, widely distributed in Turkey and Balkans [1]. It is also known as "autumn crocus" or "meadow saffron". Plants belonging to Colchicaceae are mainly used in pharmaceutical applications, thanks to therapeutic, anti-inflammatory, and antitumoral activities [2] attributed to the presence of colchicinoids (alkaloids), such as colchicine and demecolcine. In this regard, colchicine is used in the treatment of gout [3] and Behcet's disease [4], while demecolcine, together with trimethyl-colchicine acid methyl ester, demonstrated anti-neoplastic activity and is particularly suitable for the treatment of leukemia [5]. Besides, bioactive compounds characterizing plants belonging to the Colchicaceae family, such as alkaloids (e.g., colchicine), have been characterized and have been widely studied because of their beneficial effects for the treatment of cirrhosis, psoriasis, and amyloidosis [6]. Interestingly, less toxic derivatives of colchicine have also been studied as anticancer and antitumoral agents. *Colchicum* spp. also contains a considerable distribution of bioactive compounds, such as polyphenols. In particular, according to the literature [7], the most abundant (poly)-phenolic compounds are lignans, flavonoids, phenolic acids, and tannins.

Notably, *Colchicum* leaves share morphological similarities (mainly looking at leaves) with other plant species, such as *Allium ursinum* L. (wild garlic); fortunately, poisoning is rare, although some accidents (with also lethal outcomes) caused by the ingestion of toxic *Colchicum* alkaloids have been described in the scientific literature [8]. In addition, some *Colchicum* species (mainly *C. autumnale* L.) may be confused for the same reasons as Crocus spp. (mainly *Crocus sativus*); despite their morphological similarity, *Colchicum* flowers are typically larger with six stamens, while Crocus flowers are smaller with three longer stamens. Another issue related to saffron (*Crocus sativus*) is the lack of knowledge by consumers about the correct shape and botanical characteristics of the products, thus leading to potential episodes of adulteration and counterfeiting procedures on the market. In fact, besides morphological analogies, some plant-tissues from *Colchicum* spp. are extremely toxic, thus potentially affecting human health.

There are previous works based on the description and characterization of *Colchicum* spp. and its alkaloid distribution (focused mainly on colchicine and demecolcine) in several Jordanian *Colchicum* species. Recently, Rocchetti and co-authors [9] profiled, for the first time, flowers, leaves, and tubers of *Colchicum szovitsii* subsp. *szovitsii*, showing a great abundance of flavonols, phenolic acids, and total alkaloids that were the main components responsible for biological activities detected. However, to the best of our knowledge, there are no comprehensive studies based on a detailed characterization of both polyphenols and alkaloids characterizing different parts (i.e., flower, leaves, tuber) of *C. triphyllum* and based on untargeted metabolomics (i.e., ultra-high-performance liquid chromatography coupled with quadrupole time-of-flight (UHPLC-QTOF) mass spectrometry). Besides, considering that, to date, no efficient methods to synthesize *Colchicum* alkaloids have been found; colchicine and other alkaloids are mainly obtained from plant sources by different extractions techniques. Therefore, in this work, infusion, maceration (using methanol and water) together with Soxhlet extraction techniques were used to promote the extraction of both polyphenols and alkaloids from *C. triphyllum*, aiming to identify the most discriminant markers of each extraction technique used. Finally, in vitro antioxidant and enzyme inhibitory assessments were carried out to investigate the biological and pharmaceutical potential of this plant species.

2. Materials and Methods

2.1. Plant Material

The plant materials of *Colchicum triphyllum*, namely flowers, leaves, and tubers, were collected at Konya in Turkey in 2019 (Konya, around Silla Dam Lake, steppes 1200 m; Collection date: 03.02.2019). The plant materials were collected and identified by botanist Dr. Evren Yildiztugay (Selcuk University, Department of Biotechnology, Konya, Turkey, Voucher number: EY-2968). In the sampling, about twenty plants were collected in the same population. The plant materials were cleaned (first, washing with tap water and then rising by distilled water), and soil and other contaminants were removed. The plant parts, namely flowers, leaves, and tubers, were carefully separated, and these plants were dried in a shaded and well-ventilated environment at the Department of Biology, Selcuk University. After drying (about 10 days), the plant materials were powdered by using a laboratory mill (Retsch, SM-200), and the powdered materials were used to obtain extracts in the same week. The powdered plant materials were stored in well-ventilated conditions in the dark (about 20 °C). We performed the analysis in about one month after sampling.

2.2. Extraction Methods

In this work, three different extraction methods using different solvents were tested. In this regard, to obtain extracts, we performed infusion, maceration, and Soxhlet extraction techniques. Regarding infusion, the plant materials (5 g) were kept in 100 mL of boiling water for 20 min and then filtered. In the maceration technique, the plant materials (5 g) were mixed with 100 mL of both methanol and water for 24 h at room temperature. In the Soxhlet technique, the plant materials (5 g) were extracted with 100 mL methanol by using a Soxhlet apparatus for 6 h. Final extracts were obtained by using a vacuum evaporator and lyophilization. Finally, each obtained extract was stored in a refrigerator until further analyses.

2.3. UHPLC-QTOF Profiling of Polyphenols and Alkaloids

The untargeted phytochemical profile of the different *C. triphyllum* extracts was investigated through ultra-high-pressure liquid chromatography (Agilent 1290 HPLC liquid chromatograph; Agilent Technologies, Santa Clara, CA, USA) coupled to a quadrupole-time-of-flight mass spectrometer (Agilent 6550 iFunnel; Agilent Technologies, Santa Clara, CA, USA). The experimental conditions for the analysis of plant extracts using untargeted metabolomics were optimized in previous works from our research group [9–11]. The mass spectrometer acquired ions in the range 50–1200 m/z in positive (ESI+) scan mode. Three technical replications were considered, with an injection volume of 6 µL. An in-house database built, combining Phenol-Explorer 3.6 with some of the most important alkaloids reported in the literature on Colchicaceae, was then used for annotation purposes, exploiting the entire isotopic profile (i.e., combining monoisotopic accurate mass, isotopic ratios, and spacing) with a mass accuracy below 5 ppm. Therefore, the approach used was based on a Level 2 of identification (i.e., putatively annotated compounds), as set out by the COSMOS Metabolomics Standards Initiative [12–14]. Afterward, Agilent Profinder B.06 software was used for post-acquisition data filtering, retaining only those compounds identified within 100% of replications in at least one condition. Thereafter, to provide more quantitative information on the different annotated compounds, polyphenols were first ascribed into classes and subclasses and then quantified using standard solutions (80/20, *v/v* methanol/water) of pure standard compounds analyzed with the same method [9]. The following phenolic classes were targeted: anthocyanins (quantified as cyanidin equivalents), flavones (quantified as luteolin equivalents), flavonols (quantified as catechin equivalents), lignans (quantified as sesamin equivalents), low-molecular-weight phenolics (quantified as tyrosol equivalents), and phenolic acids (quantified as ferulic acid equivalents). Finally, a calibration curve of sanguinarine (Sigma grade, Sigma–Aldrich, S. Louis, MO, USA) was used to estimate the total alkaloid content. The results were finally expressed as mg equivalents/g dry matter.

2.4. In Vitro Antioxidant Capacity and Inhibitory Potential

For in vitro antioxidant capacity, different test systems, including radical quenching, reducing power, phosphomolybdenum, and ferrous ion chelating, were employed. The details of the methods are described in our earlier papers [15,16]. The results were reported as mg Trolox Equivalents (TE)/g extract and ethylenediaminetetraacetic acid (EDTA) equivalents (for ferrous ion chelating; mg EDTAE/g extract). For enzyme inhibitory activities, key enzymes for global health problems were selected, namely, α-amylase and α-glucosidase, acetylcholinesterase (AChE), butyrylcholinesterase (BChE), and tyrosinase and the inhibitory activities were compared to standard drugs (acarbose for amylase and glucosidase; galantamine for AChE and BChE; kojic acid for tyrosinase). All assays were performed considering three technical replications.

2.5. Statistical Analysis and Chemometrics

A one-way analysis of the variance (ANOVA) was performed considering data from each assay and using the software PASW Statistics 26.0 (SPSS Inc., Chicago, IL. USA), followed by a Duncan's post hoc test ($p > 0.05$). Pearson's correlations ($p < 0.05$; two-tailed) were also calculated using PASW Statistics 26.0. Afterward, the metabolomics-based dataset exported from Mass Profiler Profession B.12.06 (Agilent Technologies, Santa Clara, CA, USA) was elaborated into a second software, namely SIMCA 13 (Umetrics, Malmo, Sweden) for supervised orthogonal projections to latent structures discriminant analysis (OPLS-DA), as previously reported in previous works from our research group [9]. Two OPLS-DA models were built; the first one highlighted the differences in the phytochemical profiles as imposed by the extraction methods, while the second model showed the differences between the three plant-organs under investigation. Finally, the variables selection method VIP (i.e., variables' importance in projection) was used to evaluate those compounds mostly affected by the different extraction methods, together with those better discriminating the plant-organs. In particular, polyphenols and alkaloids showing a VIP score > 1 have been considered as marker compounds.

3. Results and Discussion

3.1. Phytochemical Profiling of the Different Extracts

To characterize the polyphenol and alkaloid composition of the different *Colchicum triphyllum* extracts we used untargeted metabolomics based on UHPLC-QTOF mass spectrometry. According to this approach, 285 compounds were putatively identified in the different matrix extracts, mainly including alkaloids, flavonoids (such as anthocyanins, flavonols, and flavones), lignans, and low-molecular-weight phenolic acids. Each compound annotated is provided in Supplementary Table S1 together with its abundance and composite mass spectra. Overall, the most abundant compounds detected when considering the metabolomic dataset were alkaloids (such as colchicine, demecolcine, and y-lumicolchicine), anthocyanins (such as petunidin 3-*O*-rutinoside and cyanidin 3-*O*-sophoroside), flavones (mainly apigenin and luteolin glucosides), and flavonols (mainly isomeric forms of kaempferol and quercetin). Regarding the class of lignans, the most abundant annotated compounds were secoisolariciresinol, pinoresinol, and its isomer, matairesinol (Supplementary Table S1). Afterward, a semi-quantitative approach based on standard compounds was used to evaluate the concentration of phytochemicals in the studied plant matrices. The results of this semi-quantitative approach are reported in Table 1.

Table 1. Semi-quantitative values for the main phenolic sub-classes and total alkaloids by ultra-high-performance liquid chromatography quadrupole time-of-flight (UHPLC-QTOF) mass spectrometry of the tested extracts together with extraction yields. Values are reported as the mean ± standard deviation ($n = 3$). The results are expressed as mg equivalents (Eq.)/g dry matter. Different superscript letters in the same column indicate significant differences ($p < 0.05$), as determined by Duncan's post-hoc test. nd = not detected.

Parts	Methods	Extraction Yield (%)	Total Alkaloids (mg Eq./g)	Anthocyanins (mg Eq./g)	Flavones (mg Eq./g)	Flavonols (mg Eq./g)	Phenolic Acids (mg Eq./g)	Lignans (mg Eq./g)	Tyrosols (mg Eq./g)
Flowers	Infusion	35.12	1.49 ± 0.49 [ab]	0.57 ± 0.05 [d]	1.98 ± 0.2 [c]	1.48 ± 0.21 [c]	0.11 ± 0.02 [a]	2.27 ± 0.29 [ab]	0.72 ± 0.35 [ab]
	Maceration-MeOH	42.6	1.77 ± 0.44 [abc]	0.78 ± 0.01 [f]	1.99 ± 0.94 [c]	1.71 ± 0.20 [d]	0.20 ± 0.00 [c]	2.67 ± 1.36 [abc]	1.09 ± 0.24 [bc]
	Maceration-Water	26.65	2.08 ± 0.02 [bcd]	0.06 ± 0.00 [b]	0.47 ± 0.08 [a]	0.14 ± 0.00 [a]	0.19 ± 0.01 [bc]	3.01 ± 0.28 [abc]	0.44 ± 0.03 [a]
	Soxhlet-MeOH	51.25	1.39 ± 0.01 [ab]	0.67 ± 0.01 [e]	2.52 ± 0.14 [c]	1.67 ± 0.01 [d]	0.15 ± 0.03 [abc]	1.28 ± 1.30 [a]	0.42 ± 0.01 [a]
Average value			1.68	2.08	1.74	1.25	0.16	2.31	0.67
Tubers	Infusion	13.44	1.86 ± 1.00 [abc]	0.005 ± 0.00 [a]	0.15 ± 0.00 [a]	nd [a]	0.50 ± 0.02 [g]	4.57 ± 0.03 [bc]	1.22 ± 0.05 [cd]
	Maceration-MeOH	28.41	3.19 ± 0.03 [cd]	nd [a]	1.06 ± 0.39 [b]	nd [a]	0.45 ± 0.05 [fg]	10.37 ± 0.72 [d]	1.84 ± 0.05 [e]
	Maceration-Water	44.02	3.46 ± 0.74 [d]	nd [a]	0.39 ± 0.24 [a]	nd [a]	0.41 ± 0.01 [e]	4.84 ± 1.41 [bc]	1.60 ± 0.01 [de]
	Soxhlet-MeOH	12.68	3.08 ± 1.14 [cd]	nd [a]	0.45 ± 0.02d [a]	nd [a]	0.35 ± 0.10 [de]	10.60 ± 2.50 [d]	2.85 ± 0.72 [f]
Average value			2.89	<0.01	0.51	nd	0.43	7.59	1.88
Leaves	Infusion	25.58	2.26 ± 0.49 [abcd]	0.06 ± 0.00 [b]	0.33 ± 0.01 [a]	0.07 ± 0.01 [a]	0.21 ± 0.01 [c]	4.77 ± 1.25 [d]	3.43 ± 0.07 [g]
	Maceration-MeOH	31.48	1.10 ± 0.54 [a]	0.05 ± 0.01 [b]	0.22 ± 0.04 [a]	0.11 ± 0.00 [a]	0.30 ± 0.04 [d]	1.65 ± 0.57 [a]	0.59 ± 0.09 [a]
	Maceration-Water	22.88	2.28 ± 0.04 [abcd]	0.13 ± 0.00 [c]	1.10 ± 0.01 [b]	0.43 ± 0.00 [b]	0.49 ± 0.06 [fg]	4.99 ± 2.75 [c]	2.48 ± 0.01 [f]
	Soxhlet-MeOH	33.61	2.79 ± 1.70 [bcd]	0.07 ± 0.01 [b]	0.56 ± 0.00 [ab]	0.09 ± 0.01 [a]	0.12 ± 0.00 [ab]	3.11 ± 0.15 [abc]	1.45 ± 0.04 [cde]
Average value			2.11	0.08	0.55	0.17	0.28	3.63	1.87

As can be observed, the three plant matrices were predominantly rich ($p < 0.05$) in alkaloids and lignans when compared to the other classes of compounds, while flavonols and anthocyanins showed the lower ($p < 0.05$) concentration; it is also interesting to notice that, when considering *C. triphyllum* tuber extracts, neither anthocyanins nor flavonols were detected. Besides, looking at *C. triphyllum* flower extracts, the maceration-water extraction method was found to promote the highest recovery of lignans and alkaloids (3.01 and 2.08 mg/g, respectively), while maceration-MeOH extraction encourages the recovery of anthocyanins, flavonols, phenolic acids, and tyrosols. Interestingly, Soxhlet extraction promoted the highest recovery of flavones (2.52 mg/g). *Colchicum* flowers are reported to be very similar from a morphological point of view to those of *Crocus Sativus L.* Overall, both plant species can be considered as a good source of (poly)-phenolic compounds In this regard, the saffron flower has been described as rich in flavonoids (such as flavonols and flavones), hydroxycinnamic acids, and lignans [10,17,18]. Regarding saffron alkaloids, Amin Mir and co-authors [19] found this class of compounds in both water and methanolic extracts of flowers. Another study by Hosseinzadeh et al. [20] based on the phytochemical screening of different *Crocus* extracts, highlighted the distribution of flavonoids (including anthocyanins) and tannins in both aqueous and ethanolic petal extracts, while alkaloids and saponins characterized the aqueous and ethanolic stigmas extracts. However, our data are difficult to compare with previously cited works, considering that in this work, both polyphenols and alkaloids were evaluated, targeting the whole flower. Recently, Jadouali and co-authors [21] evaluated the total phenolic content of different flower parts of Moroccan *Crocus sativus* L., showing a value of 54.59 mg gallic acid equivalents (GAE)/g for the whole saffron flower.

Considering the lack in the literature of similar comprehensive phytochemical screening on *C. triphyllum* extracts, we compared our findings with a previous work focused on a different species, namely *C. szovitsii*. In particular, the most abundant polyphenols characterizing *C. szovitsii* plant extracts were flavonols, phenolic acids, and tyrosols equivalents [9], while anthocyanins and flavanols were found to be less abundant. Another interesting result was obtained when comparing the extraction efficiency of polyphenols between the two *Colchicum* species and using Soxhlet-MeOH extractions. In particular, Soxhlet-MeOH promoted a better extraction of flavonols and flavones in tubers of *C. szovitsii* when compared to *C. triphyllum*, being 1.58 and 0.81 mg/g vs. not detectable values and 0.45 mg/g, respectively. Interestingly, the same extraction method (i.e., Soxhlet-MeOH) promoted a better recovery of lignans and tyrosols in *C. tryphillum*, being 10.60 and 2.85 mg/g, respectively. Regarding the alkaloids putatively annotated in *C. triphyllum* extracts, tubers obtained by water maceration showed the highest content (3.46 mg/g). This aspect is worthy of interest, considering that the colchicinoids (mainly colchicine and derivatives) are among the highly poisonous water-soluble alkaloids detected in flowers and seeds of *Colchicum* genus [22]. Regarding *C. szovitsii* leaf extracts, flavonols were better extracted by exploiting Soxhlet (21.95 mg/g), while phenolic acids and tyrosol equivalents by infusion (3.52 and 3.68 mg/g, respectively). On the other hand, *C. triphyllum* was revealed to be a great source of lignans (possessing a potential estrogenic activity), recording an average value of 3.63 mg/g when considering all the extraction methods tested (Table 1). Finally, concerning the total alkaloid content, *C. triphyllum* and *C. szovitsii* leaf extracts showed a comparable content, being 2.79 and 2.65 mg/g when considering Soxhlet and water-maceration, respectively.

3.2. Multivariate Statistical Discrimination of the Different Extraction Methods

Considering specifically the extraction type, a supervised multivariate statistical approach was carried out to find the compounds allowing the discrimination between the different methods. In more detail, an OPLS-DA (orthogonal projection to latent structures discriminant analysis) was depicted, followed by the variables importance in projection (VIP) method, to select those compounds mostly affected by the different extraction methods. As can be observed in Figure 1, a clear separation based on alkaloids and polyphenols content was achieved. In particular, each extraction method provided a differential phytochemical profile, although water-maceration extracts were different from the others. The model parameters were more than acceptable, being R^2Y (cum) (goodness-of-fit) = 0.97 and Q^2

(cum) (goodness-of-prediction) = 0.91. Besides, the model was cross-validated and showed neither suspect nor strong outliers. Afterward, the variables selection method VIP was exploited to find those metabolites mostly influenced by the extraction method used. Hence, 26 compounds were those possessing a VIP score > 1.2, including flavonoids (7 compounds), followed by low-molecular-weight phenolics (7 compounds), phenolic acids (6 compounds), and lignans (5 compounds). The list containing the remaining VIP compounds (1.2 < VIP score < 1) is reported in Supplementary Table S1. Interestingly, only one alkaloid, namely colchiceine, possessed a VIP score > 1.2 (i.e., 1.27). Regarding polyphenols, the higher VIP score was recorded for the lignan 1-acetoxypinoresinol, mainly found in tubers and leaves but not in flowers, followed by the caffeoylquinic acid isomers (VIP score = 1.39), whose presence was only detected in flowers treated with mac-MeOH. Furthermore, the lignans sesaminol, sesamolin, and episesaminol (VIP score = 1.36) were also better extracted with the infusion method. Some studies established that ethanol and methanol plant and fruit extracts can provide a better recovery of polyphenols when compared to water as the solvent, but the efficiency is strictly related to the extraction time [20]. Moreover, changes in temperature and the solvent-mixtures chosen could enhance and/or reduce the extraction efficiency. The Soxhlet method is a well-established technique requiring a smaller quantity of solvent compared to maceration, but according to literature, it is not suggested to promote the extraction of thermolabile compounds [23].

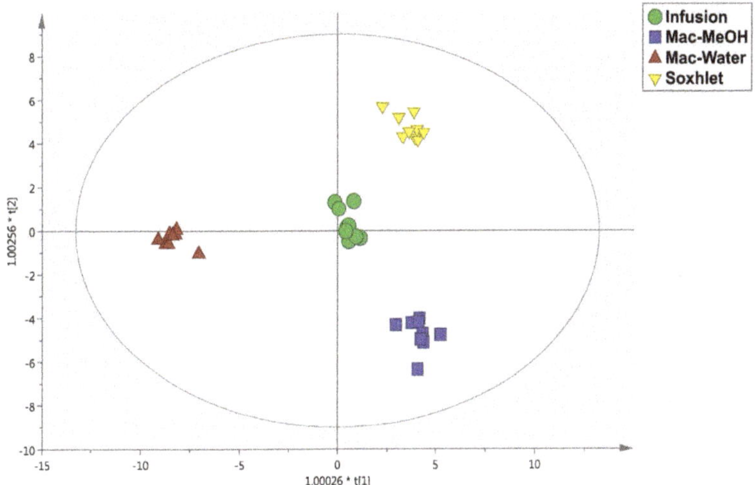

Figure 1. Orthogonal projections to latent structures discriminant analysis (OPLS-DA) score plot built according to polyphenol and alkaloid profiling and considering the different extraction methods as class membership criteria.

Thereafter, a second OPLS-DA model was built to check the major differences between the organs under investigation. The OPLS-DA score plot is reported in Figure 2.

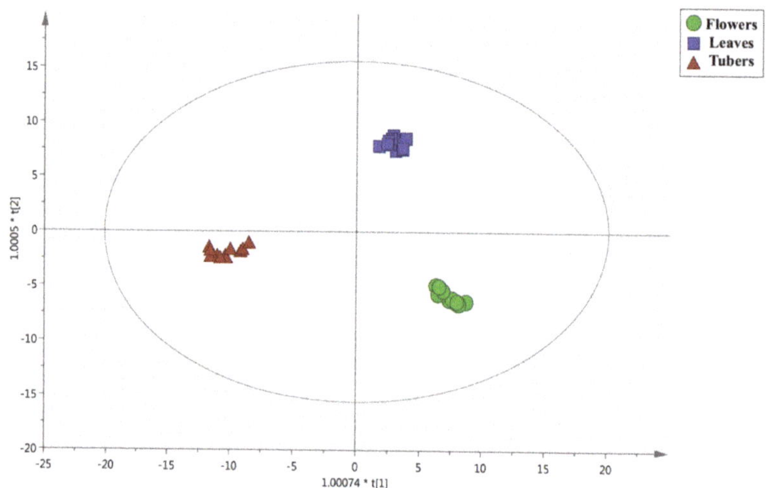

Figure 2. Orthogonal projections to latent structures discriminant analysis (OPLS-DA) score plot built according to polyphenol and alkaloid profiling and considering the different plant organs (i.e., flowers, leaves, and tubers) as class membership criteria.

As can be observed from Figure 2, the second latent vector t[2] provided a clear discrimination between tubers and the other *C. triphyllum* extracts, while the first latent vector t[1] revealed a more exclusive phytochemical profile for the leaf extracts. The goodness of the prediction model built on phenolics and alkaloids was confirmed by inspecting the goodness of fit and prediction, being 0.97 and 0.91, respectively. Finally, to check those compounds allowing the discrimination of the different organs, the VIP method was exploited, and the VIP markers can be found in Supplementary Table S1. Overall, 78 compounds were characterized by a VIP score > 1, being 34 flavonoids, 17 alkaloids, 11 phenolic acids, 11 lower-molecular-weight phenolics, and 5 lignans. In particular, the highest VIP scores were recorded for three alkaloids, namely androbine (1.51), autumnaline (1.46), and colchicoside (1.42). Looking at polyphenols, the most discriminant compounds highlighted by the VIP selection method were genistin (1.39), phlorin (1.35), and pelargonidin 3-*O*-glucoside (1.34). Overall, androbine was a specific marker of leaf extracts; autumnaline was detected in both tubers and leaves), while colchicoside was a marker of both tubers and flowers. Regarding polyphenols, genistin and pelargonidin 3-*O*-glucoside were detected mainly in flower extracts, while phlorin was a specific marker detected in tubers (Supplementary Table S1). Overall, our findings revealed a higher discrimination potential of alkaloids when compared to polyphenols. To date, more than 150 structurally elucidated alkaloids have been described for Colchicaceae genera [24]. In particular, eight distinct structural types of alkaloids characterize the Colchicaceae family, being phenethylisoquinolines (e.g., autumnaline), homoproaporphines (e.g., jolantine), homoaporphines (e.g., merobustine), androcymbines (e.g., szovitsidine), colchicines (e.g., colchicine), allocolchicines (e.g., jerusalemine), lumicolchicines (e.g., γ-lumicolchicine), and homoerythrinans (e.g., taxodine). Therefore, our findings (Supplementary Table S1) are completely following with the typical alkaloid composition reported for Colchicaceae plants. Regarding the polyphenols annotated, few comprehensive works based on high-resolution mass spectrometry are available in the literature. In previous work, Toplan et al. [1] reviewed the importance of *Colchicum* species in modern therapy together with its significance in Turkey, showing that these plants are very abundant in alkaloids, followed by polyphenols (mainly phenolic acids and flavonoids). However, looking to the different *Colchicum* species, few compounds are reported, namely ferulic acid, vanillin, luteolin, coumaric acid, caffeic acid, and 3,4-dihiydroxibenzaldehyde. Therefore, further works are required to better explore and give more insight on the phenolic composition of different *Colchicum* species.

3.3. In Vitro Antioxidant Capacity of the Tested Extracts

To date, the scientific evidence on the biological and pharmacological activities of extracts and bioactive compounds deriving from plants is very strong [25–31]. In the last years, many studies have explored the association between the ingestion of bioactive compounds and decreased risk of non-communicable diseases. Besides, considering that many non-nutrients with putative health benefits are reducing agents, hydrogen donors, singlet oxygen quenchers, or metal chelators, measurement of antioxidant activity using in vitro assays have become very popular over recent decades. In the current study, we investigated the antioxidant capacities of *C. triphyllum* flower, tuber, and leaf extracts by measuring their total in vitro antioxidant capacity (phosphomolybdenum assay), DPPH (2,2-diphenyl-1-picrylhydrazyl) and [2,2′-azinobis-(3-ethylbenzothiazoline-6-sulfonate)] (ABTS) scavenging capacity, ferric and cupric reduction activity (FRAP and CUPRAC), and metal chelating activity. However, as widely suggested in the recent scientific literature [32], these colorimetric and in vitro methods present many pitfalls and should be used as screening tools, thus supported by a comprehensive LC-MS characterization and quantification of those antioxidant compounds likely responsible of the activity observed. Based on the experimental findings (Table 2), it was observed that all of the studied extracts were characterized by potential health-promoting properties. Radical scavengers can prevent free radical-induced macromolecules or tissue damage by directly neutralizing free radicals and accepting or donating electron(s) to eliminate the unpaired condition of the radical [33]. Among the different *C. triphyllum* extracts tested in this study, we found that leaf extracts were the most active DPPH scavengers, recording an average value (when considering all the extraction method) of 45.22 mg TE/g. On the other hand, the methanolic tuber extracts obtained using the maceration method showed a similar activity, being 46.17 mg TE/g (Table 2). Regarding ABTS assay, the maceration method (based on methanol as extraction solvent) was found to promote the highest activity in tubers and leaves (being 50.93 and 52.55 mg TE/g, respectively). Overall, intriguing differences were outlined in the different organ-specific extracts when considering the cupric reduction activity (CUPRAC). In this regard, *C. triphyllum* leaf methanolic extracts showed higher values than flowers and tubers, recording 109.63 mg TE/g (using maceration method) and 123.34 mg TE/g (using Soxhlet method). Finally, regarding the other assays, FRAP ranged from 25.02 (infused flowers) up to 70.80 mg TE/g (leaves extracted using Soxhlet), metal chelating from 1.89 (tubers extracted using Soxhlet) up to 33.88 mg EDTAE/g (infused tubers), and total in vitro antioxidant capacity (phosphomolybdenum method) from 0.33 (infused tubers) up to 1.52 mmol TE/g (water macerated leaves). Therefore, taken together, our findings demonstrate that *C. triphyllum* leaf extracts were the most active as radical scavengers. This finding was in line with a previous work [9] focusing on another novel cultivar from Turkey, namely *Colchicum szovitsii subsp. szovitsii*.

Table 2. In vitro antioxidant activities of the tested extracts. Values are reported as the mean ± S.D. TE: Trolox equivalent; EDTAE: ethylenediaminetetraacetic acid equivalent. Different superscript letters in the same column indicate significant differences ($p < 0.05$), as determined by Duncan's post-hoc test.

Parts	Methods	Phosphomolybdenum (mmol TE/g)	DPPH (mg TE/g)	ABTS (mg TE/g)	CUPRAC (mg TE/g)	FRAP (mg TE/g)	Metal Chelating (mg EDTAE/g)
Flowers	Infusion	0.51 ± 0.04 [b]	31.69 ± 0.45 [c]	27.47 ± 1.57 [c]	35.00 ± 0.60 [a]	25.02 ± 0.34 [a]	20.90 ± 3.06 [e]
	Maceration-MeOH	0.46 ± 0.07 [ab]	33.99 ± 0.79 [cd]	20.53 ± 1.91 [a]	41.52 ± 1.13 [c]	25.71 ± 1.78 [a]	14.40 ± 0.55 [c]
	Maceration-water	1.10 ± 0.15 [d]	24.34 ± 0.60 [b]	19.26 ± 0.64 [a]	37.00 ± 0.22 [ab]	38.88 ± 0.67 [d]	21.53 ± 1.41 [ef]
	Soxhlet-MeOH	0.74 ± 0.09 [c]	34.63 ± 0.27 [cd]	24.48 ± 1.15 [b]	58.31 ± 0.89 [f]	30.72 ± 0.64 [b]	16.04 ± 0.41 [cd]
Average value		0.70	31.16	22.93	42.96	30.08	18.22
Tubers	Infusion	0.33 ± 0.02 [a]	12.14 ± 10.18 [a]	30.38 ± 0.38 [d]	38.38 ± 0.63 [b]	28.92 ± 0.74 [b]	33.88 ± 0.39 [g]
	Maceration-MeOH	0.72 ± 0.02 [c]	46.17 ± 1.15 [f]	50.93 ± 0.90 [h]	63.12 ± 1.52 [g]	46.16 ± 0.52 [e]	7.48 ± 0.27 [b]
	Maceration-water	0.77 ± 0.05 [c]	29.12 ± 1.05 [bc]	45.27 ± 0.60 [f]	44.75 ± 0.32 [d]	35.34 ± 0.71 [c]	23.99 ± 1.14 [f]
	Soxhlet-MeOH	0.85 ± 0.07 [c]	38.19 ± 0.91 [de]	38.79 ± 0.31 [e]	50.65 ± 2.06 [e]	35.85 ± 0.06 [c]	1.89 ± 0.25 [a]
Average value		0.67	31.40	41.34	49.22	36.57	16.81
Leaves	Infusion	1.25 ± 0.09 [e]	44.91 ± 0.35 [f]	47.90 ± 0.62 [g]	68.13 ± 0.45 [g]	60.66 ± 0.66 [f]	32.89 ± 2.20 [g]
	Maceration-MeOH	1.38 ± 0.05 [e]	45.49 ± 0.74 [f]	52.55 ± 1.65 [h]	109.63 ± 2.06 [l]	70.04 ± 3.09 [h]	22.26 ± 0.45 [ef]
	Maceration-water	1.52 ± 0.02 [f]	42.85 ± 1.09 [ef]	43.95 ± 1.34 [f]	84.71 ± 1.74 [i]	65.92 ± 1.39 [g]	15.74 ± 2.33 [cd]
	Soxhlet-MeOH	1.35 ± 0.10 [e]	47.65 ± 0.38 [f]	55.73 ± 0.36 [i]	123.34 ± 2.24 [m]	70.80 ± 0.88 [h]	18.25 ± 0.29 [d]
Average value		1.37	45.22	50.03	96.45	66.67	22.28

3.4. Enzyme Inhibitory Activity of the Tested Extracts

To investigate the enzyme inhibitory capacity of the different *C. triphyllum* extracts, the cholinesterases AChE and BChE, together with the enzymes tyrosinase, α-amylase, and α-glucosidase were considered. The results obtained using the above-mentioned enzymes are shown in Table 3.

Table 3. Enzyme inhibitory effects of the tested extracts. Values are reported as the mean ± S.D. Different superscript letters in the same column indicate significant differences ($p < 0.05$), as determined by Duncan's post-hoc test. GALAE: Galatamine equivalent; KAE: Kojic acid equivalent; ACAE: Acarbose equivalent; nd: not detected.

Parts	Methods	AChE Inhibition (mg GALAE/g)	BChE Inhibition (mg GALAE/g)	Tyrosinase Inhibition (mg KAE/g)	α-amylase Inhibition (mmol ACAE/g)	α-glucosidase Inhibition (mmol ACAE/g)
Flowers	Infusion	nd [a]	nd [a]	nd [a]	0.14 ± 0.01 [ab]	nd [a]
	Maceration-MeOH	3.41 ± 0.12 [d]	nd [a]	104.09 ± 3.66 [c]	0.66 ± 0.01 [ef]	2.69 ± 0.02 [b]
	Maceration-Water	nd [a]	nd [a]	nd [a]	0.27 ± 0.00 [c]	nd [a]
	Soxhlet-MeOH	4.09 ± 0.13 [f]	nd [a]	102.52 ± 0.56 [c]	0.61 ± 0.06 [d]	2.73 ± 0.01 [c]
Average value		1.87	nd	51.65	0.42	1.35
Tubers	Infusion	nd [a]	nd [a]	nd [a]	0.13 ± 0.00 [a]	nd [a]
	Maceration-MeOH	4.80 ± 0.02 [h]	3.91 ± 0.30 [b]	118.70 ± 0.85 [d]	0.73 ± 0.01 [g]	nd [a]
	Maceration-Water	nd [a]	nd [a]	nd [a]	0.18 ± 0.01 [b]	nd [a]
	Soxhlet-MeOH	5.11 ± 0.05 [i]	7.42 ± 1.48 [c]	118.61 ± 1.41 [d]	0.64 ± 0.01 [de]	nd [a]
Average value		2.48	2.83	59.32	0.42	nd
Leaves	Infusion	2.45 ± 0.16 [b]	nd [a]	nd [a]	0.14 ± 0.002 [ab]	nd [a]
	Maceration-MeOH	3.62 ± 0.16 [e]	nd [a]	125.78 ± 0.13 [f]	0.71 ± 0.05 [fg]	2.75 ± 0.01 [d]
	Maceration-Water	2.95 ± 0.08 [c]	nd [a]	4.10 ± 0.34 [b]	0.24 ± 0.00 [c]	nd [a]
	Soxhlet-MeOH	4.56 ± 0.09 [g]	nd [a]	121.78 ± 0.57 [e]	0.73 ± 0.02 [g]	2.76 ± 0.01 [d]
Average value		3.39	nd	62.91	0.45	1.37

According to literature, the enzyme cholinesterase is a significant therapeutic target to alleviate the deterioration of cholinergic neurons in the brain and the loss of neurotransmission, i.e., one of the major causes of Alzheimer's disease [34]. Overall, AChE inhibition activity was detected in each *C. triphyllum* leaf extract analyzed (Table 3). In particular, the highest activity values were recorded for tuber extracts (obtained by maceration-MeOH) and leaf extracts (obtained through Soxhlet-MeOH), being 4.80 and 4.56 mg galatamine equivalent (GALAE)/g, respectively. No activity against AchE was observed when considering infusion and maceration-water techniques for flowers and tubers extracts. Considering BChE, only methanolic maceration (3.91 mg GALAE/g) and Soxhlet-MeOH of *C. triphyllum* tubers (7.42 mg GALAE/g) were able to produce effective inhibitory activity against this enzyme. Regarding the enzyme tyrosinase, it plays an important role in the melanogenesis and enzymatic browning, so its inhibitors can be considered as anti-browning compounds in food and agriculture industries and as depigmentation agents in the cosmetic and medicinal industries [35]. In our experimental conditions, we found that both Soxhlet and maceration (using methanol as extraction solvent) were able to produce extracts characterized by a strong tyrosinase inhibitory ability. In particular, the highest activity was recorded for *C. triphyllum* leaves using methanolic maceration (i.e., 125.78 mg KAE/g). Finally, regarding two of the most important enzymes from a nutritional standpoint, namely α-amylase and α-glucosidase, we found a low α-amylase inhibition potential for all the tested extracts, from 0.13 (for infused tubers) up to 0.73 mmol acarbose equivalent (ACAE)/g (for leaves and tubers, using Soxhlet and methanolic maceration, respectively). Interestingly, the α-glucosidase inhibition activity was observed only for flower and leaf extracts, with the maximum average activity recorded for leaves (i.e., 1.37 mmol ACAE/g). These two enzymes are potential targets in producing lead compounds for the management of diabetes. Therefore, further research on *C. triphyllum* extracts could open a new perspective for the management of health-related and metabolic problems.

Regarding a possible comparison between different *Colchicum* species, there are few works available in the literature to make a realistic discussion. That is why we decided to compare our findings with a previous work focused on the chemical characterization of *Colchicum szovitsii* [9]. Overall, the authors showed that *C. szovitsii* possessed a strong tyrosinase inhibitory action, mainly when considering methanolic macerated leaf extracts (i.e., 116.92 mg KAE/g), thus confirming our findings. Similar trends were observed for the remaining tested enzymes. Interestingly, when considering the existing literature on saffron (*Crocus sativus*), similar activities have been reported on different saffron extracts [36], with crocetin, dimethylcrocetin, and safranal the mainly responsible for acetylcholinesterase inhibitions, as revealed by both molecular docking and in vitro enzymatic studies. Besides, Menghini and co-authors [37] showed that saffron stigmas are characterized by both AChE and BChE inhibitions (i.e., 2.51 and 3.44 mg GALAE/g, respectively) and inhibition towards amylolytic enzymes, such as α-amylase (0.44 mmol ACAE/g) and α-glucosidase (6.34 mmol ACAE/g). The same authors also showed that these values are lower when considering the combination of *C. sativus* tepals and anthers.

3.5. Correlations

Pearson's correlation coefficients were then calculated to check the contribution of polyphenols and alkaloids to the antioxidant and enzymatic related properties observed when considering the different *C. triphyllum* extracts. A summarizing table reporting all of the correlation coefficients for each part analyzed (i.e., flowers, tubers, and leaves) is reported in Supplementary Table S1. Overall, a positive and significant ($p < 0.05$) correlation coefficient was found between total alkaloids and in vitro total antioxidant capacity (phosphomolybdenum method) when considering *C. triphyllum* tuber extracts, with a correlation coefficient of 0.673. Besides, no significant correlations coefficients were recorded for leaves, flowers, and tubers when considering total alkaloids and enzymatic assays. Regarding polyphenols, we found significant correlations mainly between anthocyanins characterizing *C. triphyllum* flower extracts and enzymatic inhibitory assays. In particular, the highest correlation coefficient was measured between anthocyanins and tyrosinase inhibition (0.740), followed by α-glucosidase (0.737), AChE (0.714), and α-amylase (0.601) inhibitions. These results are consistent with recent papers reporting that flavonoids (mainly anthocyanins, flavonols, and flavones) are one of the most important classes of natural enzyme inhibitors [38–43]. Interestingly, negative and significant correlation coefficients were found when considering enzymatic assays and anthocyanins characterizing tuber extracts. In this regard, only two anthocyanins were putatively annotated in tuber extracts, namely Vitisin A and Petunidin 3,5-O-diglucoside, likely characterized by no inhibitory potential. Regarding leaf extracts, no significant correlation coefficients were observed (Supplementary Table S1). Therefore, our findings suggested that a clear matrix-effect mainly due to specific anthocyanins is conceivable, when looking at the correlation coefficients observed. Further works based on in silico studies are strongly required to confirm our findings.

Regarding the in vitro antioxidant potential, anthocyanins, flavonols, and flavones characterizing *C. triphyllum* flower extracts showed significant correlations with DPPH radical scavenging ability, recording values of 0.974, 0.972, and 0.880, respectively. Looking at the tuber extracts, we found that lignans, anthocyanins, and flavones were the most interesting classes in terms of correlation potential (Supplementary Table S1). In this regard, flavones were strongly ($p < 0.01$) correlated to FRAP (0.882), CUPRAC (0.870), ABTS (0.758), and DPPH (0.744), while lignans recorded the highest correlation coefficient for CUPRAC (0.775). Interestingly, the anthocyanins characterizing tuber extracts showed negative correlations coefficients ($p < 0.01$) with each antioxidant assay (Supplementary Table S1). Overall, it is also important to emphasize the significant correlation coefficients recorded with enzymatic assays, when considering lignans and tyrosols extracted from tubers. Regarding lignans, *C. triphyllum* tubers were mainly characterized by isomeric forms of matairesinol (Supplementary Table S1); in the last years, the interest on lignans has risen because of their estrogenic-like effects but also when considering the potential inhibitory activity on several enzymes, as revealed by in silico and molecular docking studies [44]. Finally, looking at the correlation results for *C. triphyllum* leaf extracts, few significant and positive correlation coefficients were recorded (Supplementary Table S1); overall, the phenolic profile of leaf extracts (in terms of flavonoids and phenolic acids), showed a good degree of correlation with the total in vitro antioxidant capacity ($0.01 < p < 0.05$). Clearly, the observed correlation values between phytochemical profiles (by UHPLC-QTOF mass spectrometry) and antioxidant/biological assays may be explained by considering a different exposure of abiotic and biotic stress factors for each plant part, thus leading to the production of different levels of secondary metabolites. However, looking to the wide diversity in both phenolics and alkaloids of *C. triphyllum* extracts, further studies based on in silico and targeted compound evaluations appear to be worthwhile, to confirm the potential of this plant for food and pharmaceutical purposes.

4. Conclusions

In this work, we focused the attention on the unexplored species *Colchicum triphyllum* Kunze, belonging to the *Colchicum* genus, considering the scarcity of comprehensive studies on its phytochemical profiling and biological properties. Therefore, this study reports the biochemical characterization of different *C. triphyllum* extracts, according to their phenolic and alkaloid compositions, together with the evaluation of their in vitro antioxidant and enzyme inhibitory activities. In this regard, UHPLC-QTOF-MS allowed us to identify 289 compounds (mainly alkaloids and flavonoids). Regarding the plant parts studied, *C. triphyllum* tuber extracts were the most important source of both alkaloids and polyphenolic compounds, being on average 2.89 and 10.41 mg/g, respectively. Multivariate statistics following metabolomics showed a higher impact of the different extraction methods (i.e., maceration, infusion, and Soxhlet) on the polyphenolic profile rather than alkaloids. Overall, *C. triphyllum leaf* extracts showed the stronger in vitro antioxidant capacity (as CUPRAC and FRAP), while each *C. triphyllum* methanolic extract analyzed was active against the enzyme tyrosinase. Strong correlations ($p < 0.01$) were also observed between the phytochemical profiles (mainly lignans and tyrosol equivalents) and the activities determined. Therefore, although characterized by the presence of toxic alkaloids (such as colchicine), this work sustained the utilization of different extraction methods to produce rich extracts in terms of (poly)-phenols and alkaloids, largely contributing to both antioxidant and other pharmacological properties, making *C. triphyllum* a promising source of drugs and whitening agents for both food, pharmaceutical, and cosmetic industries. However, further studies are strongly recommended to understand better the toxicity and bioavailability of the putatively identified phytochemicals, aimed at replacing synthetic antioxidants and enzyme inhibitors.

Supplementary Materials: The following are available online at http://www.mdpi.com/2304-8158/9/4/457/s1, Table S1: Dataset contained all the putatively annotated compounds by means of UHPLC-QTOF mass spectrometry, together with VIP markers from OPLS-DA and Pearson's correlation coefficients between phytochemical profiles and biological assays for each plant part.

Author Contributions: Conceptualization, G.R., G.Z., and M.A.O.; Methodology, G.R., B.S., and G.Z.; Resources, G.Z., L.L., E.Y., and D.M.; Writing—original draft preparation, B.S., G.R., G.Z., and G.A.; Writing—review and editing, G.R., D.M., G.Z., and L.L. All authors have read and agreed to the published version of the manuscript.

Funding: This research received no external funding.

Conflicts of Interest: The authors declare no conflict of interest.

References

1. Toplan, G.G.; Gurer, C.; Mat, A. Importance of Colchicum species in modern therapy and its significance in Turkey. *J. Fac. Pharm.* **2016**, *46*, 129–144.
2. Alali, F.Q.; Tawaha, K.; El-Elimat, T. Determination of (–)-demecolcine and (–)-colchicine content in selected Jordanian Colchicum species. *Int. J. Pharm. Sci. Res.* **2007**, *62*, 739–742.
3. Terkeltaub, R.A.; Furst, D.E.; Bennett, K.; Kook, K.A.; Crockett, R.S.; Davis, M.W. High versus low dosing of oral colchicine for early acute gout flare: Twenty-four-hour outcome of the first multicenter, randomized, double-blind, placebo-controlled, parallel-group, dose-comparison colchicine study. *Arthritis Rheum.* **2010**, *62*, 1060–1068. [CrossRef] [PubMed]
4. Sakane, T.; Takeno, M. Novel approaches to Behcet's disease. *Expert Opin. Investig. Drugs* **2000**, *9*, 1993–2005. [CrossRef]
5. Cifuentes, M.; Schilling, B.; Ravindra, R.; Winter, J.; Janik, M.E. Synthesis and biological evaluation of B-ring modi ed colchicine and iso colchicine analogs. *Bioorg. Med. Chem. Lett.* **2006**, *16*, 2761–2764. [CrossRef]
6. Cocco, G.; Chu, D.C.; Pandolfi, S. Colchicine in Clinical Medicine. A Guide for Internists. *Eur. J. Intern. Med.* **2010**, *21*, 503–508. [CrossRef]
7. Ondra, P.; Válka, I.; Vičar, J.; Sütlüpinar, N.; Šimánek, V. Chromatographic determination of constituents of the genus Colchicum (Liliaceae). *J. Chromatogr. A* **1995**, *704*, 351–356. [CrossRef]
8. Brvar, M.; Ploj, T.; Kozelj, G.; Mozina, M.; Noc, M.; Bunc, M. Case report: Fatal poisoning with *Colchicum autumnale*. *Crit. Care* **2004**, *8*, R56–R59. [CrossRef]
9. Rocchetti, G.; Senizza, B.; Zengin, G.; Okur, M.A.; Montesano, D.; Yildiztugay, E.; Lobine, D.; Mahomoodally, M.F.; Lucini, L. Chemical Profiling and Biological Properties of Extracts from Different Parts of *Colchicum Szovitsii* Subsp. *Szovitsii*. *Antioxidants* **2019**, *8*, 632. [CrossRef]
10. Senizza, B.; Rocchetti, G.; Ghisoni, S.; Busconi, M.; De Los Mozos Pascual, J.; Fernandez, A.; Lucini, L.; Trevisan, M. Identification of phenolic markers for saffron authenticity and origin: An untargeted metabolomics approach. *Food Res. Int.* **2019**, *126*, 108584. [CrossRef]
11. Mohamed, M.B.; Rocchetti, G.; Montesano, D.; Ali, S.B.; Guasmi, F.; Grati-Kamoun, N.; Lucini, L. Discrimination of Tunisian and Italian extra-virgin olive oils according to their phenolic and sterolic fingerprints. *Food Res. Int.* **2018**, *106*, 920–927. [CrossRef]
12. Fellah, B.; Rocchetti, G.; Senizza, B.; Giuberti, G.; Bannour, M.; Ferchichi, A.; Lucini, L. Untargeted metabolomics reveals changes in phenolic profile following *in vitro* large intestine fermentation of non-edible parts of *Punica granatum* L. *Food Res. Int.* **2020**, *128*, 108807. [CrossRef] [PubMed]
13. Salek, R.M.; Neumann, S.; Schober, D.; Hummel, J.; Billiau, K.; Kopka, J.; Correa, E.; Reijmers, T.; Rosato, A.; Tenori, L.; et al. COordination of Standards in MetabOlomicS (COSMOS): Facilitating integrated metabolomics data access. *Metabolomics* **2015**, *11*, 1587–1597. [CrossRef]
14. Schrimpe-Rutledge, A.C.; Codreanu, S.G.; Sherrod, S.D.; McLean, J.A. Untargeted metabolomics strategies-Challenges and emerging directions. *J. Am. Soc. Mass Spectrom.* **2016**, *27*, 1897–1905. [CrossRef]
15. Zengin, G. A study on *in vitro* enzyme inhibitory properties of *Asphodeline anatolica*: New sources of natural inhibitors for public health problems. *Ind. Crops Prod.* **2016**, *83*, 39–43. [CrossRef]

16. Tufa, T.; Damianakos, H.; Zengin, G.; Graikou, K.; Chinou, I. Antioxidant and enzyme inhibitory activities of disodium rabdosiin isolated from *Alkanna sfikasiana* Tan, Vold and Strid. *S. Afr. J. Bot.* **2019**, *120*, 157–162. [CrossRef]

17. Mykhailenko, O.; Kovalyov, V.; Goryacha, O.; Ivanauskas, L.; Georgiyants, V. Biologically active compounds and pharmacological activities of species of the genus Crocus: A review. *Phytochemistry* **2019**, *162*, 56–89. [CrossRef]

18. Moratalla-López, N.; Bagur, M.J.; Lorenzo, C.; Martínez-Navarro, M.E.; Salinas, M.R.; Alonso, G.L. Bioactivity and Bioavailability of the Major Metabolites of Crocus sativus L. Flower. *Molecules* **2019**, *24*, 2827. [CrossRef] [PubMed]

19. Amin Mir, M.; Parihar, K.; Tabasum, U.; Kumari, E. Estimation of alkaloid, saponin and flavonoid, content in various extracts of *Crocus sativa*. *J. Med. Plant Res.* **2016**, *4*, 171–174.

20. Hosseinzadeh, H.; Younesi, H.M. Antinociceptive and anti-inflammatory effects of *Crocus sativus* L. stigma and petal extracts in mice. *BMC Pharmacol.* **2002**, *2*, 7.

21. Jadouali, S.M.; Atifi, H.; Mamouni, R.; Majourhat, K.; Bouzoubaâ, Z.; Laknifli, A.; Faouzi, A. Chemical characterization and antioxidant compounds of flower parts of Moroccan crocus sativus L. *J. Saudi Soc. Agric. Sci.* **2019**, *18*, 476–480. [CrossRef]

22. Suica-bunghez, I.; Ion, R.; Teodorescu, I.; Sorescu, A.; Stirbescu, R.; Stirbescu, N. Fitochemical and antioxidant characterization of autumn crocus (*Colchicum autumnale*) flowers and tubers plant extracts. *J. Sci. Arts* **2017**, *3*, 539–546.

23. Lapornik, B.; Prošek, M.; Golc Wondra, A. Comparison of extracts prepared from plant by-products using different solvents and extraction time. *J. Food Eng.* **2005**, *71*, 214–222. [CrossRef]

24. Larsson, S.; Rønsted, N. Reviewing Colchicaceae Alkaloids – Perspectives of Evolution on Medicinal Chemistry. *Curr. Top. Med. Chem.* **2014**, *14*, 274–289. [CrossRef]

25. Borrelli, F.; Borbone, N.; Capasso, R.; Montesano, D.; De Marino, S.; Aviello, G.; Aprea, G.; Masone, S.; Izzo, A.A. Potent relaxant effect of a *Celastrus paniculatus* extract in the rat and human ileum. *J. Ethnopharmacol.* **2009**, *122*, 434–438. [CrossRef] [PubMed]

26. Montesano, D.; Rocchetti, G.; Putnik, P.; Lucini, L. Bioactive profile of pumpkin: An overview on terpenoids and their health-promoting properties. *Curr. Opin. Food Sci.* **2018**, *22*, 81–87. [CrossRef]

27. Rocchetti, G.; Pellizzoni, M.; Montesano, D.; Lucini, L. Italian *Opuntia ficus-indica* cladodes as rich source of bioactive compounds with health-promoting properties. *Foods* **2018**, *7*, 24. [CrossRef]

28. Pandey, K.B.; Rizvi, S.I. Plant polyphenols as dietary antioxidants in human health and disease. *Oxid. Med. Cell. Logev.* **2009**, *2*, 897484. [CrossRef]

29. Saucedo-Pompa, S.; Torres-Castillo, J.A.; Castro-López, C.; Rojas, R.; Sánchez-Alejo, E.J.; Ngangyo-Heya, M.; Martínez-Ávila, G.C.G. Moringa plants: Bioactive compounds and promising applications in food products. *Food Res. Int.* **2018**, *111*, 438–450. [CrossRef]

30. de Morais Cardoso, L.; Pinheiro, S.S.; Martino, H.S.D.; Pinheiro-Sant'Ana, H.M. Sorghum (Sorghum bicolor L.): Nutrients, bioactive compounds, and potential impact on human health. *Crit. Rev. Food Sci. Nutr.* **2017**, *57*, 372–390. [CrossRef]

31. Borbone, N.; Borrelli, F.; Montesano, D.; Izzo, A.A.; De Marino, S.; Capasso, R.; Zollo, F. Identification of a new sesquiterpene polyol ester from *Celastrus paniculatus*. *Planta Med.* **2007**, *73*, 792–794. [CrossRef]

32. Granato, D.; Shahidi, F.; Wrolstad, R.; Kilmartin, P.; Melton, L.D.; Hidalgo, F.J.; Miyashita, K.; Camp, J.V.; Alasalvar, C.; Ismail, A.B.; et al. Antioxidant activity, total phenolics and flavonoids contents: Should we ban *in vitro* screening methods? *Food Chem.* **2018**, *264*, 471–475. [CrossRef]

33. Ahmadinejad, F.; Geir Møller, S.; Hashemzadeh-Chaleshtori, M.; Bidkhori, G.; Jami, M.-S. Molecular mechanisms behind free radical scavengers function against oxidative stress. *Antioxidants* **2017**, *6*, 51. [CrossRef]

34. Silman, I.; Sussman, J.L. Acetylcholinesterase: 'classical' and 'non-classical' functions and pharmacology. *Curr. Opin. Pharmacol.* **2005**, *5*, 293–302. [CrossRef]

35. Zolghadri, S.; Bahrami, A.; Hassan Khan, M.T.; Munoz-Munoz, J.; Garcia-Molina, F.; Garcia-Canovas, F.; Saboury, A.A. A comprehensive review on tyrosinase inhibitors. *J. Enzym. Inhib. Med. Chem.* **2019**, *34*, 279–309. [CrossRef] [PubMed]

36. Geromichalos, G.D.; Lamari, F.N.; Papandreou, M.A.; Trafalis, D.T.; Margarity, M.; Papageorgiou, A.; Sinakos, Z. Saffron as a source of novel acetylcholinesterase inhibitors: Molecular docking and *in vitro* enzymatic studies. *J. Agric. Food Chem.* **2012**, *60*, 6131–6138. [CrossRef] [PubMed]

37. Menghini, L.; Leporini, L.; Vecchiotti, G.; Locatelli, M.; Carradori, S.; Ferrante, C.; Zengin, G.; Recinella, L.; Chiavaroli, A.; Leone, S.; et al. *Crocus sativus* L. stigmas and byproducts: Qualitative fingerprint, antioxidant potentials and enzyme inhibitory activities. *Food Res. Int.* **2018**, *109*, 91–98. [CrossRef] [PubMed]

38. Rocchetti, G.; Giuberti, G.; Busconi, M.; Marocco, A.; Trevisan, M.; Lucini, L. Pigmented sorghum polyphenols as potential inhibitors of starch digestibility: An *in vitro* study combining starch digestion and untargeted metabolomics. *Food Chem.* **2020**, *312*, 126077. [CrossRef]

39. Rocchetti, G.; Giuberti, G.; Gallo, A.; Bernardi, J.; Marocco, A.; Lucini, L. Effect of dietary polyphenols on the *in vitro* starch digestibility of pigmented maize varieties under cooking conditions. *Food Res. Int.* **2018**, *108*, 183–191. [CrossRef] [PubMed]

40. Tan, Y.; Chang, S.K.C. Digestive enzyme inhibition activity of the phenolic substances in selected fruits, vegetables and tea as compared to black legumes. *J. Funct. Foods* **2017**, *38*, 644–655. [CrossRef]

41. Martinez-Gonzalez, A.I.; Díaz-Sánchez, Á.G.; de la Rosa, L.A.; Bustos-Jaimes, I.; Alvarez-Parrilla, E. Inhibition of α-amylase by flavonoids: Structure activity relationship (SAR). *Spectrochim. Acta A* **2019**, *206*, 437–447. [CrossRef] [PubMed]

42. Xiao, J.; Kai, G.; Yamamoto, K.; Chen, X. Advance in Dietary Polyphenols as α-Glucosidases Inhibitors: A Review on Structure-Activity Relationship Aspect. *Crit. Rev. Food Sci. Nutr.* **2013**, *53*, 818–836. [CrossRef] [PubMed]

43. Takahama, U.; Hirota, S. Interactions of flavonoids with α-amylase and starch slowing down its digestion. *Food Funct.* **2018**, *9*, 677–687. [CrossRef] [PubMed]

44. Mocan, A.; Zengin, G.; Crisan, G.; Mollica, A. Enzymatic assays and molecular modeling studies of *Schisandra chinensis* lignans and phenolics from fruit and leaf extracts. *J. Enzym. Inhib. Med. Chem.* **2016**, *31*, 200–210. [CrossRef] [PubMed]

Article

ELISA and Chemiluminescent Enzyme Immunoassay for Sensitive and Specific Determination of Lead (II) in Water, Food and Feed Samples

Long Xu [1,2,†], Xiao-yi Suo [3,†], Qi Zhang [1,3], Xin-ping Li [3], Chen Chen [1] and Xiao-ying Zhang [1,2,3,*]

[1] College of Biological Science and Engineering, Shaanxi University of Technology, Hanzhong 723000, China; xu_lon@163.com (L.X.); 13259850610@163.com (Q.Z.); cchen2008@yahoo.com (C.C.)

[2] Centre of Molecular and Environmental Biology, University of Minho, Department of Biology, Campus de Gualtar, 4710-057 Braga, Portugal

[3] College of Veterinary Medicine, Northwest A&F University, Yangling 712100, China; suoxiaoyi@nwafu.edu.cn (X.-y.S.); lxp67cqu@163.com (X.-p.L.)

* Correspondence: zhang@bio.uminho.pt

† The first two authors contributed equally to this work.

Received: 18 January 2020; Accepted: 5 March 2020; Published: 8 March 2020

Abstract: Lead is a heavy metal with increasing public health concerns on its accumulation in the food chain and environment. Immunoassays for the quantitative measurement of environmental heavy metals offer numerous advantages over other traditional methods. ELISA and chemiluminescent enzyme immunoassay (CLEIA), based on the mAb we generated, were developed for the detection of lead (II). In total, 50% inhibitory concentrations (IC_{50}) of lead (II) were 9.4 ng/mL (ELISA) and 1.4 ng/mL (CLEIA); the limits of detection (LOD) were 0.7 ng/mL (ic-ELISA) and 0.1 ng/mL (ic-CLEIA), respectively. Cross-reactivities of the mAb toward other metal ions were less than 0.943%, indicating that the obtained mAb has high sensitivity and specificity. The recovery rates were 82.1%–108.3% (ic-ELISA) and 80.1%–98.8% (ic-CLEIA), respectively. The developed methods are feasible for the determination of trace lead (II) in various samples with high sensitivity, specificity, fastness, simplicity and accuracy.

Keywords: lead (II); ELISA; monoclonal antibody (mAb); isothiocyanobenzyl-EDTA (ITCBE); chemiluminescent enzyme immunoassay (CLEIA)

---

1. Introduction

Environmental pollution from heavy metals is a worldwide issue. Lead has been widely used in the nuclear industry, glass manufacturing, battery industry, pipe industry, cosmetics industry, toy industry and paint industry [1]. Lead can be accumulated in the environment, as it cannot be rendered harmless through a chemical or bioremediation process. Plant leaves and roots are prone to accumulate toxic metals and can therefore be used for environmental monitoring, as a tool for assessing soil-contamination levels [2].

The major sources of lead exposure include piped drinking water, soldering from canned foods, beverages and traditional medicines. When indirectly ingested through contaminated food or inhalation, lead enters the food chain from the soil, water, deposition from the air, containers or dishes, and/or from food-processing equipment. Lead primarily accumulates in blood, soft tissues, bone and neurons, and this accumulation may cause behavioral changes, cognitive obstacles, blindness, encephalopathy, kidney failure and death. Children are more susceptible and vulnerable to lead due to its impact on the nervous system, as well as on development and behavioral performance [3]. Nowadays, lead pollution has become increasingly serious because of its excessive usage. The recent water contamination of lead in Flint, Michigan, remains a topical issue in public health. The decreased intelligence of children is directly positively correlated with blood lead and bone lead levels [4]. Although regulatory authorities have established safe levels of lead in foods (Table 1), the consensus is that there is no safe level of lead.

The spectroscopy methods for the detection of lead (II) included atomic absorption spectrometry (AAS) [5], atomic fluorescence spectrometry (AFS) [6] and multiple collectors inductively coupled plasma mass spectrometry (MC-ICP-MS) [7]. An AAS was used to detect Pb^{2+} in food with detection limit of 6 ng/mL [8]. These methods are sensitive and accurate, but costly and require intricate equipment and highly qualified technicians, making them unsuitable for onsite detection [9]. Recently, several sensors based on fluorophores, organic molecules and gold nanoparticles [10] have been reported to detect lead ions. A biomimetic sensor was applied in detecting Pb^{2+} in water, with a limit of detection of 9.9 ng/mL [11]. Lead (II) fluorescent sensors detections show high sensitivity, but require fluorophores and/or quenchers. Furthermore, the background signal could lead to serious interference due to its high fluorescence intensity [12]. Electrochemical sensors need tailor-made tactical materials and biological molecules, require skillful design and lengthy sample preparation and lack sufficient specificity. Therefore, the prospects of applying these sensors are limited [13,14].

Immunoassays have been applied for heavy-metal detection (e.g., cadmium, lead, chromium, uranium and mercury) [15], as they are quick, inexpensive, easy to perform, and highly sensitive and selective. ELISA and gold immunochromatographic assay (GICA) have been applied for detection of lead ions in water samples [16,17]. Chemiluminescent enzyme immunoassay (CLEIA), which has been widely used in pesticide and veterinary drug residue analysis, uses the energy generated by chemical reactions to excite luminescence, eliminating the need for external light sources. As a pilot attempt, this study aimed to develop CLEIA and the most commonly used ELISA for Pb^{2+} analysis in water, food and feed samples, to better address the current rapid and sensitive need on Pb^{2+} detection in environment and food contamination.

Table 1. Permitted maximum amount of lead.

	Fresh Vegetable (mg/kg)	Cereals (mg/kg)	Fresh Fruits (mg/kg)	Mushroom (mg/kg)	Beans (mg/kg)	Livestock and Poultry Meat (mg/kg)	Livestock and Poultry Gut (mg/kg)	Fish (mg/kg)	Salt (mg/kg)	Drinking Water (mg/L)
CAC [18]	0.1	0.2	0.1	-	0.2	0.1	0.5	0.3	2	0.01
EFSA [19]	0.1	0.2	0.1	0.3	0.2	0.1	0.5	0.3	-	0.01
CFDA [20]	0.1	0.2	0.1	1	0.2	0.2	0.5	0.5	2	0.01
FSANZ [21]	0.1	0.2	0.1	-	0.2	0.1	0.5	0.5	-	0.05

Notes: CAC: Codex Alimentarius Commission; CFDA; Chinese Food and Drug Administration; EFSA: European Food Safety Authority; FSANZ: Food Standards Australia New Zealand.

2. Materials and Methods

2.1. Ethics Statement

All experimental animal protocols were reviewed and approved by the Ethics Committee of Shaanxi University of Technology for the Use of Laboratory Animals.

2.2. Chemicals and Reagents

Isothiocyanobenzyl-EDTA (ITCBE) was purchased from Dojindo (Kyushu, Japan). N, N′-dicyclohexylcarbodiimide (DCC), N-hydroxysuccinimide (NHS), dimethyl formamide (DMF), 3, 3′, 5, 5′-tetramethylbenzidine (TMB) and luminol were purchased from Solarbio (Beijing, China). HAT medium, keyhole hemocyanin (KLH) and bovine serum albumin (BSA) were purchased from Sigma (St. Louis, MO, USA). Goat anti-mouse IgG-HRP was purchased from Thermo (Waltham, MA, USA). $Pb(NO_3)_2$, $HgSO_4$, $3CdSO_4 \cdot 8H_2O$, $Cr_2(SO_4)_3 \cdot 6H_2O$, $CuSO_4$, $CoCl_2 \cdot 6(H_2O)$, $NiSO_4 \cdot 6H_2O$, $ZnSO_4 \cdot 7H_2O$ and $FeSO_4 \cdot 7H_2O$ were purchased from Sinopharma chemical reagent (Shanghai, China). OriginPro 8.1 (OriginLab, Northampton, MA, USA) was used for processing the analytical data.

2.3. Synthesis of Artificial Antigens of Lead

The ITCBE was conjugated to lead ions, BSA or KLH, using the DCC/NHS ester method. Briefly, equimolar amounts (0.06 mmol) of ITCBE, NHS and DCC were dissolved in 200 µL of DMF, and the same amount of lead nitrate was added to the mixture and stirred overnight. After centrifugation of the solution at 13,400× g for 10 min, the supernatant was added dropwise to 40 mg of BSA or KLH dissolved in 3 mL of 0.13 M $NaHCO_3$ (pH 8.3), under stirring. After reaction for 4 h and centrifugation, the supernatant was dialyzed in phosphate buffered saline (PBS; 0.01 M; pH 7.4) for 4 days, with daily change of buffer.

UV spectra of lead (II)-ITCBE, BSA and lead (II)-ITCBE-BSA were tested at a wavelength ranging from 200 to 400 nm.

2.4. Production of Monoclonal Antibody

Four female BALB/C mice were immunized subcutaneously with 100 µg of lead (II)-ITCBE-KLH emulsified with an equal volume of Freund's complete adjuvant. In the next two sequential booster immunizations, 50 µg of immunogen emulsified with the same volume of incomplete Freund's adjuvant was given to each mouse, in the same way, at 2-week intervals. The fourth injection was administered intraperitoneally without adjuvant. Three days after the final booster injection, the mice were killed. Their spleen cells were removed and fused with mouse SP2/0 myeloma cells, using 50% PEG 4000 (w/v) as fusion agent. The mixture was spread in 96-well culture plates supplemented with hypoxanthine–aminopterin–thymidine (HAT) medium containing 20% fetal calf serum and peritoneal macrophages as feeder cells from BALB/C mice. The plates were incubated at 37 °C, with 5% CO_2. After about 2 weeks, the supernatants were screened by an indirect competitive ELISA, using lead (II)-ITCBE-BSA as coating antigen. ITCBE, lead ions and lead (II)-ITCBE were tested as competitors. The hybridomas which were positive to lead (II)-ITCBE-BSA and negative to ITCBE-BSA were subcloned three times, using the limiting dilution method. Stable antibody-producing clones were expanded and cryopreserved in liquid nitrogen. Antibodies were collected and subjected to purification by ammonium sulfate precipitation. The purified mAb was stored at −20 °C, in the presence of 50% glycerol.

2.5. Indirect Competitive ELISA

The 96-well microtiter plates were coated with lead (II)-ITCBE-BSA conjugation (1 µg/mL, 100 µL/well) in carbonate buffer (CBS, 0.05 M, pH 9.6), and then incubated overnight at 4 °C. The plates were washed three times with PBST (PBS containing 0.05% Tween-20), using an automated plate washer, and blocked with blocking buffer (2% BSA in PBS, 200 µL/well) for 2 h, at 37 °C. After washing, diluted mAbs (stock concentration: 3.5 mg/mL, 1:32 000 dilution, 50 µL/well) were added to lead ions standard solutions (0.2, 1, 2, 5, 10, 20, 50, 100 and 200 ng/mL) or samples (50 µL/well) and incubated for 40 min, at 37 °C. After washing three times, the plates were incubated with goat anti-mouse IgG-HRP (stock concentration: 1.5 mg/mL, 1:8000, 100 µL/well), at 37 °C, for 40 min. Then, the washed plates were added with the substrate solution (TMB+H_2O_2, 100 µL/well). After 10 min of incubation, H_2SO_4 (2 M, 50 µL/well) was added, and the absorbance was measured at 450 nm. Normalized calibration curves were constructed in the form of (B/B_0) × 100(%) vs. log C (lead ions) (where B and B_0 were the absorbance of the analyte at the standard point and at zero concentration of the analyte, respectively.

2.6. Cross-Reactivity

The specificity of the mAb was investigated by cross-reactivity (CR). Different metal ions, including Hg^{2+}, Cu^{2+}, Ni^{2+}, Zn^{2+}, Cd^{2+}, Fe^{2+}, Co^{2+}, Mg^{2+} and Ca^{2+} (in the form of their soluble chloride, nitrate, carbonate or sulfate salts), were analyzed. The standard solutions of cross-reacting chemicals were prepared in the concentration range of 0.001–1000 ng/mL. CR (%) = [IC_{50} for lead ions]/ [IC_{50} for competing chemical] × 100 (%).

2.7. Indirect Competitive CLEIA

The optimal concentrations of lead (II)-ITCBE-BSA and anti-lead antibody were selected, using ELISA, by checkerboard titration. The indirect competitive CLEIA was described as follows: 100 μL/well of lead (II)-ITCBE-BSA (1 μg/mL) in 0.05 M CBS (pH 9.6) was coated on the 96-well polystyrene microtiter plates and incubated at 4 °C overnight. The following day, the plate was washed three times, using PBST, and blocked with 2% BSA in PBS (200 μL per well), at 37 °C, for 2 h. After a further washing step, 50 μL of diluted mAb (stock concentration: 3.5 mg/mL, 1:32 000 dilution) and 50 μL of lead ions standard solution were added to each well and incubated at 37 °C, for 40 min. Lead ions standard solution was prepared by diluting with PBS at a series of concentrations (0.2, 0.5, 1, 2, 5, 10, 20, 50, 100 and 200 ng/mL). After washing with PBST, the plates were incubated, and goat anti-mouse IgG-HRP (stock concentration: 1.5 mg/mL, 1:8000, 100 μL per well) was added and incubated at 37 °C, for 40 min. Finally, 100 μL of substrate solution prepared freshly was added into each well and incubated for 5 min, in the dark. Then chemiluminescence intensity was monitored on Synergy H1. The standard curve was evaluated by plotting chemiluminescence intensity against the logarithm of each concentration and fit to a logistic equation, using OriginLab 8.1 program.

2.8. Graphite Furnace Atomic Absorption Spectrometry (GFAAS)

The operating parameters of the GFAAS system were as follows: lead hollow lamp current 30 mA, wavelength 283.3 nm, shielding gas (Ar) flow rate 1500 mL/min, carrier gas (Ar) flow rate 500 mL/min, and ashing temperature and time were 450 °C and 9 s. The atomization temperature, heating rate and heating time were 2250 °C, 2200 °C/s and 3 s, respectively. The carrier solution was HNO_3 (5.0%, *v/v*). The calibration curve for lead ions was constructed with standards of 0, 0.1, 0.2, 0.4, 0.6, 0.8, 1.0, 1.4, 1.8, 2.4 and 3.0 μg/L.

2.9. Sample Preparation and Spiked Experiment

Spiked samples were used to examine the assay accuracy and precision.

Water samples, including ultrapure water, tap water and river water, were collected from different sites in Yangling, Shaanxi province, China. Water samples (100 mL) were added with Pb standard solution (1 mg/mL) at the final concentration of 100, 200 and 500 ng/mL. Ultrapure water and tap water were analyzed without any dilution and sample preparation. The river water was filtrated with a 0.45 μm nylon membrane filter and adjusted to pH 7.0 before analysis.

Milk samples were collected from the local market. Milk samples (100 mL) were added with Pb standard solution (1 mg/mL) at the final concentration of 100, 200 and 500 ng/mL. Then the samples were boiled to remove the denatured protein and fat, and then an equal volume of acetate buffer solution (0.1 M, pH 5.7) was added for precipitation. After being maintained at room temperature for 2 h, the mixture was centrifuged at 13,400× *g* for 10 min. The pH of the supernatants was adjusted to 7.0 with 1 M NaOH and diluted with pure water for analysis.

Chicken, rice and feed samples (1.0 g) were homogenized and added with Pb standard solution (1 mg/mL) at the final amounts of 100, 200 and 500 ng. Then the samples were extracted by acid leach method. The samples were soaked with 20% HNO_3, overnight, at room temperature, followed by boiling until fully dissolved. After cooling, the solution was centrifuged, and the supernatant was adjusted to a pH value of 7.0 with 1 M NaOH and diluted with pure water for further analysis.

2.10. Pretreatment of Samples for GFAAS

Water samples (10 mL) were added with Pb standard solution (1 mg/mL) at the final amounts of 1, 2 and 5 μg. Then the samples were mixed with 50% HCl (1 mL), 0.8 mL of a solution containing $KBrO_3$ (0.1 M) and KBr (0.084 M). After reaction for 15 min, an appropriate amount of hydroxylamine hydrochloride/sodium chloride (both at a concentration of 120 g/L) solution was added until the yellow color disappeared. The solution was further diluted with pure water, to 200 mL, and determined by GFAAS.

Chicken, rice and feed samples were pretreated, using a microwave-assisted acid-digestion procedure. Samples (1.0 g) were homogenized and added with Pb standard solution (1 mg/mL) at the final amounts of 100, 200 and 500 ng. Then the samples were transferred into polytetrafluoroethylene (PTFE) flasks, and then HNO_3 (8 mL) and H_2O_2 (2 mL) were added to each flask and kept for 15 min, at room temperature. The flasks were sealed and subjected to microwave digestion. Finally, the samples were diluted with pure water, to 200 mL, for GFAAS detection.

3. Results

3.1. Characterization of the Artificial Antigen and the Monoclonal Antibody

Lead (II)-ITCBE, BSA and Lead (II)-ITCBE-BSA spectra were recorded from 200 to 400 nm. BSA exhibits a characteristic ultraviolet absorption peak at 229 and 278 nm, and lead (II)-ITCBE-BSA exhibits a characteristic ultraviolet absorption peak at 215 nm. The shift of the ultraviolet absorption peak proved that the artificial antigen synthesis was successful (see Figure 1).

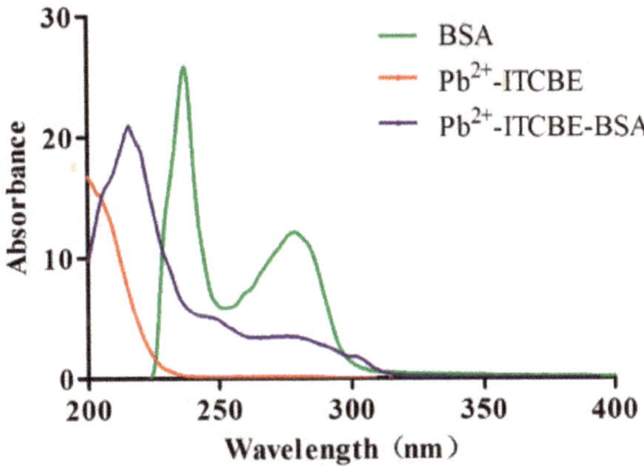

Figure 1. UV absorbance spectra of lead (II)-ITCBE, BSA and lead (II)-ITCBE-BSA.

The anti-lead mAb was purified from mice ascites, using ammonium sulfate precipitation and protein G column affinity chromatography with an obtained concentration of 3.5 mg/mL. The isotype of mAb was IgG1 with a kappa light chain.

3.2. Development of ic-ELISA

Sensitivity of ELISA was determined under optimal conditions. In the representative competitive inhibition curve for lead ions (see Figure 2), the regression curve equation of the anti-lead mAb was $Y = -0.352X + 1.195$ ($R^2 = 0.990$, $n = 3$), with an IC$_{50}$ value of 9.4 ng/mL and limit of detection (IC$_{10}$ value) of 0.7 ng/mL. The ELISA could be used for Pb^{2+} detection with a linear range from 1 to 100 ng/mL.

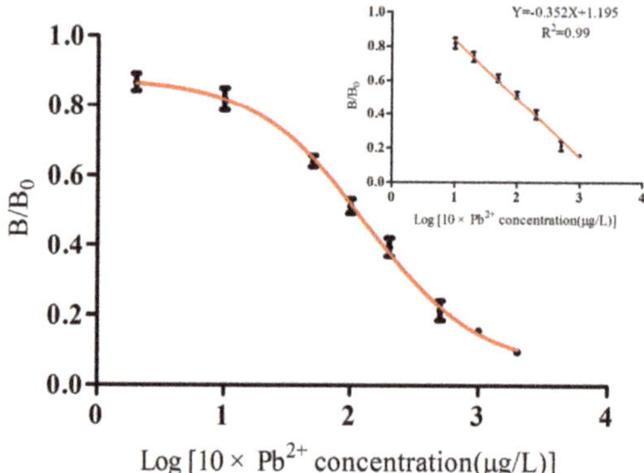

Figure 2. Standard curve of the competitive ELISA for lead ions.

3.3. Cross-Reactivity

The obtained mAb did not recognize the other eight common metal ions (see Table 2).

Table 2. Cross-reactivity of anti-lead IgG with other metal ions ($n = 3$).

Compounds	IC_{50} (µg/L)	Cross-Reactivity (%)
Pb^{2+} -ITCBE	9.4	100
Hg^{2+} -ITCBE	$>1\times10^3$	<0.943
Cd^{2+} -ITCBE	$>1\times10^3$	<0.943
Cr^{3+} -ITCBE	$>1\times10^3$	<0.943
Cu^{2+} -ITCBE	$>1\times10^3$	<0.943
Co^{2+} -ITCBE	$>1\times10^3$	<0.943
Ni^{2+} -ITCBE	$>1\times10^3$	<0.943
Zn^{2+} -ITCBE	$>1\times10^3$	<0.943
Fe^{2+} -ITCBE	$>1\times10^3$	<0.943

3.4. Chemiluminescence Immunoassay

The sensitivity of ic-CLEIA was determined under optimal conditions. The representative competitive inhibition curve (see Figure 3) revealed the regression curve equation of $Y = -0.319X + 0.862$ ($R^2 = 0.992$, $n = 3$), with IC_{50} value of 1.4 ng/mL, the limit of detection (IC_{10} value) of 0.1 ng/mL and the linear range from 0.2 to 50 ng/mL.

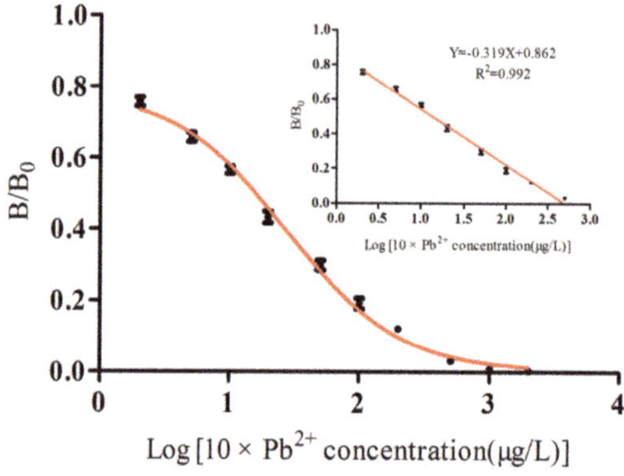

Figure 3. Standard curve of the competitive CLEIA for lead ions.

3.5. GFAAS Analysis of Pb^{2+}

The sensitivity of GFAAS was determined under optimal conditions. The regression curve equation was $Y = 2.857X - 0.020$ ($R^2 = 0.999$, $n = 3$; see Figure 4). The linearity ranged from 0 to 3.0 µg/L. The limit of quantification was 0.86 µg/L.

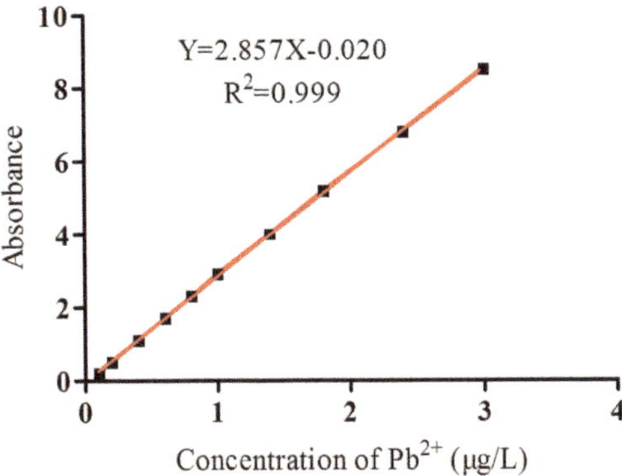

Figure 4. Standard curve of the GFAAS for lead ions.

3.6. Precision and Recovery in Sample Test

The spiked chicken, rice, chicken feed, rat feed, milk and tap water samples containing different concentrations of lead ions (100, 200 and 500 ng/g, respectively) were detected by using the proposed ic-ELISA and ic-CLEIA, respectively, and both methods showed high recoveries and low coefficients of variation (see Table 3). The recovery of the spiked samples suggested that the CLEIA is suitable as a rapid and reliable method to detect lead ions in several matrices.

Table 3. Recovery ratio of Pb^{2+} from different samples ($n = 4$).

Samples	Spiked (ng/g)	ELISA Mea-sured (ng/g)	Recovery (%)	RSD (%)	CLEIA Mea-sured (ng/g)	Recovery (%)	RSD (%)	AAS Mea-sured (ng/g)	Recovery (%)	RSD (%)
Chicken	0	<LOD	–	—	<LOD	–	—	<LOD	–	—
	100	91.0	91.0	10.1	88.2	88.2	16.0	92.1	92.1	8.9
	200	197.5	98.7	11.4	182.2	91.1	2.5	196.4	98.2	0.2
	500	442.3	88.5	15.0	487.4	97.5	4.3	482.5	96.5	0.3
Rice	0	<LOD	–	—	<LOD	–	—	<LOD	–	—
	100	82.1	82.1	5.5	98.8	98.8	12.1	90.2	90.2	9.1
	200	189.1	94.6	8.4	176.4	88.2	9.3	204.0	102.0	4.7
	500	405.2	81.0	1.7	419.3	83.9	11.7	521.2	104.2	0.2
Chicken feed	0	<LOD	–	—	<LOD	–	—	<LOD	–	—
	100	83.8	83.8	6.8	91.3	91.3	6.8	83.3	83.3	8.7
	200	216.5	108.3	12.1	164.5	82.2	12.0	209.2	104.6	0.3
	500	430.8	86.1	6.8	429.9	85.9	10.6	495.7	99.1	0.2
Rat feed	0	<LOD	–	—	<LOD	–	—	<LOD	–	—
	100	83.0	83.0	9.6	82.0	82.0	11.0	80.3	80.3	1.4
	200	171.9	85.9	8.5	175.2	87.6	8.2	195.5	97.7	0.3
	500	410.5	82.1	9.6	453.5	90.7	10.9	522.5	104.5	1.6
Milk	0	<LOD	–	—	<LOD	–	—	<LOD	–	—
	100	94.5	94.5	12.0	85.6	85.6	12.3	88.01	88.01	8.0
	200	175.0	87.5	10.5	160.3	80.1	2.3	194.6	97.3	0.6
	500	486.7	97.3	14.1	463.2	92.6	8.4	456.5	91.3	0.3
Tap water	0	<LOD	–	—	<LOD	–	—	<LOD	–	—
	100	107.7	107.7	10.2	90.9	90.9	8.7	94.2	94.2	9.3
	200	188.3	94.1	9.8	172.8	86.4	5.0	187.4	93.7	0.5
	500	513.0	102.6	3.6	426.3	85.3	4.2	524.5	104.9	0.2

3.7. Comparison of ELISA, CLEIA and GFAAS Results for Lead (II) in Samples

The linear regression curves of ELISA (see Figure 5a) and CLEIA (see Figure 5b) showed good correlation coefficients square of 0.962 and 0.972, respectively, as compared to GFAAS, indicating that the two methods developed could achieve reliable and accurate determination of lead (II) ions in samples.

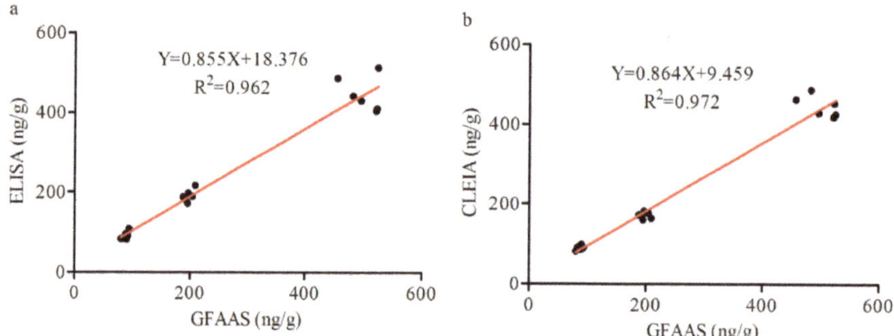

Figure 5. Correlation of ELISA and CLEIA to GFAAS on lead ions analysis.

4. Discussion

The small size and simple structure of heavy metal ions result in poor immunogenicity; as such, they are classified as incomplete antigen. To generate complete antigens for immunological assays, a highly effective bifunctional chelating agent (ITCBE) was selected to connect the lead ion and the carrier protein, which has a large relative molecular mass, reduced toxicity and enhanced immunogenicity [22,23]. The mAb we obtained is superior in sensitivity and specificity as compared to the mAb generated by using the conjugation of lead and S-2-(4-aminobenzyl) diethylenetriamine penta-acetic acid (DTPA) as immunogen, which was applied in ELISA with a limit of detection of 11.6 ng/mL and cross-activity less than 3% [24].

Sample pretreatment is a primary factor for enriching heavy metals and minimizing matrix interference in practical application, as in real detection conditions, lead ions often bind tightly to larger molecules, such as proteins, carbohydrates and colloids [17]. Several methods have been used for heavy-metal sample pretreatment. Microwave digestion method was often used to extract heavy metals from solid samples, including chicken, fish, feces and soil, with high accuracy and recovery rate, but it has limitations in real-time and high throughput detection [25]. A recent study on extracting lead ions in skin-whitening cosmetics, using microwave digestion coupled with plasma atomic emission spectrometry, showed a detection limit of 3.8 µg/kg [26]. The dry ash method is usually used to enrich heavy metals from food and plant samples; however, it demonstrated low accuracy, low recovery rate and high blank value, and it is not suitable for food containing highly volatile inorganic salt [27]. Dry ash extraction has been used for GFAAS-based lead measurement from green vegetables with obtained recovery ranging from 67% to 103% [28]. To better separate the lead ions, we used the acid leach method to enriched lead ions in samples, and have achieved high recovery and a good variable coefficient (Table 3). Furthermore, the acid leach method is easy to operate and has no loss of element, as compared to the other methods, such as microwave digestion, wet digestion and dry ash.

5. Conclusions

In this study, a monoclonal antibody against lead (II) was raised by immunizing Balb/c mouse and hybridoma technique. The LOD of ic-ELISA and ic-CLEIA were 0.7 and 0.1 ng/mL, respectively. The ic-ELISA and ic-CLEIA demonstrated low coefficient of variation. Compared to GFAAS, the two developed methods showed a wide detection range, and the ic-CLEIA showed even more sensitivity compared to the ic-ELISA.

Collectively, ic-ELISA and ic-CLEIA were developed for handy, sensitive and specific detection of lead (II) ions in water, food and feed.

Author Contributions: Conceptualization, L.X., X.-y.S. and X.-y.Z.; data curation, C.C.; formal analysis, Q.Z.; investigation, L.X. and X.Y.S.; methodology, L.X. and X.Y.S.; resources, X.-p.L. and X.-y.Z.; validation, L.X. and X.-y.S.; writing—original draft, L.X. and X.-y.S.; writing—review and editing, X.-y.Z. All authors have read and agreed to the published version of the manuscript.

Funding: This research was funded by National Natural Science Foundation of China, grant numbers 31572556, 31873006; the Key Program for International S&T Cooperation Project of Shaanxi Province, grant number 2017KW-ZD-10; and the Incubation Project on State Key Laboratory of Biological Resources and Ecological Environment of Qinba Areas, grant number SLGPT2019KF04-04.

Conflicts of Interest: The authors declare no potential conflicts of interest with respect to the research, authorship and publication of this article.

References

1. Molera, J.; Pradell, T.; Salvadó, N.; Vendrell-Saz, M. Interactions between Clay Bodies and Lead Glazes. *J. Am. Ceram. Soc.* **2001**, *84*, 1120–1128. [CrossRef]

2. Szczyglowska, M.; Bodnar, M.; Namiesnik, J.; Konieczka, P. The Use of Vegetables in the Biomonitoring of Cadmium and Lead Pollution in the Environment. *Crit. Rev. Anal. Chem.* **2014**, *44*, 2–15. [CrossRef] [PubMed]

3. Sang Yong, E.; Young-Sub, L.; Seul-Gi, L.; Mi-Na, S.; Byung-Sun, C.; Yong-Dae, K.; Ji-Ae, L.; Myung-Sil, H.; Ho-Jang, K.; Yu-Mi, K. Lead, Mercury, and Cadmium Exposure in the Korean General Population. *J Korean Med. Sci.* **2018**, *33*, e9.

4. Wasserman, G.A.; Factor-Litvak, P.; Liu, X.; Todd, A.C.; Graziano, J.H. The Relationship Between Blood Lead, Bone Lead and Child Intelligence. *Child Neuropsychol.* **2003**, *9*, 22–34. [CrossRef]

5. Oliveira de, T.M.; Peres, J.A.; Felsner, M.L.; Justi, K.C. Direct determination of Pb in raw milk by graphite furnace atomic absorption spectrometry (GF AAS) with electrothermal atomization sampling from slurries. *Food Chem.* **2017**, *229*, 721–725. [CrossRef]

6. da Silva, D.L.F.; da Costa, M.A.P.; Silva, L.O.B.; dos Santos, W.N.L. Simultaneous determination of mercury and selenium in fish by CVG AFS. *Food Chem.* **2019**, *273*, 24–30. [CrossRef]

7. Chen, K.y.; Fan, C.; Yuan, H.l.; Bao, Z.a.; Zong, C.l.; Dai, M.n.; Ling, X.; Yang, Y. High-Precision In Situ Analysis of the Lead Isotopic Composition in Copper Using Femtosecond Laser Ablation MC-ICP-MS and the Application in Ancient Coins. *Spectrosc. Spect. Anal.* **2013**, *33*, 1342–1349.

8. Da Col, J.A.; Domene, S.M.A.; Pereira-Filho, E.R. Fast Determination of Cd, Fe, Pb, and Zn in Food using AAS. *Food Anal. Methods* **2009**, *2*, 110–115. [CrossRef]

9. Wei, C.; Shunbi, X.; Jin, Z.; Dianyong, T.; Ying, T. Immobilized-free miniaturized electrochemical sensing system for Pb^{2+} detection based on dual Pb^{2+}-DNAzyme assistant feedback amplification strategy. *Biosens. Bioelectron.* **2018**, *117*, 312–318.

10. Wang, X.Y.; Niu, C.G.; Guo, L.J.; Hu, L.Y.; Wu, S.Q.; Zeng, G.M.; Li, F. A Fluorescence Sensor for Lead (II) Ions Determination Based on Label-Free Gold Nanoparticles (GNPs)-DNAzyme Using Time-Gated Mode in Aqueous Solution. *J. Fluoresc.* **2017**, *27*, 643–649. [CrossRef]

11. Chu, W.; Zhang, Y.; Li, D.; Barrow, C.J.; Wang, H.; Yang, W. A biomimetic sensor for the detection of lead in water. *Biosens. Bioelectron.* **2015**, *67*, 621–624. [CrossRef] [PubMed]

12. Liu, C.W.; Huang, C.C.; Chang, H.T. Highly Selective DNA-Based Sensor for Lead(II) and Mercury(II) Ions. *Anal. Chem.* **2009**, *81*, 2383–2387. [CrossRef] [PubMed]

13. Tang, S.r.; Lu, W.; Gu, F.; Tong, P.; Yan, Z.; Zhang, L. A novel electrochemical sensor for lead ion based on cascade DNA and quantum dots amplification. *Electrochim. Acta* **2014**, *134*, 1–7. [CrossRef]

14. Zhang, H.; Jiang, B.; Xiang, Y.; Su, J.; Chai, Y.; Yuan, R. DNAzyme-based highly sensitive electronic detection of lead via quantum dot-assembled amplification labels. *Biosens. Bioelectron.* **2011**, *28*, 135–138. [CrossRef] [PubMed]

15. Xiang, J.J.; Zhai, Y.f.; Tang, Y.; Wang, H.; Liu, B.; guo, C.W. A competitive indirect enzyme-linked immunoassay for lead ion measurement using mAbs against the lead-DTPA complex. *Environ. Pollut.* **2010**, *158*, 1376–1380. [CrossRef] [PubMed]

16. Tang, Y.; Zhai, Y.F.; Xiang, J.J.; Wang, H.; Guo, C.W. Colloidal gold probe-based immunochromatographic assay for the rapid detection of lead ions in water samples. *Biosens. Bioelectron.* **2010**, *158*, 2074–2077. [CrossRef]

17. Mandappa, I.M.; Ranjini, A.; Haware, D.J.; Manonmani, H.K. Immunoassay for lead ions: analysis of spiked food samples. *J. Immunoassay* **2014**, *35*, 1–11. [CrossRef]

18. Codex Alimentarius Commission (CSC). Codex general standard for contaminants and toxins in food and feed. *Codex stan* **1995**, *193*, 229–234.

19. The commission of the European Community. Commission regulation (EC) NO 1881/2006 setting maximum levels for certain contaminants in foodstuffs. *OJEC* **2006**, *364*, 5–24.

20. National food safety standard for maximum levels of contaminants in foods. Available online: http://www.nhc.gov.cn/ewebeditor/uploadfile/2013/01/20130128114248937.pdf. (accessed on 29 January 2013).

21. Australia New Zealand Standard 1.4.1 contaminants and natural toxicants. Available online: https://www.foodstandards.gov.au/code/Documents/Sched%2019%20Contaminant%20MLs%20v157.pdf (accessed on 1 March 2016).

22. Perrin, C.L.; Kim, Y.J. Symmetry of Metal Chelates. *Inorg. Chem.* **2000**, *39*, 3902–3910. [CrossRef]

23. Love, R.A.; Villafranca, J.E.; Aust, R.M.; Nakamura, K.K.; Jue, R.A.; Major, J.G.; Radhakrishnan, R.; Butler, W.F. How the anti-(metal chelate) antibody CHA255 is specific for the metal ion of its antigen: X-ray structures for two Fab'/hapten complexes with different metals in the chelate. *Biochemistry* **1993**, *32*, 10950–10959. [CrossRef] [PubMed]

24. Zhu, X.; Hu, B.; Lou, Y.; Xu, L.; Yang, F.; Yu, H.; Blake, D.A.; Liu, F. Characterization of monoclonal antibodies for lead chelate complexes: applications in antibody-based assays. *J. Agric. Food Chem.* **2007**, *55*, 4993–4998. [CrossRef] [PubMed]

25. Safari, Y.; Karimaei, M.; Sharafi, K.; Arfaeinia, H.; Moradi, M.; Fattahi, N. Persistent sample circulation microextraction combined with graphite furnace atomic absorption spectroscopy for trace determination of heavy metals in fish species marketed in Kermanshah, Iran and human health risk assessment. *J. Sci. Food Agric.* **2017**, *98*, 2915–2924. [CrossRef] [PubMed]

26. Alqadami, A.A.; Mu, N.; Abdalla, M.A.; Khan, M.R.; Alothman, Z.A.; Wabaidur, S.M.; Ghfar, A.A. Determination of heavy metals in skin-whitening cosmetics using microwave digestion and inductively coupled plasma atomic emission spectrometry. *IET Nanobiotechnol.* **2017**, *11*, 597–603. [CrossRef]

27. Hadiani, M.R.; Farhangi, R.; Soleimani, H.; Rastegar, H.; Cheraghali, A.M. Evaluation of heavy metals contamination in Iranian foodstuffs: Canned tomato paste and tomato sauce (ketchup). *Food Addit. Contam. Part B* **2014**, *7*, 74–78. [CrossRef]

28. Baxter, M.J.; Burrell, J.A.; Crews, H.M.; Massey, R.C.; McWeeny, D.J. A procedure for the determination of lead in green vegetables at concentrations down to 1 µg/kg. *Food Addit. Contam.* **1989**, *6*, 341–349. [CrossRef]

 *foods*

Article

Aromatic Characterization of Mangoes (*Mangifera indica* L.) Using Solid Phase Extraction Coupled with Gas Chromatography–Mass Spectrometry and Olfactometry and Sensory Analyses

Haocheng Liu, Kejing An, Siqi Su, Yuanshan Yu, Jijun Wu, Gengsheng Xiao and Yujuan Xu *

Sericulture & Agri-Food Research Institute Guangdong Academy of Agricultural Sciences/Key Laboratory of Functional Foods, Ministry of Agriculture and Rural Affairs/Guangdong Key Laboratory of Agricultural Products Processing, Guangzhou 510610, China; AnsisHC@163.com (H.L.); ankejing@gdaas.cn (K.A.); lijun@gdaas.cn (S.S.); yuyuanshan@gdaas.cn (Y.Y.); wujijun@gdaas.cn (J.W.); gshxiao@yahoo.com.cn (G.X.)
* Correspondence: xuyujuan@gdaas.cn; Tel.: +86-136-0901-1905

Received: 14 December 2019; Accepted: 5 January 2020; Published: 9 January 2020

Abstract: Mangoes (*Mangifera indica* L.) are wildly cultivated in China with different commercial varieties; however, characterization of their aromatic profiles is limited. To better understand the aromatic compounds in different mango fruits, the characteristic aromatic components of five Chinese mango varieties were investigated using headspace solid-phase microextraction (HS-SPME) coupled with gas chromatography-mass spectrometry-gas chromatography-olfactometry (GC-MS-O) techniques. Five major types of substances, including alcohols, terpenes, esters, aldehydes, and ketones were detected. GC-O (frequency detection (FD)/order-specific magnitude estimation (OSME)) analysis identified 23, 20, 20, 24, and 24 kinds of aromatic components in Jinmang, Qingmang, Guifei, Hongyu, and Tainong, respectively. Moreover, 11, 9, 9, 8, and 17 substances with odor activity values (OAVs) ≥1 were observed in Jinmang, Qingmang, Guifei, Hongyu, and Tainong, respectively. Further sensory analysis revealed that the OAV and GC-O (FD/OSME) methods were coincided with the main sensory aromatic profiles (fruit, sweet, flower, and rosin aromas) of the five mango pulps. Approximately 29 (FD ≥ 6, OSME ≥ 2, OAV ≥ 1) aroma-active compounds were identified in the pulps of five mango varieties, namely, γ-terpinene, 1-hexanol, hexanal, terpinolene trans-2-heptenal, and *p*-cymene, which were responsible for their special flavor. Aldehydes and terpenes play a vital role in the special flavor of mango, and those in Tainong were significantly higher than in the other four varieties.

Keywords: mango; volatile compounds; frequency detection (FD); order-specific magnitude estimation (OSME); odor activity value; sensory analysis

---

1. Introduction

Mango (*Mangifera indica* L.), a native crop of South Asia, is a member of the *Anacardiaceae* family [1] and has been historically grown for more than 4000 years [2], thereby earning the title "king of fruits". It is recognized as one of the most popular fruits around the world, with the highest rates of production, marketing, and consumption [3,4]. Among tropical fruits, mango is the second most common crop involved in international trade, following banana. Global mango production has been estimated at 50.65 million tons, with China being the second largest mango-growing country, 2017 production reaching close to 4.94 million tons [5]. Common mango cultivars in China featuring specific regional characteristics include Guifei, Hongyu Jinmang, Qingmang, and Tainong.

Aroma is a major factor that influences the quality and consumer acceptance of mango products. Investigating various aromatic components would improve our understanding and facilitate controlling critical quality parameters that could influence mango processing. Hundreds of compounds have been characterized in various mango cultivars, which mainly include aldehydes, alcohols, esters, ketones, and terpenes [6]. However, previous studies have mainly focused on the volatiles in various mangoes in China [7,8], and a few volatiles were detected because of their odor threshold. Thus, scientific information relating to aromatic constituents as well as sensory characteristics of various mango cultivars is limited. Therefore, an in-depth investigation is required to identify the volatile or aromatic components of various mango cultivars in China.

To determine the aromatic components of mango, simultaneous solvent-assisted flavor evaporation (SAFE), solid phase microextraction (SPME), and distillation and extraction (SDE) have been employed in food aroma extraction [9–12]. The procedures of SDE and SAFE isolate aromatic compounds from food matrices using organic solvents [13]; however, these methods are highly laborious, time-consuming, and entail preconcentration of extracts. Unlike well-established protocols, SPME has been extensively used in the preparation of volatile and semi-volatile compounds from various types of samples [14]. This technique was developed more than two decades ago and is a rapid, simple, sensitive, and solvent-free technique for the analysis of volatile organic compounds (VOCs) [15]. Coupling of the methods of aroma extraction with the GC-MS/O technique, in particular, headspace solid-phase microextraction (HS-SPME) for extraction, together with detection frequencies (FD) and order-specific magnitude estimation (OSME) for GC-MS/O, creates a highly reliable method of identifying potent odorants.

Thus, to fully understand the aromatic compounds present in typical mango varieties, this study conducted the following studies: (1) identification and quantification of volatiles of different types of mangoes by HS-SPME-GC-MS; (2) discrimination of the major aroma-active compounds in various types of mangoes using combined GC-O detection (FD, OSME) and odor activity value (OAV); and (3) validation of the sensory differences using quantitative descriptive analysis (QDA). In summary, objectives of this investigation were to reveal the major aromatic compounds of mangoes in China.

2. Materials and Methods

2.1. Samples Preparation

This study used seven mature samples of the cultivars Jinmang (JM), Qingmang (QM), Hongyu (HY), Guifei (GF), and Tainong (TN), which were purchased from the regional market in Guangzhou (133.35° N, 23.12° E), Guangdong Province (April 2019). The samples were shipped to the laboratory and kept at 25 °C until complete maturity (2 days). Full maturity and maturity of the mangoes were based on fruit color (green to yellow-orange or red, except for the green lawn, Table S1), odor (sweet scent), and hardness (pulp hardness index changed from 5.28 to 4.32 N). Finally, the mango samples with the same maturity were washed, and the peeled pulp was immediately frozen in liquid nitrogen, and then stored at −80 °C for further studies.

2.2. Chemicals

Humulene, 2-penten-1-ol, 2-hexen-1-ol, (E)-3-hexen-1-ol, 1-hexanol, p-cymen-8-ol, 2-vinyloxy)-ethanol, 3-methyl-1-butanol, 1,3,8-p-menthatriene, allo-ocimene, 2-carene, α-phellandrene, 3-carene, terpinolene, 1,3-cyclohexadiene, 1-methyl-4-(1-methylethyl), isovaleraldehyde, 3-hexenal, 2,4-dimethyl-benzaldehyde, heptanal, *trans*-2-heptenal, trans-2-pentenal, 1-nonanal, decanal, citral, 1-penten-3-one, 2-cyclohepten-1-one, 3-methylcyclohex-3-en-1-one, 6-methyl-5-hepten-2-one, ethyl-propionate, ethyl cyclopropanecarboxylate, ethyl crotonate, isoamyl acetate, tetraethyl orthosilicate, ethyl butyrate, and γ-octanoic were purchased from TCI (Tokyo, Japan); and linalool, α-pinene, β-pinene, β-myrcene, D-limonene, p-cymene, β-ocimene, hexanal, γ-terpinene, and *trans*-2-hexenal were purchased from Sigma (St. Louis, MO, USA). All of the chemical standards were of GC quality.

2.3. HS-SPME-GC-MS

Based on previous studies, the optimized SPME experimental conditions were established [16–18]. Approximately 5.0 g of the juice with 1.5 g of NaCl were blended in a 15 mL vial tightly capped with a PTFE-silicon septum at a stirring speed of 80 rpm. The flavor compounds in mango pulp are formed during equilibrium and during the extraction process. Therefore, the extraction temperature was set at 40 °C. After the vial containing the sample was equilibrated at 40 °C for 10 min on a heating platform agitation, the pretreated (conditioned at 270 °C for 30 min) SPME fiber (50/30 μm DVB/Carboxen/PDMS, Supelco, Bellefonte, PA, USA) was then inserted into the headspace, and extraction was performed for 30 min with continued heating and agitation. Afterward, the fiber was withdrawn and instantly introduced to the GC for desorption and analysis.

GC-MS analysis was performed on an Agilent 7890 (Agilent Technologies, Palo Alto, CA, USA) gas chromatography and Agilent 5977 mass selective detector. Samples were separated using both HP-5 and DB-WAX (both 30 m × 0.25 mm i.d., 0.25 mm film thickness, Agilent Technologies, Palo Alto, CA, USA). Helium was used as carrier gas at a flow rate of 1.7 mL/min, and the GC inlet was set in the split-less mode. The injector temperature was 250 °C. The temperature program was from 40 °C (2 min hold) to 160 °C at 4 °C/min and finally raised to 280 °C at 50 °C/min. Then, electron ionization mode (EI) was used with a 70 eV ionization energy. The ion source temperature was 230 °C, and the mass range was from m/z 35 to 450. The volatile compounds were determined by authentic standards, retention indices (RI), and NIST 14.0 library. The retention indices (RIs) of compounds were determined via sample injection with a homologous series of straight-chain alkanes (C6–C30) (Sigma Aldrich, St. Louis, MO, USA).

2.4. Identification of Aroma-Active Compounds by GC-O

The odorant compounds were analyzed using a sniffing port (ODP3, Gerstel, Germany) coupled with a GC-MS (7890B–5977B, Agilent Technologies, Inc.). Upon exiting the capillary column, the effluents were divided to a ratio of 3:1 (by volume) into a sniffing port as well as an MS detector using an Agilent capillary flow technique. The transfer line directed to the GC-O sniffing port was set at a temperature of 270 °C. The GC-MS settings were similar to those described earlier. Aroma extraction was conducted by four highly skilled personnel (in an alternate order of 50 min intervals) using reference compounds. All personnel were extensively trained on the GC-O technique for at least 90 h.

Frequency analysis was conducted by four trained sensory panelists (i.e., two males and two females). Retention time and odor quality, together with substance detection, were recorded. Frequency analysis was performed in duplicate by every panelist. Odorants with an FD ≥2 (determined by at least two analysts) were considered to have potential aroma activity [19].

The OSME reflected the aromatic intensity of the stimulus that was based on a five-point scale that ranged from 0 to 5, where 0 = none, 3 = moderate, and 5 = extreme. Every sample was sniffed thrice by each panelist, and then the average aromatic intensity values were calculated. If the panelists did not utilize a similar attribute for an aroma that was eluted by GC, the analysis was repeated, and only the descriptors used for the same aroma were included in the analysis [20].

2.5. Quantitative Analysis of Aromatic Compounds

Quantitative data on the identified compounds were gathered by calculating their relative quantitative correction factors (RQCFs) with the "single-point correction method", which is similar to the standard addition method. Similar GC conditions as described above were used in GC-MS analysis, with solvent delay time set at 3 min to prevent the solvent in the standard solutions to reach the filament. The method of obtaining RQCFs consisted of the following: 5.0 g of mango pulp was analyzed using SPME-GC-MS, resulting in an ion peak area for each identified compound. A similar volume of mango pulp with defined amounts of various authentic and internal standards (to avoid run-to-run variations) was then analyzed to generate a new quantifying ion peak for each detected compound and quantitative correction factor for the internal standard. The RQCF of each volatile was generated using the following equation:

$$f_i' = \frac{f_{wi}}{f_{ws}} = \frac{m_i/A_i}{m_s/A_s} = \frac{A_s m_i}{A_i m_s},$$
(1)

where f_i' was the RQCF of a detected compound (i); m_s and m_i were the respective known contents of authentic (i) and internal standard (s); A_s was the peak area of the quantifying ions of (s); and A_i was the peak area of the quantifying ions of (i) before and after the addition of the standard solution to the juice sample.

To determine the amounts of the identified volatiles in mango pulp, approximately 5.0 g of juice per volume containing a similar amount of internal standard as that of the calculated RQCF was prepared and used in GC-MS analysis. The concentration was computed using the equation:

$$m_i = f_i' \times A_i \times \frac{m_s}{A_s}$$
(2)

where m_i was the amount of compound (i); m_s was the known amount of (s), A_i and A_s were the respective peak areas of the quantifying ions of detected compound (i) and internal (s) standards; and f_i' was the RQCF of (i). The peak area of the quantifying ion of every component in selected ion chromatograms was assessed in triplicate, and the average value was computed. Then, the concentration of every identified volatile in mango pulp is described in nanograms per milliliter of juice [19].

2.6. Odor Activity Value (OAV)

OAV pertains to the concentration of the odor divided by its threshold in water. Compounds with an OAV ≥ 1 were considered as major contributors to the aromatic profile of each sample [21].

2.7. Sensory Panel and Aroma Profile Analysis

Aromatic profiling was performed using descriptive sensory analysis, as earlier described [22]. The mango pulps were analyzed by a highly skilled panel of 10 members consisting of five males and five females. Prior to quantitative descriptive analysis, 50 mL of mango pulp was placed in a 100 mL cubage of a plastic cup with a Teflon lid, which was handed over to a panelist in the laboratory without peculiar smell at a temperature of 25 °C. Then, the panelists assessed the aromatic profile of the mango pulp using three preliminary sessions (each spent approximately 2 h), until all of them attained a consensus as to the degree of aromatic flavor. Then, the organoleptic characteristic descriptors were assessed using eight sensory features (i.e., "overall aroma", "tropical fruit", "citrus", "floral", "fresh", "rosin", "honey/sweet", and "green") to evaluate aroma negative and positive mango pulp features. The descriptors were described as the following odors: linalool for the "floral" descriptor, β-phellandrene for "citrus", butyl acetate for the "fruity" descriptor, β-damascenone for the "honey/sweet" descriptor, phenylacetaldehyde for "fresh", (E)-2-hexenal for "green and grassy", and terpinolene for "rosin". The complete profile of each sample was randomly assessed in triplicate for each treatment. The assessors then rated the odor intensities a seven-point scale ranging from 0 to 3, with 0 as not perceivable, 1 as weak, 2 as significant, and 3 as strong. The results were then averaged for every odor note and plotted using a spider web diagram.

3. Results

3.1. Sensory Analysis of Five Mango Cultivars

Figure 1 shows that the five mango pulps had similar aromatic intensities, and these can be divided into six aroma attributes, including tropical fruit, flower aroma, sweet aroma, green grass aroma, green melon aroma, wood aroma, and rosin aroma, although to different extents. SPSS was used to distinguish differences between the mango samples through sensory evaluation scores (Table S2). In most cases, the five mango pulps exhibited significant differences in intensity of fruit aroma ($p < 0.05$), with some exceptions for "tropical fruit" and sweetness between JM and TN, tropical fruit between GF and QM, green between JM and HY, melon between GF and TN, and wood between JM and GF, where no significant difference was observed ($p > 0.05$). In general, innate differences among the mango varieties significantly influenced the intensities of most of the key sensory attributes.

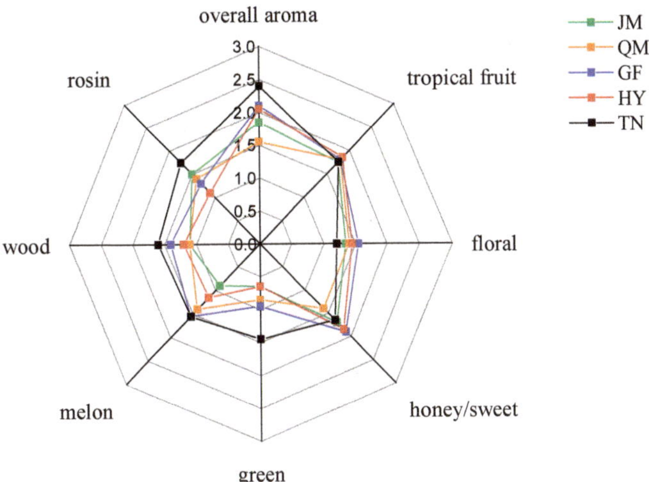

Figure 1. Spider plot for flavor attributes of five varieties mango samples. Jinmang—JM, Qingmang—QM, Hongyu—HY, Guifei—GF, and Tainong—TN.

The overall (decreasing) order of flavor scores was: TN > GF > HY > JM > QM. The intensity of fruit aroma was highest in both GF and QM, followed by TN and JM. The intensity of flower aroma was highest in both GF and HY, followed by QM and JM. TN had the lowest. GF and HY had the highest intensity of sweet aroma, followed by TN, QM, and JM. TN had the highest intensity of green grass aroma, followed by GF and QM. The intensity of green grass aroma in JM and HY was almost identical. The intensities of green melon aroma in QM, GF, and TN were markedly higher than those of JM and HY. TN had the highest intensity of green melon aroma. In addition, the intensity of wood aroma in GF and TN was higher than in JM, QM, and HY. TN had the highest intensity of wood aroma.

In general, GF had the strongest flower and sweet aromas, HY had the strongest fruit aroma, and TN had the strongest green grass, green melon, wood, and Rosin aromas. According to Bonneau [9], different mango varieties significantly influenced the intensities of key sensory attributes of mango, which was due to the different amounts of aroma-active components in mango.

3.2. Comparasion of the GC-MS Results of Five Mango Cultivars

The MS and RI results preliminarily identified 47 volatile compounds (Figure 2). There were 25, 24, 24, 29, and 23 volatile compounds in JM, QM, GF, HY, and TN, respectively. These were generally composed of alcohols, alkenes, aldehydes, esters, ketones, and ethers. Most of the volatiles found in this study were similar to those in the findings of previous studies [1,6,23]. Six compounds, namely, *p*-cymen-8-ol, 2-(vinyloxy)ethanol, 1-methyl-4-(1-methylethyl)-1,3-cyclohexadiene, 3-methyl butanal, 2-cyclohepten-1-one, and tetraethyl orthosilicate, were first observed in the volatile composition of mango, although those were not the major contributors to the mango aroma. This difference might be caused by habitat, maturity conditions, and aromatic extraction method.

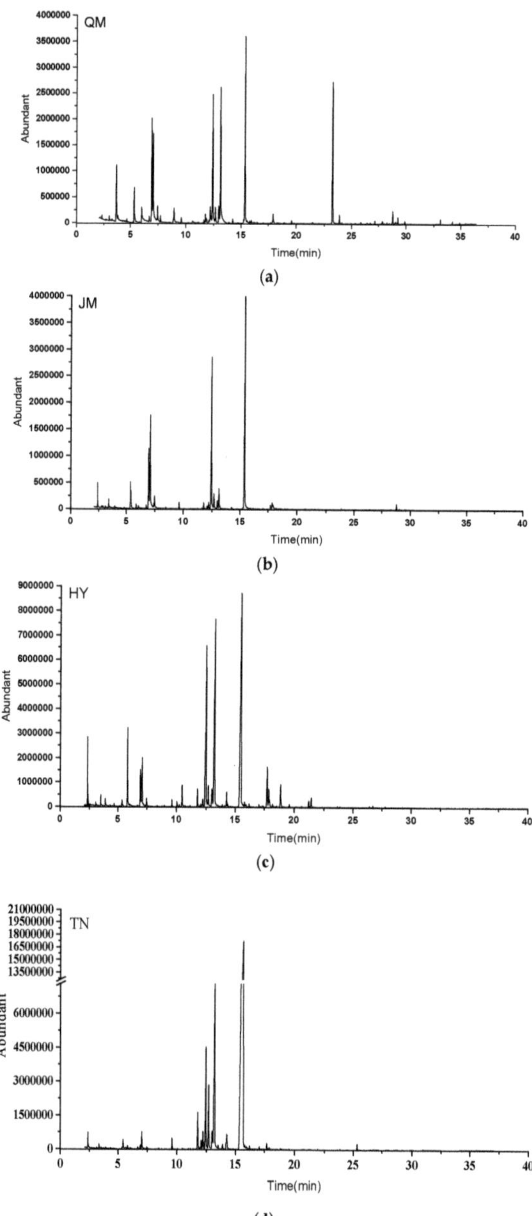

Figure 2. *Cont.*

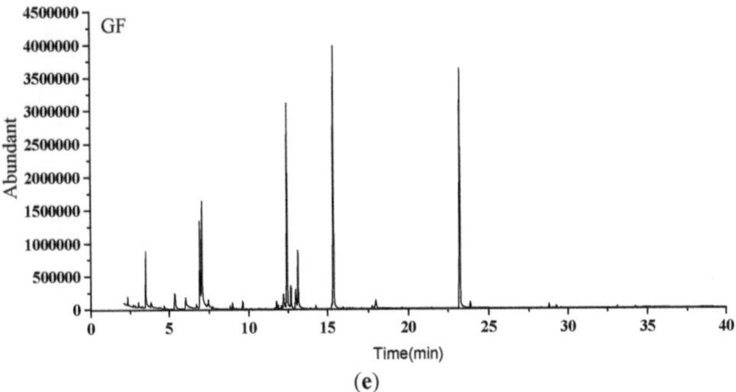

Figure 2. The representative Total ion chromatograms of five varieties mango samples. **a:** QM, **b:** JM, **c:** HY, **d:** TN, **e:** GF.

Figure 3 shows that the number of alkenes in the five mangoes was much higher than that of the other kinds of volatile compounds, indicating that alkenes were the most important volatile substances. In addition, the highest number of alcohols, terpenes, and aldehydes were found in JM, TN, and HY, respectively. The number of ketones in QM and HY was higher than in other mango varieties, whereas the number of esters in GF was higher than in other mangoes. Moreover, only nine volatile components, including one alcohol, one aldehyde, and seven terpenes, were identified in the five kinds of mango pulp. This showed that the volatiles of the five mango cultivars varied.

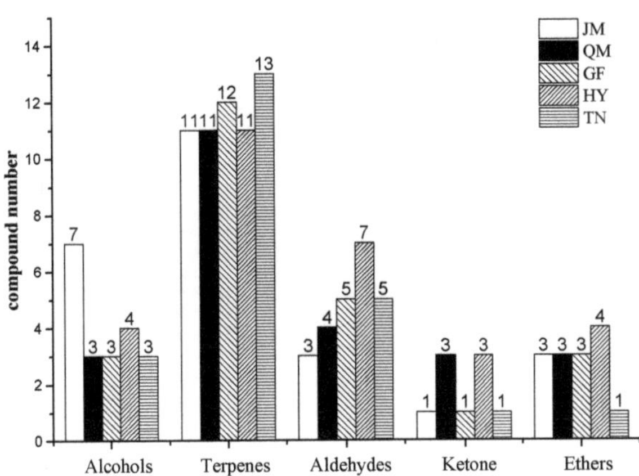

Figure 3. Comparisons of the numbers of volatile compounds detected in different mango samples

Full quantifications using standards were conducted for the volatile substances in the five mango varieties (Table 1 and Figure 4). In JM, the total amount of volatiles was 539.29 µg/kg; the content of terpinolene was the highest (147.07 µg/kg); then came 3-carene (103.63 µg/kg), 1-hexanol (55.83 µg/kg), e-3-hexen-1-ol (41.50 µg/kg), isoamyl alcohol (38.87 µg/kg), and *p*-cymene (27.43 µg/kg). Isoamyl alcohol (38.87 µg/kg) and 2-(vinyloxy) ethanol (3.49 µg/kg) were the unique volatile components. In QM, the total amount of volatiles was 662.92 µg/kg; the content of (e)-2-hexenal was the highest (180.60 µg/kg); then came e-3-hexen-1-alcohol (132.56 µg/kg), 3-carene (60.36 µg/kg), and terpinolene (59.42 µg/kg). The unique volatile components in QM were ethyl crotonate (9.83 µg/kg), e-2-nonenal (7.60 µg/kg), and 3-methyl-3-cyclohexene-1-one (0.62 µg/kg). In GF, the total amount of volatiles was 381.12 µg/kg; (e)-2-hexenal (84.03 µg/kg), trans-3-hexen-1-ol (76.82 µg/kg), and terpinolene (43.53 µg/kg) were the main volatile substances; the unique volatile component was isovaleraldehyde (1.77 µg/kg). In HY, the total amount of volatiles was 400.50 µg/kg; (e)-2-hexenal (85.64 µg/kg), hexanal (75.54 µg/kg), (e)-2-heptenal (65.65 µg/kg), 3-carene (27.67 µg/kg), and terpinolene (24.90 µg/kg) were the main volatile substances; HY is the most special kind of mango. Its unique volatile components were β-caryophyllene (2.18 µg/kg), (e)-2-heptenal (65.65 µg/kg), *p*-cymen-8-ol (12.23 µg/kg), ethyl propionate (6.39 µg/kg), 3-hexen-1-ol (6.59 µg/kg), humulene (2.93 µg/kg), β-caryophyllene (2.18 µg/kg), 2-cyclohepten-1-one (0.66 µg/kg), (e,z)-2,6-nonadienal (0.66 µg/kg), and ethyl cyclopropanecarboxylate (0.12 µg/kg). In the TN pulp, the total amount of volatiles was 1279.68 µg/kg; the content of terpinolene was the highest (811.61 µg/kg), followed by *p*-cymene (132.96 µg/kg), 3-hexenal (44.13 µg/kg), 3-carene (70.92 µg/kg), and phellandrene (30.17 µg/kg), with 3-hexenal being the unique volatile component. Based on the above analysis, the contents of volatile substances in the five cultivars of mangoes were different, and the total content of volatile substances in TN was the highest. And all five kinds of mangoes have unique volatile substances with individual aromatic components; especially for ruby, further analyses by GC-O are needed.

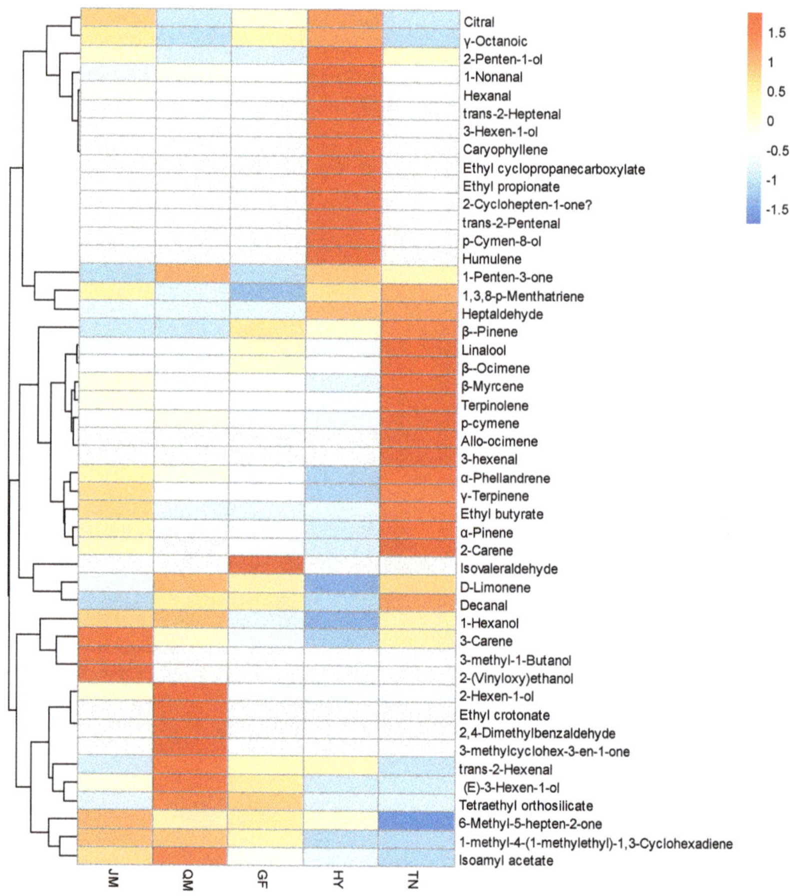

Figure 4. Hierarchical clustering of the all compounds in different mango samples.

3.3. Identification of Key Aromatic Compounds in the Pulps of Five Mango Cultivars

Not all volatile substances in mango contribute to its aroma, and thus aromatic intensity identification was conducted to determine whether the high content or unique volatile components in mango contribute to its overall aroma. The frequency-of-detection (FD) method requires the evaluator to smell the same aromatic substance and record the peak time and aromatic properties of the aromatic substance. The more times the aromatic substance was detected, the greater the contribution to the overall aroma. The intensity (OSME) method was a GC-O detection method that was used to evaluate flavor contribution based on the odor intensity of the aroma substance. After the aromatic substances are separated by GC-MS, the method directly describes the change in odor intensity (measured) and the frequency of aroma attributes. It is considered to be the simplest and most effective GC-O analysis method because it is less time-consuming and less demanding on assessors. Thirty-three characteristic aromatic components were identified in the five mango pulps by GC-O combined with FD and OSME methods, including three alcohols, 10 aldehydes, 11 terpenes, three esters, and four ketones. The differences in aromatic components in the pulp of five mango cultivars were not significant.

Further analysis, the FD analysis identified 23 aromatic substances (FD ≥ 2) in JM. According to the FD statistical data of each substance in the Table 2, one of the substances (FD = 8) was identified in all of the tests: γ-terpinene with citrus aroma. The substances with FD = 7 were phellandrene (citrus-like aroma) and γ-octanoic (floral, violet aroma). The substances with FD = 6 were ethyl cyclopropanecar-boxylate (fruit aroma) and terpinolene (rosin aroma). In QM, 18 aromatic substances with FD ≥2 were identified. According to the FD statistical data of each substance in the table, three kinds of substances (FD = 8) were identified in all of the tests, including 1-penten-3-one, 3-hexenal, and terpinolene. 1-penten-3-one had mushroom-like aroma, 3-hexenal had a grassy aroma, and terpinolene had a rosin-like aroma. The substance with FD = 7 was β-pinene, which has a green-grass-type aroma. The substances with FD = 6 were p-cymene (citrus and green aroma) and γ-terpinene (citrus and lemon-like aroma). In GF, 18 aromatic substances with FD ≥2 were identified. According to the FD statistical data of each substance in Table, 2 substances with FD = 8 were identified in all of the tests: ethyl cyclopropanecarboxylate (fruit aroma) and γ-terpinene (citrus and lemon aroma). The substances with FD = 7 were 3-hexenal (green grass aroma) and terpinolene (rosin-like aroma). The substance with FD = 6 was p-cymene (citrus and green aroma). In HY, 23 aromatic substances with FD ≥2 were identified. The substances with FD = 8 were ethyl cyclopropanecarboxylate (fruit aroma). The substances with FD = 7 were cis-2-penten-1-ol (grass and tea aroma), 3-hexenal (green grass aroma), and terpinolene (rosin-like). The substance with FD = 6 was p-cymene (citrus and green aroma). In TN, there were 24 aromatic substances with FD ≥ 2. The substances with FD = 8 were terpinolene and γ-terpinene, which mainly had a rosin citrus-like aroma. The substances with FD = 6 were ethyl cyclopropanecarboxylate, β-myrcene, 2-carene, phellandrene, and p-cymene. They had a fruity rosin and sweet aroma.

The OSME analysis showed that the aromatic intensities of the five mango pulps had significant differences (Figure 5). There were 23 kinds of aromatic components in JM, and the substances with the aromatic intensity greater than 2 were phellandrene (4.1), ocimene (2.33), γ-terpinene (2.3), and terpinolene (2.3). This suggests an overall aromatic profile of citrus, green grass, and flowers. There were 19 kinds of aromatic components in QM, and the substances with an aromatic intensity greater than 2 were 1-hexanol (2.12), methyl-3-methylcyclohex-3-en-1-one (2.67), β-phellandrene (2.1), and decanal (3) showing an overall aromatic profile of lemon, sweet, and green grass. Seventeen aromatic components were identified in GF, and the substances with aroma intensities greater than 2 were 1-hexanol (2.13), phellandrene (2.5), terpinolene (2.33), and p-menthol (2.25) showing an overall aroma profile of citrus and lemon, fresh green leaves, and green grass. There were 22 aromatic components in HY, and the substances with aromatic intensity greater than 2 were phellandrene (3), ocimene (2.5), α-pinene (2.2), γ-terpinene (2.5), terpinolene (2.2), and 1,3,8-p-menthatriene (2.2) showing an overall aroma profile of flower, citrus, green grass, and pine wood. TN had 22 aromatic components, including hexanal (2.2), 2-carene (2.2), phellandrene (3.53), p-cymene (8.7), γ-terpinene (2.5), terpinolene (2.8), and decanal (3) with aromatic intensity values greater than 2. This has an overall aroma profile of sweet, fruit, green grass, and pine wood.

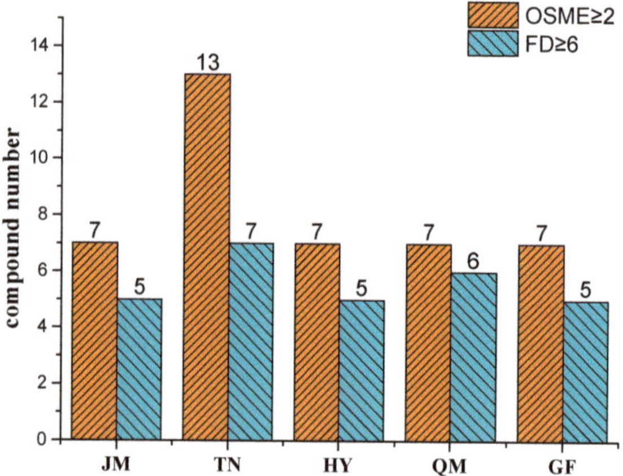

Figure 5. Identification of key aromatic compounds by frequency detection (FD) ≥6 and order-specific magnitude estimation (OSME) ≥2 in five mango pulps.

There were no significant differences in the numbers of overall aromatic substances identified using the two methods (FD and OSME). The number of substances with an FD ≥6 identified in TN by the FD method was the highest, followed by QM, JM, HY, and GF. The highest number of aromatic components with intensity values >2 was also found in TN, followed by HY and JM. QM and GF had the same number of aromatic substances. Therefore, the identification results obtained using FD and OSME were similar. These were highly consistent in identifying key aromatic substances. 2-Cyclohepten-1-one, ethyl cyclopropanecarboxylate, 1,3,8-*p*-menthatriene, and citral were identified for the first time and regarded as a characteristic aroma substance in mangoes, regardless of analytical method used (FD or OSME), although 1,3,8-*p*-menthatriene and citral were previously detected in lychee fruits [16]. In contrast to previous findings of aromatic substances in mangoes, some potent odorants, such as linalool and isoamyl acetate, were not detected, which might be due to differences in varieties, storage conditions, and extraction techniques. According to sensory analysis, the main aromatic profiles (fruit, sweet, flower, and rosin aromas) of pulp from the five mango cultivars were similar to those identified using the FD and intensity methods, indicating that the intensity method combined with the FD method can accurately illustrate the characteristic aromatic components with high or low intensity.

3.4. Odor Activity Values (OAVs) for the Pulps of Five Mango Cultivars

Aromatic analysis techniques such as the FD and intensity methods can effectively analyze the major aromatic compounds in mango pulp, but these cannot accurately reflect the contributions of individual components to the overall aroma characteristics [24]. Therefore, OAVs may be a more accurate scale to evaluate the contribution of volatile substances to fully consider the interactions between the food matrix and aromatic substances. The main component of mango pulp is water, and the calculation of the OAVs of each substance was based on the results of accurate quantitative analysis performed in this study, and the aroma threshold of each compound in water was previously reported [25,26].

According to the literature [19], substances with OAVs >1 contribute to the overall aroma of the sample. Table 3 and Figure 6 show the results of OAV analysis of pulp from five mango cultivars. There were 25 characteristic aromatic components with an OAV ≥ 1 in pulps of five mango cultivars, including two alcohols, seven aldehydes, three esters, 11 terpenes, and two ketones. Terpenes (44%) and aldehydes (28%) were the main aromatic components of mango, of which γ-terpinene had the highest OAV (3.04–10.04), followed by β-phellandrene (2.41–3.41), hexanal (1.10–16.97), and 1-nonanal (5.37–56.2), which were also considered as major aroma-active compounds in Australian mango cultivars. In contrast, although alcohols were the predominant component of all substances (Table 1), these showed minimal contribution (8%), due to their relatively high odor threshold. For instance, the OAV of the highest concentration of (e)-3-hexen-1-ol was only within the range of 0.38–1.21. In addition, 2-cyclohepten-1-one, ethyl cyclopropanecarboxylate, 1,3,8-*p*-menthatriene, and citral were first identified to be useful in aroma activity in mango based on OVA, which coincides with the results of GC-O.

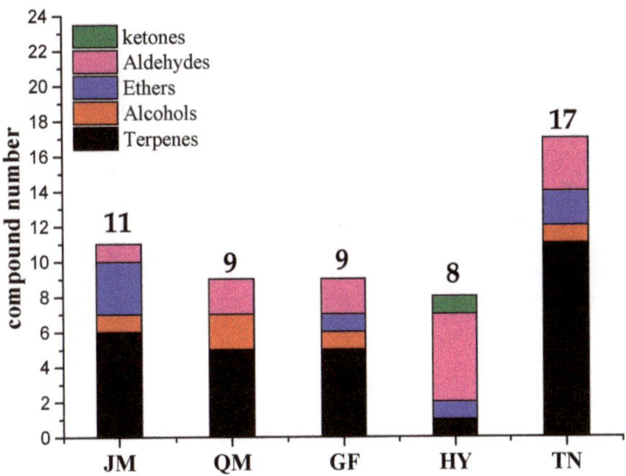

Figure 6. Identification of key aromatic compounds by OVA ≥ 1 in five mango pulps.

OAV values of characteristic aromatic substances were also different in different varieties of mango. Table 3 shows 11 substances with OAV ≥1 in JM. The OAV of γ-terpinene was the largest (7.70), followed by 1-hexanol (6.20), γ-octanoic (2.66), phellandrene (2.41), and ethyl cyclopropanecar-boxylate (2.17). In QM, nine aromatic substances with OAV ≥1 were identified, of which 1-nonanal (9.73) had the largest OAV, followed by 1-hexanol (6.58), γ-terpinene (3.37), *p*-cymene (1.83), and β-phellandrene (1.55). GF has nine substances with OAV ≥1, among which 1-nonanal (6.07) has the largest OAV, followed by γ-terpinene (3.04), 1-hexanol (2.55), γ-octanoic (2.36), decanal (1.41), and *p*-cymene (1.05). HY has eight substances with OAV ≥1, among which 1-nonanal (56.2) has the largest OAV value, followed by 1-hexanal (16.79), trans-2-heptenal (5.05), 2-cyclohepten-1-one (4.71), and γ-octanoic (3.76). HY has 17 substances with OAV ≥1, among which γ-terpinene (10.44) has the largest OAV, followed by *p*-cymene (7.19), terpinolene (5.80), 1-nonanal (5.73), 1-hexanol (5.13), and β-phellandrene (4.31).

3.5. Comparison of GC-O(FD/OSME) and OAV Aroma-Active Compounds

The joint analysis revealed 29 components (FD ≥ 6, OSME ≥ 2, OAV ≥ 1) as aroma-active compounds in the pulps of five mango cultivars (Figure 7). A total of 28 components were detected by GC-O (FD/OSME), whereas 25 substances were detected only by OVA. Compounds with high OAVs, such as 1-nonanal (5.73–56.20), ethyl butyrate (1.56–5.40), and heptanal (1.65–1.83), were not detected using GC-O (FD/OSME). Among the components discriminated by all the panelists in GC-O (FD/OSME), the contributions of 2-penten-1-ol, β-pinene, 3-methylcyclohex-3-en-1-one, and 6-methyl-5-hepten-2-one to the overall mango aroma were limited, as their OAVs were ≤0.1. The discrepancies between the two assessments mainly resulted from differences in application principles [19]. The calculations of the OAVs were based on the odor threshold in water instead of the food matrix. For the actual food matrix, the release of aroma is promoted or inhibited by interactions of volatiles with food components [27–29]. Therefore, OAV identification may not precisely match the actual results generated using GC-O (FD/OSME). However, biological variations, including respiratory rate and receptor state, may lead to errors in aromas based on GC-O (FD/OSME). This explains why the synergetic use of the two methods is strongly recommended for the identification of aroma-active compounds. In this study, the results of OAV coincided with those of GC-O (FD/OSME) to a certain extent, and the 29 key contributors to the five Chinese mango pulps were thus identified. These included 1-penten-3-one, 2-vyclohepten-1-one, 2-penten-1-ol, hexanal, ethyl cyclopropanecarboxylate, *trans*-2-hexenal, 3-hexenal, 1-hexanol, heptanal, α-pinene, β-pinene, e-2-heptenal, 3-methylcyclohex-3-en-1-one, 6-methyl-5-hepten-2-one, β-myrcene, ethyl butyrate, 2-carene, β-phellandrene, 3-carene, *p*-cymene, d-limonene, (E)-beta-ocimene, γ-terpinene, terpinolene, 1-nonanal, 1,3,8-*p*-menthatriene, citral, decanal, and γ-octanoic. These were all recognized as potent and major aroma contributors to mango pulp flavor. Further investigation showed that 1-hexanol, γ-terpinene, β-phellandrene, terpinolene, ethyl cyclopropanecarboxylate, and γ-octanoic were aroma-active compounds in JM; β-phellandrene, *p*-cymene, d-limonene, decanal, 1-hexanol, 1-nonanal were the most important aromatic substances in QM; 1-nonanal, 2-cyclohepten-1-one, 1,3,8-*p*-menthatriene, hexanal, 2-cyclohepten-1-one, and γ-octanoic were the most important aromatic substances in HY; and 1-hexanol, γ-terpinene, γ-octanoic, β-phellandrene, and 1-nonanal were the most important aromatic substances in GF. TN was significantly higher than in the other four aromatic substances. β-myrcene, 2-carene, β-phellandrene, 3-carene, *p*-cymene, d-limonene, γ-terpinene, terpinolene, 1,3,8-*p*-menthatriene, and decanal were the most important aromatic substances in HY.

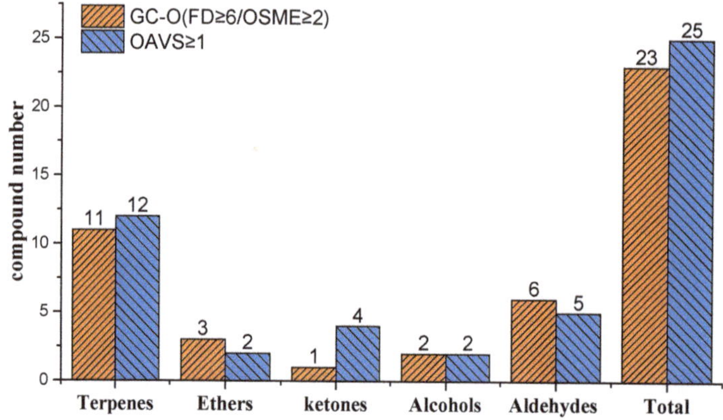

Figure 7. Comparison of GC-O (FD/OSME) and OAVs for aroma-active compounds.

115

Table 1. Volatile compounds identified in five different cultivars of Chinese mango samples using SPME-GC/MS.

Code	Compounds	RI [1]		Identification [2]	RQCF (fi') [3]	Concentration (µg/kg Mango Must) [4]									
		HP-5	Reference			JM	RSD (%)	QM	RSD (%)	GF	RSD (%)	HY	RSD (%)	TN	RSD (%)
						Alcohols									
A1	2-Penten-1-ol	761	769	RI, Std, MS,	1.25	1.92 [b5]	0.02	- [8]	-	-	-	5.45 [a]	0.22	1.87 [c]	0.07
A2	(E)-3-Hexen-1-ol	844	852	RI, Std, MS,	0.52	41.50 [c]	0.03	132.56 [a]	0.12	76.82 [b]	0.12	-	-	-	-
A3	2-Hexen-1-ol	852	852	RI, Std, MS,	1.09	7.08 [b]	0.23	35.53 [a]	0.11	-	-	-	-	-	-
A4	3-Hexen-1-ol	854	857	RI, Std, MS,	4.3	-	-	-	-	-	-	6.59	0.11	-	-
A5	1-Hexanol	856	865	RI, Std, MS,	2.01	55.83 [b]	0.08	59.24 [a]	0.13	22.91 [d]	0.10	7.71 [e]	0.12	46.14 [c]	0.17
A6	isoamyl alcohol	921	910	RI, Std, MS,	0.74	38.87	0.03	-	-	-	-	-	-	-	-
A7	p-Cymen-8-ol	945	ND	RI, Std, MS,	11.08	-	-	-	-	-	-	12.23	0.04	-	-
A8	2-(Vinyloxy) ethanol	1012	ND	RI, Std, MS,	3.34	3.49	0.06	-	-	-	-	-	-	-	-
A9	Linalool	1103	1098	RI, Std, MS,	0.80	-	-	-	-	1.92 [b]	0.07	-	-	6.90 [a]	0.2
						Terpenes									
B1	α-Pinene	918	916	RI, Std, MS,	0.36	4.05 [b]	0.24	1.76 [d]	0.17	1.93 [c]	0.12	1.01 [e]	0.17	7.38 [a]	0.04
B2	Allo-ocimene	947	950	RI, Std, MS,	2.56	-	-	-	-	-	-	-	-	0.08	0.12
B3	β–Pinene	958	946	RI, Std, MS,	2.10	-	-	-	-	2.33 [b]	0.08	1.34[c]	0.17	3.54 [a]	0.01
B4	β-Myrcene	973	990	RI, Std, MS,	0.43	6.28 [b]	0.24	3.83 [d]	0.34	4.5 [c]	0.07	1.41 [e]	0.17	22.94 [a]	0.05
B5	2-Carene	981	995	RI, Std, MS,	0.22	1.60 [b]	0.31	0.67 [c]	0.14	0.54 [d]	0.22	-	-	4.44 [a]	0.05
B6	Humulene	984	990	RI, Std, MS,	0.81	-	-	-	-	-	-	2.93	0.03	-	-
B7	α-Phellandrene	985	989	RI, Std, MS,	0.50	16.84 [b]	0.24	10.84 [c]	0.01	6.90 [d]	0.07	-	-	30.17 [a]	0.04
B8	3-Carene	990	1107	RI, Std, MS,	0.25	103.63 [a]	0.24	60.36 [c]	0.23	43.32 [d]	0.07	27.67 [e]	0.17	70.92 [b]	0.01
B9	D-Limonene	1007	1026	RI, Std, MS,	0.10	12.4 [d]	0.24	32.54 [a]	0.23	26.2 [c]	0.07	2.56 [e]	0.17	28.87 [b]	0.04
B10	p-cymene	1005	1016	RI, Std, MS,	1.11	27.43 [c]	0.08	33.85 [b]	0.14	19.44 [d]	0.06	15.57 [e]	0.09	132.96 [a]	0.04
B11	β-Caryophyllene	1018	ND	RI, Std, MS,	0.64	-	-	-	-	-	-	2.18	0.17	-	-
B12	1,3-Cyclohexadiene, 1-methyl-4-(1-methylethyl)	1021	1030	RI, Std, MS,	0.18	4.96 [a]	0.07	4.84 [b]	0.02	3.42 [c]	0.06	-	-	-	-
B13	γ-Terpinene	1036	1045	RI, Std, MS,	0.68	15.44 [b]	0.16	6.73 [c]	0.19	6.08 [d]	0.26	1.10 [e]	0.07	20.88 [a]	0.25
B14	β-Ocimene	1050	1060	RI, Std, MS,	1.52	-	-	-	-	6.97 [b]	0.08	-	-	9.11 [a]	0.04
B15	Terpinolene	1066	1065	RI, Std, MS,	0.19	147.07 [b]	0.24	59.42 [c]	0.23	43.5 [d]	0.07	24.90 [e]	0.17	811.60 [a]	0.04
B16	1,3,8-p-Menthatriene	1086	1110	RI, Std, MS,	1.37	11.92 [c]	0.18	3.79 [d]	0.19	-	-	13.58 [b]	0.16	16.38 [a]	0.14
						Aldehydes									
C1	Hexanal	792	800	RI, Std, MS,	1.30	4.94 [b]	0.07	-	-	-	-	75.54 [a]	1.50	-	-
C2	Isovaleraldehyde	837	ND	RI, Std, MS,	2.47	-	-	-	-	1.77	0.17	-	-	-	-
C3	trans-2-Hexenal	840	844	RI, Std, MS,	2.20	8.38 [c]	0.02	180.60 [a]	0.26	84.03[b]	0.15	85.64 [b]	1.50	4.11 [d]	0.26
C4	3-hexenal	857	847	RI, Std, MS,	11.55	-	-	-	-	-	-	-	-	44.13	0.12
C5	Heptanal	888	899	RI, Std, MS,	0.07	-	-	-	-	-	-	1.65 [b]	1.50	1.83 [a]	0.44
C6	trans-2-Heptenal	940	957	RI, Std, MS,	1.35	-	-	-	-	-	-	65.65	1.50	-	-

Table 1. *Cont.*

Code	Compounds	RI [1]		Identification [2]	RQCF (fi') [3]	Concentration (µg/kg Mango Must) [4]									
		HP-5	Reference			JM	RSD (%)	QM	RSD (%)	GF	RSD (%)	HY	RSD (%)	TN	RSD (%)
C7	1-Nonanal	1080	1140	RI, Std, MS,	0.45	-	-	1.46 [b]	0.16	0.91 [c]	0.28	8.43 [a]	1.50	0.86 [c]	0.26
C8	(E,Z)-2,6-Nonadienal	1152	1152	RI, Std, MS,	0.64	-	-	-	-	-	-	0.66	0.10	-	-
C9	E-2-Nonenal	1162	1160	RI, Std, MS,	1.87	-	-	7.60	0.09	-	-	-	-	-	-
C10	Decanal	1175	1205	RI, Std, MS,	6.05	-	-	8.48 [c]	0.06	8.45 [c]	0.15	-	-	12.46 [a]	0.04
C11	Citral	1236	1242	RI, Std, MS,	0.02	0.18 [b]	0.28	-	-	0.09 [c]	0.05	1.23 [a]	0.36	-	-
					ketones										
D1	1-Penten-3-one	652	687	RI, Std, MS,	0.05	-	-	0.52 [a]	0.56	-	-	0.47 [b]	0.28	0.31 [c]	0.22
D2	2-Cyclohepten-1-one	673	ND	RI, Std, MS,	1.50	-	-	-	-	-	-	0.66	0.14	-	-
D3	3-methylcyclohex-3-en-1-one	967	986	RI, Std, MS,	3.15	-	-	0.62	0.17	-	-	-	-	-	-
D4	6-Methyl-5-hepten-2-one	969	985	RI, Std, MS,	0.42	0.39 [a]	0.05	0.28 [c]	0.22	0.30 [b]	0.25	0.23 [d]	0.32	-	-
					Ethers										
E1	Ethyl propionate	765		RI, Std, MS,	1.10	-	-	-	-	-	-	6.39	0.11	-	-
E2	Ethyl cyclopropanecarboxylate	808	755	RI, Std, MS,	0.26	-	-	-	-	-	-	0.12	0.55	-	-
E3	Ethyl crotonate	876	802	RI, Std, MS,	1.60	-	-	9.83	0.05	-	-	-	-	-	-
E4	Isoamyl acetate	879	876	RI, Std, MS,	1.37	5.31 [b]	0.09	7.45 [a]	0.22	2.21 [c]	0.19	1.28 [d]	0.14	-	-
E5	Tetraethyl orthosilicate	923	880	RI, Std, MS,	0.06	-	-	0.12 [a]	0.54	0.09 [b]	0.93	-	-	-	-
E6	Ethyl butyrate	978	ND	RI, Std, MS,	0.84	1.17 [b]	0.06	-	-	-	-	-	-	1.80 [a]	0.23
E7	γ-Octanoic	1222	994	RI, Std, MS,	2.48	18.60 [b]	0.21	-	-	16.49 [c]	0.20	26.32 [a]	0.11	-	-

[1] Retention index of volatile compounds on HP-5 columns according to equation proposed [30]; reference: comparing linear retention indices (LRI) on columns (HP-5) in the literature. "ND" not detected in literature. [2] Method of identification: RI, retention index (HP-5) in agreement with literature value; Std, confirmed by authentic standards; MS, mass spectrum comparisons using NIST14 library; [3] RQCF (fi') equals the ratio of quantitative factor of identified components standards to that of internal standard (ethyl hexanoate). [4] Concentrations are expressed in nanograms per milliliter of mango must, with ethyl hexanoate as the internal standard, and data listed are the means of three assays ± RSDs (%); all RSDs were <15%. [5] Values in total data with different letters are significantly different ($p < 0.05$). [8] "-" not detected in samples.

Table 2. GC-O identified aroma-active compounds in mango samples with the method of aromatic intensity and frequency.

No.	Compounds	RI (Calculate) [A]		RI (Reference) [B]		Aroma description [C]	Identification [D]	Aromatic Intensity [E]										Frequency [F]				
		HP-5	Wax	HP-5	Wax			JM	RSD (%)	TN	RSD (%)	HY	RSD (%)	QM	RSD (%)	GF	RSD (%)	JM	TN	HY	QM	GF
1	1-Penten-3-one	652	ND	687 [6]	ND	mushroom	R, S, M, O	0	0	2.25	10.00	1.67	5.99	1.33	11.28	0	0	0	5	4	8	0
2	2-Cyclohepten-1-one	673	ND	ND	ND	coffee-like	R, S, M, O	0	0	0	0	2.00	0.50	0	0	0	0	0	0	5	0	0
3	2-Penten-1-ol	761	1316	769 [31]	1305 [6]	grassy, tea	R, S, M, O	2.00	0.01	1.25	0.80	1.00	1.00	0	0	0	0	4	2	7	0	0
4	Hexanal	792	1061	800 [32]	1088 [32]	grass, tallow, fat	R, S, M, O	1.83	0.49	2.2	0.91	2.00	1.00	0	0	1.80	0.56	5	2	2	0	2
5	Ethyl cyclopropanecarboxylate	808	ND	ND	ND	fruity	R, S, M, O	1.50	0.23	1.30	0.20	1.30	0.90	0	2.11	0.56	0	6	6	8	0	8
6	trans-2-Hexenal	840	1202	844 [32]	1192 [32]	green, leaf	R, S, M, O	1.50	0.23	1.67	6.95	1.67	3.78	2.00	0.50	2.00	0.50	3	5	4	5	4
7	3-hexenal	857	ND	857 [6]	ND	grass	R, S, M, O	1.27	0.67	0	0	1.36	3.78	1.60	8.13	1.43	0.45	3	1	7	8	7
8	3-Hexen-1-ol, (E)-	844	1356	852 [31]	1343 [31]	grassy, tea	R, S, M, O	1.00	0.01	0	0	0	0	1.33	0.75	1.75	5.71	4	2	0	2	3
9	1-Hexanol	856	1360	865 [32]	1362 [32]	resin, flower, green	R, S, M, O	1.60	0.28	1.83	2.73	1.23	0.56	2.12	0.34	2.13	0.67	5	2	4	4	3
10	Heptanal	888	1174	899 [6]	1184 [32]	fruity	R, S, M, O	0	0	1.5	0.67	1.67	2.34	1.00	1.00	0	0	0	5	4	0	0
11	α-Pinene	918	1032	916 [31]	1056 [23]	pine-like	R, S, M, O	1.10	0.34	1.80	10.00	2.20	20.00	1.00	1.00	1.5	0.67	2	4	2	2	2
12	trans-2-Heptenal	940	1300	957 [31]	1307 [31]	grassy, irritant	R, S, M, O	0	0	0	0	0	0	1.25	1.60	1.56	0.9	0	2	1	3	3
13	Allo-ocimene	947	1125	ND	1135 [1]	floral	R, S, M, O	0	0	0	0	1.00	1.00	0	0	0	0	0	0	3	0	0
14	β-Pinene	958	1113	946 [23]	1108 [23]	grass	R, S, M, O	0	0	1.56	0.80	1.23	1.50	1.67	0.60	0	0	0	5	4	7	0
15	3-methylcyclohex-3-en-1-one	967	ND	986 [1]	ND	grass	R, S, M, O	0	0	0	0	0	0	2.67	2.25	0	0	0	1	0	3	0
16	6-Methyl-5-hepten-2-one	969	ND	985 [23]	ND	fruit, grass	R, S, M, O	1.67	0.32	2.00	0.50	1.33	4.51	1.56	0.78	1.75	3.43	2	4	2	2	3
17	β-Myrcene	973	1137	990 [32]	1158 [32]	light balsam, wood	R, S, M, O	1.80	1.20	2.00	0.50	0	0	0	0	0	0	4	6	0	0	0
18	Ethyl butyrate	978	1070	994 [31]	1076 [31]	fruity, apple	R, S, M, O	1.23	5.79	1.67	6.95	0	0	0	0	0	0	3	5	0	0	0
19	2-Carene	981	ND	995 [23]	ND	sweet, rosin	R, S, M, O	1.89	0.98	2.20	2.27	0	0	1.54	0.56	1.23	0.44	5	6	0	4	2
20	β-Phellandrene	985	1166	989 [31]	1171 [31]	citrus like	R, S, M, O	4.10	1.34	3.53	6.67	3.00	0.33	2.10	0.90	2.50	0.67	7	6	2	5	4
21	3-Carene	990	1148	1007 [31]	1153 [31]	sweet, rosin	R, S, M, O	1.50	0.87	2.00	0.50	1.14	0.34	1.45	0.76	1.23	1.56	3	4	2	2	2
22	p-Cymene	1005	1275	1026 [32]	1274 [32]	citrus, green	R, S, M, O	1.00	0.29	8.70	0.45	1.67	6.95	2.00	15.0	1.80	6.67	5	6	6	6	6
23	D-Limonene	1028	1205	1030 [32]	1208 [32]	citrus-like, sweet	R, S, M, O	0	0	2	0.5	0	0	2.00	0.50	0	0	0	5	0	4	0
24	(E)-beta-ocimene	1050	1241	1048 [33]	1250 [33]	floral and green	R, S, M, O	2.33	0.01	0	0	2.50	0.40	0	0	0	0	3	0	3	0	0
25	γ-Terpinene	1057	1249	1060 [33]	1245 [33]	citrus, lemon	R, S, M, O	2.33	0.02	2.50	1.60	2.33	1.72	1.67	6.95	2.00	2.50	8	8	5	6	8
26	Terpinolene	1066	1275	1065 [31]	1276 [31]	rosin-like	R, S, M, O	2.30	0.02	2.28	2.19	2.20	3.18	1.33	0.75	2.33	9.31	6	8	7	8	7

Table 2. *Cont.*

No.	Compounds	RI (Calculate) [A]		RI (Reference) [B]		Aroma description [C]	Identification [D]	Aromatic Intensity [E]										Frequency [F]				
		HP-5	Wax	HP-5	Wax			JM	RSD (%)	TN	RSD (%)	HY	RSD (%)	QM	RSD (%)	GF	RSD (%)	JM	TN	HY	QM	GF
27	1-Nonanal	1080	1328	1104 [31]	1349 [31]	cucumber-like	R, S, M, O	1.67	0.26	1.50	3.33	0	0	0	0	1.40	7.14	5	4	0	0	3
28	1,3,8-p-Menthatriene	1086	ND	1110 [34]	ND	minty-like	R, S, M, O	1.00	0.01	2.00	0.50	2.20	0.45	0	0	2.25	4.44	5	4	4	0	4
29	(E,Z)-2,6-Nonadienal	1152	1460	1152 [31]	1469 [6]	fresh, green, cucumber	R, S, M, O	0	0	0	0	1.67	0.60	0	0	0	0	0	0	2	0	0
30	E-2-Nonenal	1162	1416	1160 [31]	1436 [31]	cucumber-like	R, S, M, O	1.20	0.01	0	0	0	0	0	0	0	0	2	0	0	0	0
31	Citral	1236	1700	1242 [34]	1679 [33]	lemon-like	R, S, M, O	2.00	0.01	0	0	0	0	0	0	0	0	2	0	0	0	0
32	Decanal	1175	1835	1205 [34]	1846 [34]	fruity, citrus, orange	R, S, M, O	0	0	3	0.33	0	0	3.00	0.33	0	0	0	2	0	2	0
33	γ-Octanoic	1222	1721	1236 [31]	1733 [31]	floral, violet	R, S, M, O	2.00	0.01	0	0	1.50	2.00	0	0	2.00	0.50	7	0	4	0	3

[A] Retention index of volatile compounds on columns (HP-5 and WAX) according to equation proposed [30]; "ND" not detected in samples. [B] RI (reference): comparing linear retention indices (LRI) on columns (HP-5 and WAX) in the literature. [C] Odor note perceived at the sniffing port. [D] Method of identification: RI, retention index (HP-5) in agreement with literature value; Std, confirmed by authentic standards; MS, mass spectrum comparisons using NIST14 library. [E] Aromatic intensity, the data listed are the means of three assays ± RSDs (%); all RSDs were <15%. [F] Aroma frequency.

Table 3. Concentrations and calculations of odor activity values (OAVs) of the important aroma-active compounds in mango samples.

No	Compounds	Threshold [a] (µg kg^{-1})	Source [b]	OAV [c]				
				JM	QM	GF	HY	TN
1	1-Penten-3-one	1.00 [35]	mango	0	0.52	0	0.47	0.31
2	2-Cyclohepten-1-one	0.14 [35]	new	0	0	0	4.71	0
3	2-Penten-1-ol	720.00 [23]	mango	<0.01	0	0	<0.01	<0.01
4	Hexanal	4.50 [23]	mango	1.10	0	0.29	16.79	0.40
5	Ethyl cyclopropanecarboxylate	0.12 [23]	new	2.17	0	0	0	1.00
6	trans-2-Hexenal	400.00 [23]	mango	0.02	0.45	0.21	0.21	0.01
7	3-hexenal	550.00 [23]	mango	0	0	0	0	0.08
8	3-Hexen-1-ol, (E)-	110.00 [23]	mango	0.38	1.21	0.70	0	0
9	1-Hexanol	9.00 [26]	mango	6.20	6.58	2.55	0.86	5.13
10	Heptanal	1.00 [23]	mango	0	0	0	1.65	1.83
11	α-Pinene	6.00 [35]	mango	0.68	0.29	0.32	0.17	1.23
12	trans-2-Heptenal	13.00 [35]	mango	0	0	0	5.05	0
13	Allo-ocimene	140.00 [35]	mango	0	0	0.07	0.04	0.10
14	β-Pinene	100.00 [23]	mango	0	0	0	<0.01	0
15	3-methylcyclohex-3-en-1-one	7.00 [36]	mango	0	0.09	0	0	0
16	6-Methyl-5-hepten-2-one	50.00 [23]	mango	<0.01	<0.01	<0.01	<0.01	0
17	β-Myrcene	20.00 [23]	mango	0.31	0.19	0.23	0.07	1.15
18	Ethyl butyrate	0.75 [23]	mango	1.56	0	0	0	2.40
19	2-Carene	4.00 [23]	mango	0.40	0.17	0.14	0	1.11
20	β-Phellandrene	7.00 [23]	mango	2.41	1.55	1.00	0	4.31
21	3-Carene	50.00 [23]	mango	2.07	1.21	0.87	0.55	1.42
22	p-Cymene	18.50 [23]	mango	1.48	1.83	1.05	0.84	7.19
23	D-Limonene	26.00 [23]	mango	0.48	1.25	1.01	0.10	1.11
24	(E)-beta-ocimene	6.70 [23]	mango	0	0	1.04	0	1.36
25	γ-Terpinene	2.00 [23]	mango	7.72	3.37	3.04	0.55	10.44
26	Terpinolene	140.00 [23]	mango	1.05	0.42	0.31	0.18	5.80
27	1-Nonanal	0.15 [23]	mango	0	9.73	6.07	56.20	5.73
28	1,3,8-p-Menthatriene	6.80 [23]	Lychee	1.75	0.56	0	2.00	2.41
29	(E,Z)-2,6-Nonadienal	4.50 [23]	mango	0.57	0	0	0	0.02
30	E-2-Nonenal	50.00 [36]	mango	0	0.15	0	0	0
31	Citral	1.00 [23]	Lychee	0.18	0	0.09	1.23	0
32	Decanal	6.00 [23]	mango	0	1.41	1.41	0	2.08
33	γ-Octanoic	7.00 [35]	mango	2.66	0	2.36	3.76	0

[a] OT odor threshold in water (ppb) found in the newly determined and taken from the literature. [b] Source: It indicates substances found in the related literature for mangoes and litchis; New: first identified to be useful in aroma activity in mango. [c] An OAV was calculated by dividing the concentration of an odorant by its orthonasal odor threshold.

4. Conclusions

Mango has a pleasing sensory quality and rich nutritional components, and thus it is essential to study the composition of flavor components in mango. A total of 47 volatile compounds were preliminarily identified by GC-MS, which were subsequently classified into alcohols, alkenes, aldehydes, esters, ketones, and ethers. The results of GC-O (FD/OSME) analysis showed that there were 23, 20, 20, 24, and 24 kinds of aromatic components in JM, QM, GF, HY, and TN, respectively. Sensory analysis indicated that the main sensory aroma profiles (fruit, sweet, flower, and rosin aromas) of the pulps of five mango cultivars were consistent with those of identified using the FD and OSME methods, indicating that the intensity method combined with the FD method could accurately reflect the characteristic aromatic components with high or low intensities. Moreover, OAV calculations indicated that there were 11 substances with OAVs ≥1 in JM, nine in QM, nine in GF, eight in HY, and 17 in TN. Analysis of OAV and GC-O(FD/OSME) identified 29 predominant aroma-active compounds (FD ≥ 6, OSME ≥ 2, OAV ≥ 1) in the pulps of five mango cultivars, which included citrus, lemon-like γ-terpinene and β-phellandrene, rosin-like terpinolene, floral, green-like 1-hexanol and γ-octanoic, and fruit-like ethyl cyclopropanecarboxylate in JM. The predominant aroma-active compounds of cucumber, fruity, floral, citrus, green-like cucumber-like β-phellandrene, p-cymene, d-limonene, decanal, 1-hexanol, and 1-nonanal were observed in QM. The predominant aroma-active compounds of minty, citrus, green, floral, violet, coffee, cucumber-like 1-nonanal, 2-cyclohepten-1-one, 1,3,8-*p*-menthatriene, hexanal, 2-cyclohepten-1-one, and γ-octanoic were detected in HY. The predominant aroma-active compounds of resin, flower, green, citrus, lemon, cucumber-like 1-hexanol, γ-terpinene, γ-octanoic, β-phellandrene, and 1-nonanal were observed in GF. Light balsam, wood, sweet, rosin, citrus, minty, fruity, citrus, orange-like β-myrcene, 2-carene, β-phellandrene, 3-carene, *p*-cymene, d-limonene, γ-terpinene, terpinolene, 1,3,8-*p*-menthatriene, and decanal were the most important aromatic substances in HY. TN was significantly higher than HY in four other aromatic substances. In addition, 2-cyclohepten-1-one, ethyl cyclopropanecarboxylate, 1,3,8-*p*-menthatriene, and citral were identified to be associated for the first time with aroma activity in mango based on OVA and GC-O(FD/OSME). Hence, this research not only revealed the aroma-active compounds in different mangoes, but also improved our understanding and control of critical aroma parameters in different mango cultivars in China.

Supplementary Materials: The following are available online at http://www.mdpi.com/2304-8158/9/1/75/s1, Table S1: Physicochemical characterization of five different cultivars of mango samples, Table S2: Statistical analysis for flavor attributes of five varieties mango samples.

Author Contributions: Conceptualization, H.L. and K.A.; Data curation and S.S., H.L.; Formal analysis, G.X. and Y.X.; Funding acquisition, Y.X.; Methodology, H.L.; Project administration, Y.Y., J.W. and Y.X.; Resources, Y.Y. and J.W.; Writing—original draft, H.L.; Writing—review & editing, Y.X. and K.A. All authors have read and agreed to the published version of the manuscript.

Funding: We thank the financial support of the National Key Research Project of China (2017YFD0400900, 2017YFD0400904); the Science and Technology Project of Guangzhou (201906010097); and Guangdong Provincial Agricultural Science and Technology Innovation and Extension Project in 2019 (2019KJ101) and Guangdong academy of agricultural sciences president foundation (201806B).

Conflicts of Interest: The authors declare no conflict of interest.

References

1. Pandit, S.S.; Chidley, H.G.; Kulkarni, R.S.; Pujari, K.H.; Giri, A.P.; Gupta, V.S. Cultivar relationships in mango based on fruit volatile profiles. *Food Chem.* **2009**, *114*, 363–372. [CrossRef]

2. Munafo, J.P., Jr.; Didzbalis, J.; Schnell, R.J.; Schieberle, P.; Steinhaus, M. Characterization of the major aroma-active compounds in mango (*Mangifera indica* L.) cultivars Haden, White Alfonso, Praya Sowoy, Royal Special, and Malindi by application of a comparative aroma extract dilution analysis. *J. Agric. Food Chem.* **2014**, *62*, 4544–4551. [CrossRef] [PubMed]

3. Dar, M.S.; Oak, P.; Chidley, H.; Deshpande, A.; Giri, A.; Gupta, V. Nutrient and flavor content of mango (*Mangifera indica* L.) cultivars: An appurtenance to the list of staple foods. In *Nutritional Composition of Fruit Cultivars*; Elsevier: Amsterdam, The Netherlands, 2016; pp. 445–467.

4. Musharraf, S.G.; Uddin, J.; Siddiqui, A.J.; Akram, M.I. Quantification of aroma constituents of mango sap from different Pakistan mango cultivars using gas chromatography triple quadrupole mass spectrometry. *Food Chem.* **2016**, *196*, 1355–1360. [CrossRef] [PubMed]

5. FAOSTAT. Available online: http://www.fao.org/faostat/en/#data/QC/visualize (accessed on 1 October 2019).

6. Pino, J.A.; Mesa, J.; MUNoz, Y.; Martí, M.P.; Marbot, R. Volatile components from mango (*Mangifera indica* L.) cultivars. *J. Agric. Food Chem.* **2005**, *53*, 2213–2223. [CrossRef] [PubMed]

7. Li, L.; Ma, X.W.; Zhan, R.L.; Wu, H.X.; Yao, Q.S.; Xu, W.T.; Luo, C.; Zhou, Y.G.; Liang, Q.Z.; Wang, S.B. Profiling of volatile fragrant components in a mini-core collection of mango germplasms from seven countries. *PLoS ONE* **2017**, *12*, e0187487. [CrossRef]

8. Ma, X.W.; Su, M.Q.; Wu, H.X.; Zhou, Y.G.; Wang, S.B. Analysis of the Volatile Profile of Core Chinese Mango Germplasm by Headspace Solid-Phase Microextraction Coupled with Gas Chromatography-Mass Spectrometry. *Molecules* **2018**, *23*, 1480. [CrossRef]

9. Bonneau, A.; Boulanger, R.; Lebrun, M.; Maraval, I.; Gunata, Z. Aroma compounds in fresh and dried mango fruit (*Mangifera indica* L. cv. Kent): Impact of drying on volatile composition. *Int. J. Food Sci. Technol.* **2016**, *51*, 789–800. [CrossRef]

10. Kung, T.L.; Chen, Y.J.; Chao, L.K.; Wu, C.S.; Lin, L.Y.; Chen, H.C. Analysis of Volatile Constituents in Platostoma palustre (Blume) Using Headspace Solid-Phase Microextraction and Simultaneous Distillation-Extraction. *Foods* **2019**, *8*, 415. [CrossRef] [PubMed]

11. Lau, H.; Liu, S.Q.; Xu, Y.Q.; Lassabliere, B.; Sun, J.; Yu, B. Characterising volatiles in tea (*Camellia sinensis*). Part I: Comparison of headspace-solid phase microextraction and solvent assisted flavor evaporation. *LWT Food Sci. Technol.* **2018**, *94*, 178–189. [CrossRef]

12. Usami, A.; Nakahashi, H.; Marumoto, S.; Miyazawa, M. Aroma evaluation of setonojigiku (*Chrysanthemum japonense var. debile*) by hydrodistillation and solvent-assisted flavor evaporation. *Phytochem. Anal. PCA* **2014**, *25*, 561–566. [CrossRef]

13. Schuh, C.; Schieberle, P. Characterization of the key aroma compounds in the beverage prepared from Darjeeling black tea: Quantitative differences between tea leaves and infusion. *J. Agric. Food Chem.* **2006**, *54*, 916–924. [CrossRef] [PubMed]

14. He, C.; Guo, X.; Yang, Y.; Xie, Y.; Ju, F.; Guo, W. Characterization of the aromatic profile in "zijuan" and "pu-erh" green teas by headspace solid-phase microextraction coupled with GC-O and GC-MS. *Anal. Methods* **2016**, *8*, 4727–4735. [CrossRef]

15. Bianchin, J.N.; Nardini, G.; Merib, J.; Dias, A.N.; Martendal, E.; Carasek, E. Screening of volatile compounds in honey using a new sampling strategy combining multiple extraction temperatures in a single assay by HS-SPME–GC–MS. *Food Chem.* **2014**, *145*, 1061–1065. [CrossRef] [PubMed]

16. Feng, S.; Huang, M.; Crane, J.H.; Wang, Y. Characterization of key aroma-active compounds in lychee (*Litchi chinensis* Sonn.). *JFDA* **2018**, *26*, 497–503. [CrossRef] [PubMed]

17. San, A.T.; Joyce, D.C.; Hofman, P.J.; Macnish, A.J.; Webb, R.I.; Matovic, N.J.; Williams, C.M.; De Voss, J.J.; Wong, S.H.; Smyth, H.E. Stable isotope dilution assay (SIDA) and HS-SPME-GCMS quantification of key aroma volatiles for fruit and sap of Australian mango cultivars. *Food Chem.* **2017**, *221*, 613–619. [CrossRef]

18. Zhao, J.-H.; Liu, F.; Pang, X.-L.; Xiao, H.-W.; Wen, X.; Ni, Y.-Y. Effects of different osmo-dehydrofreezing treatments on the volatile compounds, phenolic compounds and physicochemical properties in mango (*Mangifera indica* L.). *Int. J. Food Sci. Technol.* **2016**, *51*, 1441–1448. [CrossRef]

19. Pang, X.; Guo, X.; Qin, Z.; Yao, Y.; Hu, X.; Wu, J. Identification of aroma-active compounds in Jiashi muskmelon juice by GC-O-MS and OAV calculation. *J. Agric. Food Chem.* **2012**, *60*, 4179–4185. [CrossRef]

20. Mastello, R.B.; Janzantti, N.S.; Monteiro, M. Volatile and odoriferous compounds changes during frozen concentrated orange juice processing. *Food Res. Int.* **2015**, *77*, 591–598. [CrossRef]

21. Munafo, J.P., Jr.; Didzbalis, J.; Schnell, R.J.; Steinhaus, M. Insights into the Key Aroma Compounds in Mango (*Mangifera indica* L. Haden) Fruits by Stable Isotope Dilution Quantitation and Aroma Simulation Experiments. *J. Agric. Food Chem.* **2016**, *64*, 4312–4318. [CrossRef]

22. Matheis, K.; Granvogl, M. Characterization of Key Odorants Causing a Fusty/Musty Off-Flavor in Native Cold-Pressed Rapeseed Oil by Means of the Sensomics Approach. *J. Agric. Food Chem.* **2016**, *64*, 8168–8178. [CrossRef]

23. Bonneau, A.; Boulanger, R.; Lebrun, M.; Maraval, I.; Valette, J.; Guichard, E.; Gunata, Z. Impact of fruit texture on the release and perception of aroma compounds during in vivo consumption using fresh and processed mango fruits. *Food Chem.* **2018**, *239*, 806–815. [CrossRef] [PubMed]

24. Song, H.; Liu, J. GC-O-MS technique and its applications in food flavor analysis. *Food Res. Int.* **2018**, *114*, 187–198. [CrossRef] [PubMed]

25. Zhao, D.; Shi, D.; Sun, J.; Li, A.; Sun, B.; Zhao, M.; Chen, F.; Sun, X.; Li, H.; Huang, M.; et al. Characterization of key aroma compounds in Gujinggong Chinese Baijiu by gas chromatography-olfactometry, quantitative measurements, and sensory evaluation. *Food Res. Int.* **2018**, *105*, 616–627. [CrossRef] [PubMed]

26. Zhu, J.; Wang, L.; Xiao, Z.; Niu, Y. Characterization of the key aroma compounds in mulberry fruits by application of gas chromatography-olfactometry (GC-O), odor activity value (OAV), gas chromatography-mass spectrometry (GC-MS) and flame photometric detection (FPD). *Food Chem.* **2018**, *245*, 775–785. [CrossRef]

27. Chalier, P.; Angot, B.; Delteil, D.; Doco, T.; Gunata, Z. Interactions between aroma compounds and whole mannoprotein isolated from Saccharomyces cerevisiae strains. *Food Chem.* **2007**, *100*, 22–30. [CrossRef]

28. Kühn, J.; Considine, T.; Singh, H. Interactions of milk proteins and volatile flavor compounds: Implications in the development of protein foods. *J. Food Sci.* **2006**, *71*, R72–R82. [CrossRef]

29. Xu, J.; He, Z.; Zeng, M.; Li, B.; Qin, F.; Wang, L.; Wu, S.; Chen, J. Effect of xanthan gum on the release of strawberry flavor in formulated soy beverage. *Food Chem.* **2017**, *228*, 595–601. [CrossRef]

30. Van den Dool, H.; Kratz, P.D. A Generalization of the Retention Index System including Linear Temperature Programmed Gas-Liquid Partition Chromatography. *J. Chromatogr.* **1963**, *11*, 463–471. [CrossRef]

31. Zhang, W.; Dong, P.; Lao, F.; Liu, J.; Liao, X.; Wu, J. Characterization of the major aroma-active compounds in Keitt mango juice:. Comparison among fresh, pasteurization and high hydrostatic pressure processing juices. *Food Chem.* **2019**, *289*, 215–222. [CrossRef]

32. Goodner, K.L. Practical retention index models of OV-101, DB-1, DB-5, and DB-Wax for flavor and fragrance compounds. *LWT Food Sci. Technol.* **2008**, *41*, 951–958. [CrossRef]

33. Babushok, V.I.; Linstrom, P.J.; Zenkevich, I.G. Retention Indices for Frequently Reported Compounds of Plant Essential Oils. *J. Phys. Chem. Ref. Data* **2011**, *40*, 043101. [CrossRef]

34. An, K.; Liu, H.; Fu, M. Identification of the cooked off-flavor in heat-sterilized lychee (*Litchi chinensis* Sonn.) juice by means of molecular sensory science. *Food Chem.* **2019**, *301*, 125282. [CrossRef] [PubMed]

35. Pino, J.A.; Mesa, J. Contribution of volatile compounds to mango (*Mangifera indica* L.) aroma. *Flavor Fragr. J.* **2006**, *21*, 207–213. [CrossRef]

36. Van Gemert, L. *Compilations of Odour Threshold Values in Air, Water and Other Media*; BACIS: Zeist, The Netherlands, 2003.

Article

Application of Near-Infrared Hyperspectral Imaging with Machine Learning Methods to Identify Geographical Origins of Dry Narrow-Leaved Oleaster (*Elaeagnus angustifolia*) Fruits

Pan Gao [1,2,†], Wei Xu [3,4,†], Tianying Yan [1,2], Chu Zhang [5,6], Xin Lv [2,3] and Yong He [5,6,*]

1   College of Information Science and Technology, Shihezi University, Shihezi 832000, China; gp_inf@shzu.edu.cn (P.G.); yantianying@163.com (T.Y.)
2   Key Laboratory of Oasis Ecology Agriculture, Shihezi University, Shihezi 832003, China; lxshz@126.com
3   College of Agriculture, Shihezi University, Shihezi 832003, China; xu_wei082@163.com
4   Xinjiang Production and Construction Corps Key Laboratory of Special Fruits and Vegetables Cultivation Physiology and Germplasm Resources Utilization, Shihezi 832003, China
5   College of Biosystems Engineering and Food Science, Zhejiang University, Hangzhou 310058, China; chuzh@zju.edu.cn
6   Key Laboratory of Spectroscopy Sensing, Ministry of Agriculture and Rural Affairs, Hangzhou 310058, China
*   Correspondence: yhe@zju.edu.cn; Tel.: +86-571-88982143
†   These two authors contributed equally to this manuscript.

Received: 21 October 2019; Accepted: 23 November 2019; Published: 27 November 2019

Abstract: Narrow-leaved oleaster (*Elaeagnus angustifolia*) fruit is a kind of natural product used as food and traditional medicine. Narrow-leaved oleaster fruits from different geographical origins vary in chemical and physical properties and differ in their nutritional and commercial values. In this study, near-infrared hyperspectral imaging covering the spectral range of 874–1734 nm was used to identify the geographical origins of dry narrow-leaved oleaster fruits with machine learning methods. Average spectra of each single narrow-leaved oleaster fruit were extracted. Second derivative spectra were used to identify effective wavelengths. Partial least squares discriminant analysis (PLS-DA) and support vector machine (SVM) were used to build discriminant models for geographical origin identification using full spectra and effective wavelengths. In addition, deep convolutional neural network (CNN) models were built using full spectra and effective wavelengths. Good classification performances were obtained by these three models using full spectra and effective wavelengths, with classification accuracy of the calibration, validation, and prediction set all over 90%. Models using effective wavelengths obtained close results to models using full spectra. The performances of the PLS-DA, SVM, and CNN models were close. The overall results illustrated that near-infrared hyperspectral imaging coupled with machine learning could be used to trace geographical origins of dry narrow-leaved oleaster fruits.

Keywords: narrow-leaved oleaster fruits; near-infrared hyperspectral imaging; geographical origin; convolutional neural network; effective wavelengths

---

1. Introduction

Narrow-leaved oleaster (*Elaeagnus angustifolia*) is a shrub-like plant of Elaeagnus, which is widely distributed from the Mediterranean region to the northern hemisphere, including in northern Russia and northwestern China. Narrow-leaved oleaster fruits contain a variety of functional health components; in particular, they contain polysaccharides, phenolic acids, and flavonoids. Therefore, narrow-leaved oleaster fruits, as a traditional medicine, are used to treat many diseases in nations and countries from

Central Asia to West Asia. As a medicine and food, the fruit of narrow-leaved oleaster fruits is not only a raw material for food industry processing but also a raw material for functional food and new drugs [1–11]. It has good prospects for development and utilization in arid and semi-arid regions of Northwest China. Its unique habitat environment and long history of planting have produced unique qualities of narrow-leaved oleaster fruits in different producing areas. The qualities of narrow-leaved oleaster fruits are different depending on their place of origin, so it is urgent to establish effective methods for identification of the place of origin of narrow-leaved oleaster fruits.

At present, different scholars have isolated the bioactive components of narrow-leaved oleaster fruits [12], studied the physical and chemical properties and antioxidant properties of narrow-leaved oleaster fruits [13], used Gas Chromatography-Mass Spectrometer (GC-MS) to analyze the components of narrow-leaved oleaster fruit oil [14], and studied the diseases of narrow-leaved oleaster fruits [15]. However, there have been few studies on differentiation of the origins of narrow-leaved oleaster fruits. It is feasible to differentiate narrow-leaved oleaster fruits from different producing areas by synthesizing external morphological and microscopic characteristics and physicochemical identification of fruit powder. Manual sorting has many drawbacks, such as involving monotonous work and strong subjectivity, and being time-consuming and difficult to quantify. Physical and chemical index testing is destructive, and requires complicated sample pretreatment, a long detection cycle, and so on. It also has higher professional requirements for testers. These methods are time-consuming and laborious and cannot achieve the goal of fast and non-destructive classification. In view of the drawbacks of traditional detection methods, many applications use hyperspectral imaging for non-destructive detection due to its advantages of non-destructive, rapid, and accurate measurement, which has broad prospects.

Near-infrared hyperspectral imaging is a chemical analysis tool that can detect different absorption frequencies of specific molecules in substances. Near-infrared hyperspectral imaging can acquire spectral and image information of samples simultaneously. It can obtain comprehensive spectral information of samples. It has the characteristics of fastness and high accuracy. Near-infrared hyperspectral imaging has been widely used in geographical origins and variety identification of food [16]. C. Ru et al. used the hyperspectral imaging method of spectral image fusion in the range of visible and near-infrared (VNIR) and shortwave infrared (SWIR) to classify the geographical origin of Rhizoma Atractylodis Macrocephalae [17]. A. Noviyanto et al. used hyperspectral imaging and machine learning to distinguish honey botanical origins [18]. S. Minaei et al. used visible-near-infrared (VIS-NIR) hyperspectral imaging combined with a machine learning algorithm to predict honey floral origins [19]. M. Puneet et al. used near-infrared hyperspectral imaging to identify six different tea products [20]. Our research team has used near-infrared hyperspectral imaging for varietal and geographical origin identification of agricultural and food materials. C. Zhang et al. used near-infrared hyperspectral imaging to identify coffee bean varieties from different locations [21]. W. Yin et al. used near-infrared hyperspectral imaging to identify geographical origins of Chinese wolfberries [22]. S. Zhu et al. used near-infrared hyperspectral imaging to identify cotton seed varieties [23]. These researchers obtained good performances and illustrated the feasibility of using near-infrared hyperspectral imaging to identify the varietal and geographical origin of agricultural and food materials.

In this study, a near-infrared hyperspectral imaging system covering the spectral range of 874–1734 nm was used. This spectral range is related to various chemical compounds. Researchers have used hyperspectral imaging at this spectral range to obtain good performances for determining contents of protein [24], oil [25], water [26], total iron-reactive phenolics, anthocyanins and tannins [27], and flavanol [28], etc. Previous studies have shown that near-infrared hyperspectral imaging can achieve target classification, but there is no relevant research on the place of origin classification of dry narrow-leaved oleaster fruits. The main purpose of this study was to detect the geographical origin of dry narrow-leaved oleaster fruits based on near-infrared hyperspectral imaging technology, combined with characteristic wavelength selection and machine learning algorithms, including deep learning, providing theoretical methods and a basis for distinguishing the different producing areas of narrow-leaved oleaster fruits.

2. Materials and Methods

2.1. Sample Preparation

Dry narrow-leaved oleaster fruits from three different geographical origins, including Miqin County, Gansu province (Gansu), China (103°4′48″ E, 38°37′12″ N); Zhongwei City, Ningxia Hui Autonomous Region (Ningxia), China (105°10′48″ E, 37°30′36″ N); and Aksu City, Xinjiang Uygur Autonomous Region (Xinjiang), China (80°17′24″ E, 41°9′00″ N), were collected. For each geographical origin, fully matured fruits were harvested in October 2018 and air-dried for consumption and trade. For each geographical origin, intact, clean, and dry narrow-leaved oleaster fruits were collected for hyperspectral image acquisition. In total, 1105, 1205, and 962 intact fruits were obtained from Gansu, Ningxia, and Xinjiang, respectively. The convolutional neural network (CNN) was trained with an independent validation set. To build discriminant models, the samples were randomly split into calibration, validation, and prediction sets. There were 539, 602, and 481 samples from Gansu, Ningxia, and Xinjiang in the calibration set, 291, 303, and 241 samples from Gansu, Ningxia, and Xinjiang in the validation set, and 275, 300, and 240 samples from Gansu, Ningxia, and Xinjiang in the prediction set, respectively. Samples of each geographical origin for hyperspectral imaging acquisition are placed and presented in Figure 1.

Figure 1. Samples of each geographical origin for hyperspectral imaging acquisition.

2.2. Hyperspectral Image Acquisition and Correction

A near-infrared hyperspectral imaging system was used to acquire hyperspectral images of single narrow-leaved oleaster fruits. This hyperspectral imaging system consisted of four major modules, including an imaging module, an illumination module, a sample motion module, and a software module. The imaging module consisted of an imaging spectrograph (ImSpector N17E, Spectral Imaging Ltd., Oulu, Finland) coupled with an InGaAs camera (Xeva 992, Xenics Infrared Solutions, Leuven, Belgium). The spectral range of the hyperspectral imaging system was 874–1734 nm, the spectral resolution 5 nm, and the number of wavebands 256. The lens for the camera was OLES22 (Spectral Imaging Ltd., Oulu, Finland). The illumination module had a 3900 light source (Illumination Technologies Inc., New York, NY, USA). The sample motion module was formed by an IRCP0076 electric displacement table (Isuzu Optics Corp., Taiwan, China) and samples were placed in the motion platform for line-scan. The software module was used to control the image acquisition and motion platform. The structure of the acquired hyperspectral image was able to be expressed as 320 pixels × L pixels × 256 (wavebands), where 320 pixels was the width of the image, the number 256 was the number of wavebands, and L pixels was the length of the image. L was manually determined during the image acquisition to ensure all samples in one plate were covered in one image.

The image quality, which was determined by the distance between the sample and the lens, the moving speed of the motion platform, and the camera exposure time, was determined by setting these parameters as 12.6 cm, 11 mm/s, and 3000 µs, respectively. In this study, intact narrow-leaved oleaster fruits were placed separately on a black plate for image acquisition. For each image, a random number of fruits was placed there (as shown in Figure 1), and there were at least twenty fruits in an image. During image acquisition, the imaging conditions and system parameters always remained. After image acquisition, the raw hyperspectral images were corrected into reflectance images according to the equation

$$I_c = \frac{I_r - I_d}{I_w - I_d}, \tag{1}$$

where I_c is the corrected image, I_r is the raw original image, I_d is the dark reference image and I_w is the white reference image.

2.3. Spectral Data Extraction

After image correction, spectral data were extracted from each narrow-leaved oleaster fruit. The hyperspectral imaging system collected reflectance spectra of the samples, and reflectance spectra were used for analysis in this study. Each single narrow-leaved oleaster fruit was defined as a region of interest (ROI). A binary image was formed of each hyperspectral image by binarizing the gray-scale image at 1119 nm, in which the narrow-leaved oleaster fruits region was '1' and the background region was '0'. The binary image was then applied to the gray-scale images at each gray-scale image to remove background information. Considering that obvious noises existed at the beginning and end of the spectra, only spectra in the range 975–1646 nm (waveband numbers 31 to 230) were studied, resulting in 200 wavelength variables in the spectral range. Pixel-wise spectra were preprocessed by wavelet transform (wavelet function Daubechies 6 with decomposition level 3) to reduce random noise and area normalization to reduce the influence of sample shape. Pixel-wise spectra within one narrow-leaved oleaster fruit were averaged to represent the sample.

2.4. Data Analysis Methods

2.4.1. Principal Component Analysis

Principal component analysis (PCA) is a widely used qualitative analysis and feature extraction method for spectral data analysis. PCA projects the original spectral data to some new principal component variables (PCs) through linear transformation. Each principal component is linearly combined with the original data. The PCs are ranked by the explained variance. The first PC (PC1) explains the largest of the total variance, followed by PC2 and PC3 and so on. In general, the first few PCs could explain most of the total variance and these few principal components with the largest variance could reflect the data information. In general, the scores of scatter plots which are obtained by projecting scores of one PC onto another PC are used to explore clusters of samples from different classes. In this study, PCA was used to explore qualitative discrimination of narrow-leaved oleaster fruit samples from Gansu, Ningxia, and Xinjiang.

2.4.2. Partial Least Squares Discriminant Analysis

The partial least squares discriminant analysis (PLS-DA) algorithm is based on the PLS regression model to discriminate the target, where the variables in the X block (spectral data) are related to the category values corresponding to the classes contained in the Y vector [29–35]. The integer values are assigned to each class. The category values can be assigned as real integer numbers or they can be formed by dummy variables (0 and 1). PLS regression is firstly conducted on X and Y and the decimal prediction results are transformed into category values according to certain rules.

2.4.3. Support Vector Machine

The support vector machine (SVM) system has been widely applied in statistics, especially for classification. The main idea of SVM is to find the most distinguishable hyperplane by maximizing the margin between the closest points in each class [34–38]. By choosing and optimizing parameters such as penalty factor and kernel function, the discriminant model established by small data samples can still produce small errors for independent test sets. In this paper, the parameter penalty coefficient C of SVM model was searched, and the optimum range was 10^{-8} to 10^8. The kernel function was a radial basis function (RBF) and the searching range of the width of the kernel function (g) was 10^{-8} to 10^8.

2.4.4. Convolutional Neural Network

The convolutional neural network has been proved as a data processing method with high efficiency and high performance for hyperspectral data analysis due to its ability to aid automatic feature learning [39]. In this study, a simplified CNN architecture based on the model proposed in [40] was designed for narrow-leaved oleaster fruit discrimination.

Figure 2 shows the CNN architecture used in this research. It consisted of two main parts. The first part included two one-dimensional convolution layers (Conv1D, represented by a box with a green background), each of which having been followed by a ReLU activation (yellow box), a one-dimension MaxPooling layer (MaxPool1D, blue box) and a batch-normalization (white box) process. The other part included a fully connected network which was constructed by three Dense layers (light red box) and a SoftMax layer (gray box). The numbers of kernels in the convolution layers were 64 and 32, respectively, with a kernel size of 3 and stride of 1 without padding. MaxPooling layers were configured with a pool size of 2 and stride of 2. The numbers of neurons in the Dense layers were defined as 512, 128, and 3, in order. The first two Dense layers were activated by the ReLU function and followed by a batch-normalization process.

The training procedure was implemented by minimizing the SoftMax Cross Entropy Loss using a stochastic gradient descent (SGD) algorithm. The learning rate was optimized and set as 0.0005. The batch size was set as 400. The train epoch was defined as 400.

2.4.5. Optimal Wavelength Selection

Extracted spectra data contain redundant and collinear information, and some of the wavelengths are uninformative. These uninformative wavelengths may result in unstable calibrations. Moreover, a large number of wavelengths for calibration may result in a complex model structure. Selecting the most informative wavelengths is an important step for further multivariate analysis.

In this study, second derivative spectra were used to select the optimal wavelengths for narrow-leaved oleaster fruits. The second derivative is a widely used spectral preprocessing method which can highlight spectral peaks and suppress background information. In second derivative spectra, the background information is quite small and close to zero, and the positive and negative peaks with greater differences among different categories of samples are manually selected as optimal wavelengths [41].

2.5. Software and Model Evaluation

In this study, PCA, PLS-DA, and SVM were executed on a Matlab R2014b (The Math Works, Natick, MA, USA), the second derivate was conducted on Unscrambler 10.1 (CAMO AS, Oslo, Norway), and the CNN model was performed on Python 3 and MXNET framework (Amazon, Seattle, WA, USA). PCA and PLS-DA was computed using leave-one-out cross validation, SVM was computed using five-fold cross validation, and CNN was computed using an independent validation set. Model performances were evaluated by their classification accuracy, which was calculated as the ratio of the number of correctly classified samples to the total number of samples.

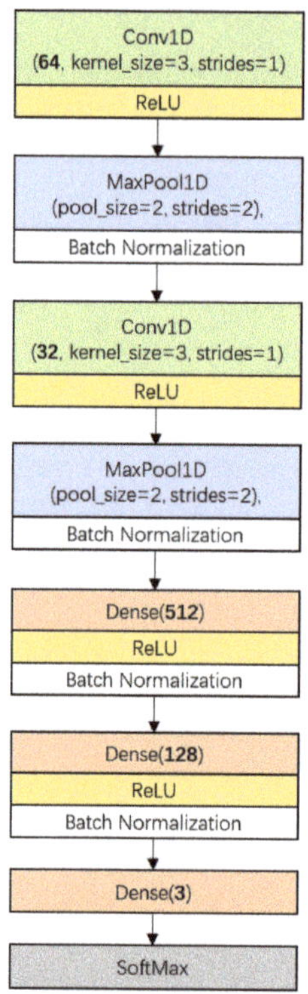

Figure 2. The proposed convolutional neural network (CNN) architecture for narrow-leaved oleaster fruit identification. Conv1D denotes 1-dimension convolution layer, ReLU (Rectified Linear Unit) is the activation function, MaxPool1D denotes 1-dimension max pooling layer, Dense denotes densely-connected neural network layer. The parameter of Conv1D which is defined as 'Channels' is the number of the kernels or filters. The parameter of Dense which is defined as 'units' is the number of the neurons.

3. Results

3.1. Spectral Profiles and Effective Wavelength Identification

Figure 3 shows the average spectra with standard deviation of each wavelength of narrow-leaved oleaster fruits from Gansu, Ningxia, and Xinjiang. Slight differences in reflectance values exist in the average spectra. The differences exist across the whole spectral ranges. However, the overlaps can be observed according to the standard deviation in Figure 3. With these overlaps, the samples from different geographical origins cannot simply be identified by observing their spectral differences. Figure 4 shows the second derivative spectra of the average spectra of narrow-leaved oleaster fruit samples from Gansu, Ningxia and Xinjiang. There are wavelengths with differences. Wavelengths corresponding to the peaks and valleys with greater differences were manually identified. As shown in Figure 4, a total of 22 wavelengths can be identified: 995, 1022, 1032, 1042, 1056, 1072, 1089, 1136, 1190, 1244, 1274, 1284, 1315, 1352, 1365, 1375, 1402, 1433, 1456, 1487, 1500, and 1632 nm. These wavelengths were selected as the effective wavelengths for geographical identification. In this study, the full spectra were used to conduct PCA for qualitative analysis of the sample cluster within one geographical origin and sample separability among different geographical origins. The full spectra were also used to build machine learning models to quantitatively assess the sample separability among different geographical origins. To reduce redundant and collinear information which are informative in full spectra, simplify the models and improve model robustness, the selected effective wavelengths were used to build machine learning models for comparison with the full-spectra-based models.

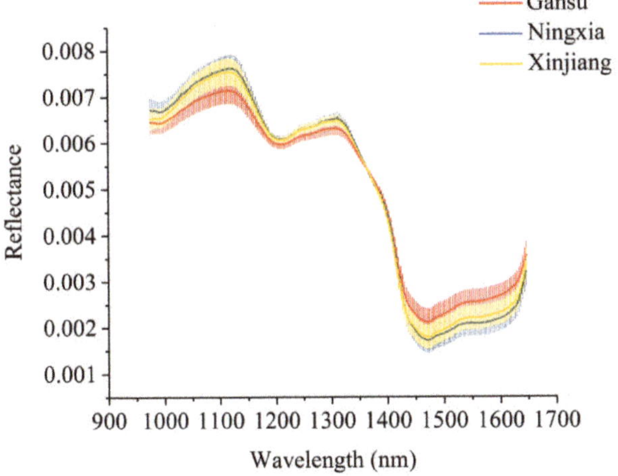

Figure 3. Average spectra with standard deviation of each wavelength of narrow-leaved oleaster fruits from Gansu, Ningxia, and Xinjiang.

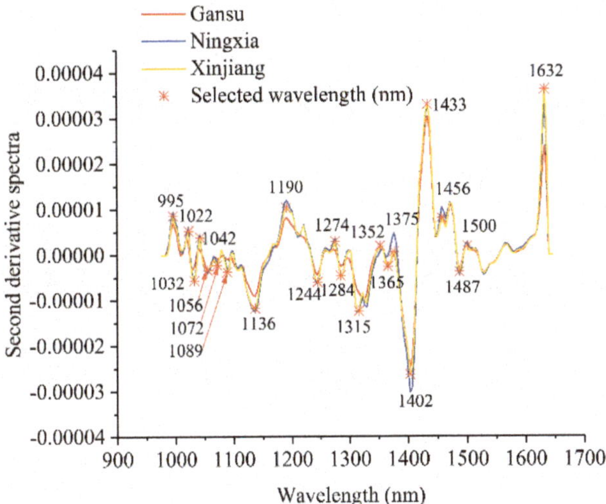

Figure 4. Effective wavelength selection using the second derivative spectra of average spectra of the samples from Gansu, Ningxia, and Xinjiang.

3.2. Principal Component Analysis

PCA was conducted to qualitatively cluster the samples in the scoring spaces. PCA was conducted on the full spectra of the calibration set, and the spectral data were centered for PCA analysis. The first three PCs explain most of the total variance, which was over 99% (PC1: 97.34%, PC2: 1.24%, PC3: 0.63%). Score scatter plots of two different PCs are shown in Figure 5. Samples from the same geographical origins are marked with the same color, as well as the confidence ellipse (confidence level at 0.95). As shown in the score scatter plot of PC1 versus PC2, samples from each geographical origin are able to cluster well. Overlaps exist among the samples from Gansu, Ningxia, and Xinjiang.

In the score scatter plot of PC1 versus PC3, samples from each geographical origin are able to cluster well. Samples from Gansu show greater overlaps with samples from the other geographical origins, and samples from Ningxia and Xinjiang are able to separate well. In the score scatter plot of PC2 versus PC3, samples from each geographical origin are able to cluster well. Samples from Gansu show greater overlaps with samples from the other geographical origins, and samples from Ningxia and Xinjiang are able to separate well. The score scatter plots in Figure 5 showed that the samples from different geographical origins are able to be well clustered and that they have great potential to be correctly identified.

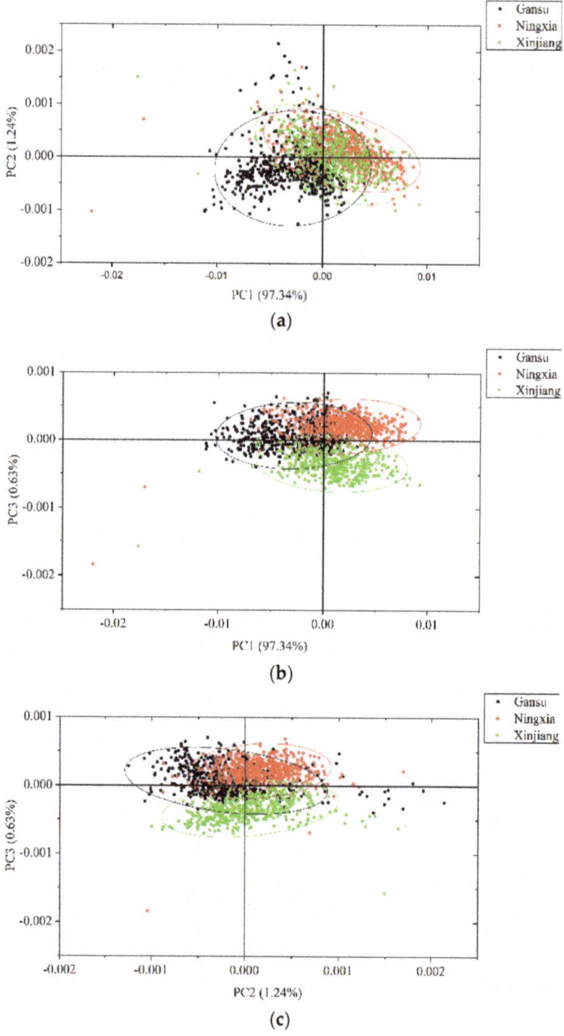

Figure 5. Principal component analysis (PCA) score scatter plots of (**a**) PC1 versus PC2; (**b**) PC1 versus PC3; and (**c**) PC2 versus PC3. The ellipse is the confidence ellipse (confidence level at 0.95).

3.3. Classification Models Using Full Spectra

PLS-DA, SVM, and CNN models were built using the full spectra. For the PLS-DA models, the category values of the samples from Gansu, Ningxia, and Xinjiang were labelled 001, 010, and 100.

For the SVM and CNN models, the category values of the samples from Gansu, Ningxia, and Xinjiang were labelled 0, 1, and 2.

The classification results of the three different models are shown in Table 1. All discriminant models obtained good performances, with the classification accuracy of the calibration, validation, and prediction sets all over 90%. For the PLS-DA model, the optimal number of latent variables (LVs) was 12, and good classification performance was obtained. Classification accuracies of the calibration, validation, and prediction sets were all over 99%. For the SVM model, the model parameters (C, g) were optimized as (100, 10,000). The classification accuracy of the calibration set was 100%, while the classification accuracy of the validation and prediction sets was found to be lower. For the CNN model, the classification accuracy of the calibration, validation, and prediction sets were determined to be all over 97%. With regard to all three models, the PLS-DA model performed the best, the CNN model obtained results quite close to and slightly worse than those for PLS-DA, and the SVM model performed the worst.

Table 1. Confusion matrix of the partial least squares discriminant analysis (PLS-DA), support vector machine (SVM) and convolutional neural network (CNN) models using full spectra.

Model	Category Values	Calibration				Validation				Prediction			
		0	1	2	Total (%)	0	1	2	Total (%)	0	1	2	Total (%)
PLS	0 *	539	0	0		291	0	0		268	0	7	
	1	0	601	1		0	303	0		0	299	1	
	2	0	0	481		0	0	241		0	0	240	
	Total (%)				99.94				100				99.02
SVM	0	539	0	0		289	0	2		224	0	51	
	1	0	602	0		0	303	0		0	300	0	
	2	0	0	481		0	0	241		0	0	240	
	Total (%)				100				99.76				93.74
CNN	0	539	0	0		289	0	2		253	0	22	
	1	1	601	0		0	303	0		0	300	0	
	2	6	0	475		4	0	237		0	0	240	
	Total (%)				99.57				99.28				97.30

* 0, 1, and 2 are the assigned category values of the samples from Gansu, Ningxia, and Xinjiang, respectively.

When using the PLS-DA model, samples from Ningxia were misclassified as samples from Xinjiang and samples from Gansu were misclassified as samples from Xinjiang; when using the SVM model, samples from Gansu were misclassified as samples from Xinjiang; and when using the CNN model, samples from Gansu and Xinjiang were misclassified as each other. The overall classification results indicated good separability among the samples from the three geographical origins. Samples from Gansu and Xinjiang were more likely to be misclassified, due to the results of the three discriminant models.

3.4. Classification Models Using Optimal Wavelengths

After effective wavelength selection, the PLS-DA, SVM, and CNN models were built using the selected effective wavelengths. The results of the three discriminant models are shown in Table 2. Good performances were obtained by the three models, with the classification accuracy of the calibration, validation, and prediction sets all over 95%. For the PLS-DA model, the optimal number of LVs was found to be 17. The classification accuracies of the calibration, validation, and prediction sets were all over 99%. For the SVM model, the model parameters (C, g) were optimized as (100, 108). The classification accuracies of the calibration, validation, and prediction sets were all over 95%. For the CNN model, the classification accuracies of the calibration, validation, and prediction sets were all over 97%.

Table 2. Confusion matrices of the PLS-DA, SVM, and CNN models using effective wavelengths.

Model	Category Values	Calibration				Validation				Prediction			
		0	1	2	Total (%)	0	1	2	Total (%)	0	1	2	Total (%)
PLS	0 *	538	0	1		291	0	0		272	0	3	
	1	1	601	0		0	303	0		0	300	0	
	2	1	0	480		0	0	241		0	0	240	
	Total (%)				99.92				100				99.63
SVM	0	539	0	0		271	0	20		238	0	37	
	1	0	602	0		0	303	0		0	300	0	
	2	2	0	479		1	0	240		1	0	239	
	Total (%)				99.88				97.49				95.34
CNN	0	539	0	0		287	0	4		263	0	12	
	1	0	602	0		0	303	0		0	299	1	
	2	4	0	477		8	0	233		5	0	235	
	Total (%)				99.75				98.56				97.79

* 0, 1, and 2 are the assigned category values of the samples from Gansu, Ningxia, and Xinjiang, respectively.

When using the PLS-DA model, samples from Gansu and Xinjiang were misclassified as each other, and one sample from Ningxia was misclassified as a sample from Gansu. When using the SVM model, it was observed that samples from Gansu and Xinjiang were misclassified as each other. When using the CNN model, samples from Gansu and Xinjiang were misclassified as each other, and one sample from Ningxia was misclassified as a sample from Xinjiang. The confusion matrices of the three models illustrate that samples from Gansu and Xinjiang were more likely to be misclassified.

The PLS-DA, SVM, and CNN models using effective wavelengths obtained similar results to those using effective wavelengths, illustrating the effectiveness of effective wavelength selection. The overall classification accuracy of all models indicates that there are great differences existing in narrow-leaved oleaster fruits from the three different geographical origins considered. As shown in Tables 1 and 2, the PLS-DA models performed slightly better than the CNN models, and the CNN models performed slightly better than the SVM models. Although differences existed in these model performances, the differences were quite small. The results illustrate that CNN models could be used for narrow-leaved oleaster fruit geographical origin identification. Moreover, the results of the discriminant models using full spectra and effective wavelengths all showed that samples from Gansu and Xinjiang were more likely to be misclassified.

4. Conclusions

In this work, near-infrared hyperspectral imaging was successfully used to identify the geographical origins of narrow-leaved oleaster fruits from Gansu, Ningxia, and Xinjiang. PCA score scatter plots showed the separability of the samples from the three geographical origins. PLS-DA, SVM, and CNN models were established using full spectra and effective wavelengths selected by second derivative spectra. The high classification accuracy, which was over 90% for models using full spectra and effective wavelengths, illustrates that the proposed method can effectively distinguish narrow-leaved oleaster fruits from different geographical origins. The performances of the models using effective wavelengths were similar to those using full spectra. Moreover, deep CNN models obtained close results to the PLS-DA and SVM models, showing good performances of deep learning for narrow-leaved oleaster fruit geographical origin detection. According to the discriminant models, samples from Gansu and Xinjiang were more likely to be misclassified. These results indicate that it would be possible to develop online systems for narrow-leaved oleaster fruit origin detection using near-infrared hyperspectral imaging and machine learning methods.

Author Contributions: Conceptualization, P.G.; data curation, P.G. and C.Z.; formal analysis, W.X.; funding acquisition, W.X.; investigation, C.Z.; methodology, C.Z., X.L., and Y.H.; project administration, P.G.; resources,

T.Y.; software, T.Y.; supervision, Y.H.; validation, X.L.; visualization, T.Y.; writing—original draft, P.G., W.X., and C.Z.; writing—review and editing, X.L. and Y.H.

Funding: This research was funded by the National Natural Science Foundation of China, grant number 61965014, and the Special Project for Scientific and Technological Innovation, grant number CXFZ201906.

Acknowledgments: The authors want to thank L.Z., a Ph.D candidate in College of Biosystems Engineering and Food Science, Zhejiang University, China, for providing help on data analysis.

Conflicts of Interest: The authors declare no conflict of interest.

References

1. Zhang, L.; Alvarez, L.V.; Bonthond, G.; Tian, C.; Fan, X. Cytospora elaeagnicola sp. nov. Associated with Narrow-leaved oleaster fruits Canker Disease in China. *Mycobiology* **2019**, *47*, 1–10. [CrossRef] [PubMed]
2. Zhanga, X.; Lia, G.; Sheng, D. Simulating the potential distribution of *Elaeagnus angustifolia* L. based on climatic constraints in China. *Ecol. Eng.* **2018**, *113*, 27–34. [CrossRef]
3. Lin, J.; Li, J.P.; Yuan, F.; Yang, Z.; Wang, B.S.; Chen, M. Transcriptome profiling of genes involved in photosynthesis in *Elaeagnus angustifolia* L. under salt stress. *Photosynthetica* **2018**, *56*, 1–12. [CrossRef]
4. Chen, X.; Yushuang, L.; Guangying, C.; Chi, G.; Shengge, L.; Huiming, H.; Tao, Y. Angustifolinoid A, a macrocyclic flavonoid glycoside from *Elaeagnus angustifolia* flowers. *Tetrahedron Lett.* **2018**, *59*, 2610–2613. [CrossRef]
5. Du, H.; Chen, J.; Tian, S.; Gu, H.; Li, N.; Sun, Y.; Ru, J.; Wang, J. Extraction optimization, preliminary characterization and immunological activities in vitro of polysaccharides from *Elaeagnus angustifolia* L. pulp. *Carbohydr. Polym.* **2016**, *151*, 348–357. [CrossRef]
6. Mcshane, R.; Auerbach, D.; Friedman, J.M.; Auble, G.T.; Shafroth, P.B.; Merigliano, M.; Scott, M.; Poff, N. Distribution of invasive and native riparian woody plants across the western USA in relation to climate, river flow, floodplain geometry and patterns of introduction. *Ecography* **2016**, *38*, 1254–1265. [CrossRef]
7. Collette, L.K.D.; Pither, J. Insect assemblages associated with the exotic riparian shrub Russian olive (Elaeagnaceae), and co-occurring native shrubs in British Columbia, Canada. *Can. Entomol.* **2016**, *148*, 316–328. [CrossRef]
8. Tredick, C.A.; Kelly, M.J.; Vaughan, M.R. Impacts of large-scale restoration efforts on black bear habitat use in Canyon de Chelly National Monument, Arizona, United States. *J. Mammal.* **2016**, *97*, gyw060. [CrossRef]
9. Khamzina, A.; Lamers, J.P.A.; Martius, C. Above- and belowground litter stocks and decay at a multi-species afforestation site on arid, saline soil. *Nutr. Cycl. Agroecosyst.* **2016**, *104*, 187–199. [CrossRef]
10. Singh, A.; Singh, N.B.; Hussain, I.; Singh, H.; Yadav, V.; Singh, S.C. Green synthesis of nano zinc oxide and evaluation of its impact on germination and metabolic activity of *Solanum lycopersicum*. *J. Biol.* **2016**, *233*, 84–94. [CrossRef]
11. Singh, A.; Singh, N.B.; Afzal, S.; Singh, T.; Hussain, I. Zinc oxide nanoparticles: A review of their biological synthesis, antimicrobial activity, uptake, translocation and biotransformation in plants. *J. Mater. Sci.* **2017**, *53*, 185–201. [CrossRef]
12. Hassanzadeh, Z.; Hassanpour, H. Evaluation of physicochemical characteristics and antioxidant properties of *Elaeagnus angustifolia* L. *Sci. Hortic.* **2018**, *238*, 83–90. [CrossRef]
13. Waili, A.; Yili, A.; Maksimov, V.V.; Mijiti, Y.; Atamuratov, F.N.; Ziyavitdinov, Z.F.; Mamadrakhimov, A.; Asia, H.A.; Salikhov, S.I. Erratum to: Isolation of Biologically Active Constituents from Fruit of *Elaeagnus angustifolia*. *Chem. Nat. Compd.* **2016**, *52*, 776. [CrossRef]
14. Wei, Q.; Wei, Y.; Wu, H.; Yang, X.; Zhang, H. Chemical Composition, Anti-oxidant, and Antimicrobial Activities of Four Saline-Tolerant Plant Seed Oils Extracted by SFC. *J. Am. Oil Chem. Soc.* **2016**, *93*, 1–10. [CrossRef]
15. Morehart, A.L. Phomopsis canker and dieback of *Elaeagnus angustifolia*. *Plant Dis.* **2015**, *64*, 66. [CrossRef]
16. Marena, M. Near-infrared spectroscopy and hyperspectral imaging: Non-destructive analysis of biological materials. *Chem. Soc. Rev.* **2014**, *43*, 8200–8214.
17. Ru, C.; Li, Z.; Tang, R. A Hyperspectral Imaging Approach for Classifying Geographical Origins of Rhizoma Atractylodis Macrocephalae Using the Fusion of Spectrum-Image in VNIR and SWIR Ranges (VNIR-SWIR-FuSI). *Sensors* **2019**, *19*, 2045. [CrossRef]

18. Noviyanto, A.; Abdulla, W.H. Honey botanical origin classification using hyperspectral imaging and machine learning. *J. Food Eng.* **2019**, *265*, 109684. [CrossRef]

19. Minaei, S.; Shafiee, S.; Polder, G.; Moghadam-Charkari, N.; Van Ruth, S.; Barzegar, M.; Zahiri, J.; Alewijn, M.; Kuś, P.M. VIS/NIR imaging application for honey floral origin determination. *Infrared Phys. Technol.* **2017**, *86*, 218–225. [CrossRef]

20. Puneet, M.; Alison, N.; Julius, T.; Guoping, L.; Sally, R.; Stephen, M. Near-infrared hyperspectral imaging for non-destructive classification of commercial tea products. *J. Food Eng.* **2018**, *238*, 70–77. [CrossRef]

21. Zhang, C.; Liu, F.; He, Y. Identification of coffee bean varieties using hyperspectral imaging: Influence of preprocessing methods and pixel-wise spectra analysis. *Sci. Rep.* **2018**, *8*, 2166. [CrossRef] [PubMed]

22. Yin, W.; Zhang, C.; Zhu, H.; Zhao, Y.; He, Y. Application of near-infrared hyperspectral imaging to discriminate different geographical origins of chinese wolfberries. *PLoS ONE* **2017**, *12*, e0180534. [CrossRef] [PubMed]

23. Zhu, S.; Zhou, L.; Gao, P.; Bao, Y.; He, Y.; Feng, L. Near-Infrared Hyperspectral Imaging Combined with Deep Learning to Identify Cotton Seed Varieties. *Molecules* **2019**, *24*, 3268. [CrossRef] [PubMed]

24. Mahesh, S.; Jayas, D.S.; Paliwal, J.; White, N.D.G. Comparison of partial least squares regression (plsr) and principal components regression (pcr) methods for protein and hardness predictions using the near-infrared (nir) hyperspectral images of bulk samples of Canadian wheat. *Food Bioprocess Technol.* **2015**, *8*, 31–40. [CrossRef]

25. Weinstock, B.A.; Janni, J.; Hagen, L.; Wright, S. Prediction of oil and oleic acid concentrations in individual corn (*Zea mays* L.) kernels using near-infrared reflectance hyperspectral imaging and multivariate analysis. *Appl. Spectrosc.* **2006**, *60*, 9. [CrossRef]

26. Sun, J.; Lu, X.; Mao, H.; Wu, X.; Gao, H. Quantitative determination of rice moisture based on hyperspectral imaging technology and bcc-ls-svr algorithm. *J. Food Process Eng.* **2016**, *40*. [CrossRef]

27. Zhang, N.; Liu, X.; Jin, X.; Li, C.; Wu, X.; Yang, S.; Ning, J.; Yanne, P. Determination of total iron-reactive phenolics, anthocyanins and tannins in wine grapes of skins and seeds based on near-infrared hyperspectral imaging. *Food Chem.* **2017**, *237*, 811–817. [CrossRef]

28. Rodríguez-Pulido, F.J.; Hernández-Hierro, J.M.; Nogales-Bueno, J.; Gordillo, B.; González-Miret, M.L.; Heredia, F.J. A novel method for evaluating flavanols in grape seeds by near infrared hyperspectral imaging. *Talanta* **2014**, *122*, 145–150.

29. Mazivila, S. Discrimination of the type of biodiesel/diesel blend (B5) using mid-infrared spectroscopy and PLS-DA. *Fuel* **2015**, *142*, 222–246. [CrossRef]

30. Botelho, B.G.; Reis, N.; Oliveira, L.S.; Sena, M.M. Development and analytical validation of a screening method for simultaneous detection of five adulterants in raw milk using mid-infrared spectroscopy and PLS-DA. *Food Chem.* **2015**, *181*, 31–37. [CrossRef]

31. Balage, J.M.; Amigo, J.M.; Antonelo, D.S.; Mazon, M.R.; e Silva, S.D. Shear force analysis by core location in Longissimus steaks from Nellore cattle using hyperspectral images—A feasibility study. *Meat Sci.* **2018**, *143*, 30–38. [CrossRef] [PubMed]

32. Melucci, D.; Bendini, A.; Tesini, F.; Barbieri, S.; Zappi, A.; Vichi, S.; Conte, L.; Gallina, T.T. Rapid direct analysis to discriminate geographic origin of extra virgin olive oils by flash gas chromatography electronic nose and chemometrics. *Food Chem.* **2016**, *204*, 263–273. [CrossRef] [PubMed]

33. Da Costa, G.B.; Fernandes, D.D.S.; Gomes, A.A.; De Almeida, V.E.; Veras, G. Using near infrared spectroscopy to classify soybean oil according to expiration date. *Food Chem.* **2016**, *196*, 539–543. [CrossRef]

34. Du, L.; Lu, W.; Cai, Z.J.; Bao, L.; Hartmann, C.; Gao, B.; Yu, L.L. Rapid detection of milk adulteration using intact protein flow injection mass spectrometric fingerprints combined with chemometrics. *Food Chem.* **2017**, *240*, 573–578. [CrossRef]

35. Schmutzler, M.; Beganovic, A.; Böhler, G.; Huck, C.W. Methods for detection of pork adulteration in veal product based on FT-NIR spectroscopy for laboratory, industrial and on-site analysis. *Food Control* **2015**, *57*, 258–267. [CrossRef]

36. Yang, H.X.; Fu, H.B.; Wang, H.D.; Jia, J.W.; Sigrist, M.W.; Dong, F.Z. Laser-induced breakdown spectroscopy applied to the characterization of rock by support vector machine combined with principal component analysis. *Chin. Phys. B* **2016**, *25*, 065201. [CrossRef]

37. Li, J.L.; Sun, D.W.; Pu, H.; Jayas, D.S. Determination of trace thiophanate-methyl and its metabolite carbendazim with teratogenic risk in red bell pepper (*Capsicum annuum* L.) by surface-enhanced Raman imaging technique. *Food Chem.* **2017**, *218*, 543–552. [CrossRef]

38. Ropodi, A.I.; Panagou, E.Z.; Nychas, G.J.E. Multispectral imaging (MSI): A promising method for the detection of minced beef adulteration with horsemeat. *Food Control* **2017**, *73*, 57–63. [CrossRef]

39. Wu, N.; Zhang, C.; Bai, X.; Du, X.; He, Y. Discrimination of Chrysanthemum Varieties Using Hyperspectral Imaging Combined with a Deep Convolutional Neural Network. *Moleclues* **2018**, *23*, 2831. [CrossRef]

40. Qiu, Z.; Chen, J.; Zhao, Y.; Zhu, S.; He, Y.; Zhang, C. Variety Identification of Single Rice Seed Using Hyperspectral Imaging Combined with Convolutional Neural Network. *Appl. Sci.* **2018**, *8*, 212. [CrossRef]

41. Zhang, C.; Feng, X.; Wang, J.; Liu, F.; He, Y.; Zhou, W. Mid-infrared spectroscopy combined with chemometrics to detect Sclerotinia stem rot on oilseed rape (*Brassica napus* L.) leaves. *Plant Methods* **2017**, *13*, 39. [CrossRef] [PubMed]

Article

Validation of a HILIC UHPLC-MS/MS Method for Amino Acid Profiling in Triticum Species Wheat Flours

Emmanouil Tsochatzis [1,*], Maria Papageorgiou [2] and Stavros Kalogiannis [3]

[1]   Department of Chemical Engineering, Aristotle University of Thessaloniki, GR-54124 Thessaloniki, Greece
[2]   Department of Food Science and Technology, International Hellenic University, Sindos Campus, P.O. Box 141, GR-57400 Thessaloniki, Greece; mariapapage@food.teithe.gr
[3]   Department of Nutritional Sciences and Dietetics, International Hellenic University, Sindos Campus, P.O. Box 141, GR-57400 Thessaloniki, Greece; kalogian@nutr.teithe.gr
*   Correspondence: tsochatzism@gmail.com; Tel.: +30-6977-441091

Received: 25 September 2019; Accepted: 12 October 2019; Published: 18 October 2019

Abstract: Amino acids are essential nutritional components as they occur in foods either in free form or as protein constituents. An ultra-high-performance (UHPLC) hydrophilic liquid chromatography (HILIC)-tandem Mass Spectrometry (MS) method has been developed and validated for the quantification of 17 amino acids (AA) in wheat flour samples after acid hydrolysis with 6 M HCl in the presence of 4% (*v/v*) thioglycolic acid as a reducing agent. The developed method proved to be a fast and reliable tool for acquiring information on the AA profile of cereal flours. The method has been applied and tested in 10 flour samples of spelt, emmer, and common wheat flours of organic or conventional cultivation and with different extraction rates (70%, 90%, and 100%). All the aforementioned allowed us to study and evaluate the variation of the AA profile among the studied flours, in relation to other quality characteristics, such as protein content, wet gluten, and gluten index. Significant differences were observed in the AA profiles of the studied flours. Moreover, AA profiles exhibited significant interactions with quality characteristics that proved to be affected based mainly on the type of grain. A statistical and multivariate analysis of the AA profiles and quality characteristics has been performed, as to identify potential interactions between protein content, amino acids, and quality characteristics.

Keywords: amino acid profiling; hydrophilic interaction chromatography (HILIC); tandem mass spectrometry; Triticum species flours; flour quality characteristics

---

1. Introduction

Amino acids (AA) are essential nutritional components present in foods either in their free form (FAA) or as protein constituents. They directly contribute to the flavor of foods as they are precursors of aroma compounds and colors formed by thermal or enzymatic reactions during production, processing, and storage of food. Hence, information on the profile and amount of free AA is highly needed in food science and nutrition studies [1–5].

Analysis of AA, in either free form or in protein profile, was highly challenging due to their structural and polarity differences. Ion exchange liquid chromatography has found wide application in AA analysis, especially through the commercial amino acid analyzers, which utilize cation-exchange chromatography followed by post-column derivatization with a chromophore or fluorophore derivatizing agent [6–8]. On the other hand, conventional reversed-phase (RP) High Pressure Liquid Chromatography (HPLC) proved to be time-consuming since most AAs are highly polar and cannot be determined without pre- or post-column derivatization [2,9,10]. Recently the advancement of hydrophilic interaction chromatography (HILIC) and new analytical columns provided alternative paths for the profiling of AA by liquid chromatography (LC). HILIC is generally known to enhance the sensitivity of electrospray ionization-mass spectrometry (ESI–MS) detection, and it is increasingly employed in the analysis of polar analysis in various matrices, [3,5,11–15].

So far, the application of HILIC-MS has found limited use in the analysis of AA in food samples. To our knowledge, three methods have been reported on the determination of FAA in liquid food matrices such as juice, beer, honey, or tea [3], ginkgo seeds [5], and fruits of Ziziphus jujuba [14]. Only recently, it was developed a HILIC-MS/MS method for the determination of either free AA or amino acids profile in high protein content food matrix (mussels) [15]. Moreover, the analytical methods that determine amino acid profile (TAA) were developed for the application in pure protein samples, such as collagen [16] or bovine serum albumin (BSA) and angiotensin I [13]. In all these methods, single ion monitoring-MS (SIM-MS) detection [5] or tandem MS [4,13] were applied. By using tandem MS detection, the separation of isobaric AA was feasible in most cases, whereas the use of single MS detection led to increased total analysis time since longer chromatographic runs were needed for the separation of all compounds. Prior to the analysis of amino acids, proteins need to be hydrolyzed in order to release their constituting AA. The most commonly applied method is hydrolysis by digestion with a strong inorganic acid [13,15–17].

Cereals are considered as one of the basic foods consumed by humans and animals. The carbohydrates they contain provide approximately 50% of the total daily calories, whereas the proteins one-third of the total protein need [17]. The composition of amino acids varies among the proteins of the different cereal grains or flours. Wheat (*Triticum* species) is the 3rd most-produced cereal worldwide. Wheat proteins are known to be low in some amino acids that are considered essential for the human diet, especially lysine (the most deficient amino acid) and threonine (the second limiting amino acid) [18–21]. On the other hand, they are rich in glutamine and proline (s), the functional amino acids in dough formation. Tetraploid Emmer wheat (*Triticum turgidum* species, dicoccum, genomes AABB), also known as emmer, faro or zea in different countries, is hulled wheat that differs from the domestic species on the fact that the ripened seed head of the wild species shatters and spreads the seed onto the ground while in the domesticated counterpart the seed head remains intact, making it easier for humans to harvest the grain. It is considered as the ancestor of bread wheat and durum wheat growing in the margin of the Mediterranean area [22]. Hexaploid spelt, *Triticum aestivum* variety spelta (genomes AABBDD) is also a hulled cereal grain with high resistance to environmental factors (diseases, stress) showing good yields under disadvantageous conditions [23]. It is suitable for organic farming and contributes to agro-diversity [24]. Spelt is becoming widely used in the growing natural food market. It has been reported that spelt protein content showed a great variation depending on the genotype [21,25,26], but it is higher than in common wheat [25,27]. The amino acid composition of the proteins from spelt differs slightly from one of the modern bread and pasta wheats [21]. It has been suggested that spelt-based products could be potentially more digestible than those from common wheat. Certain ancient wheats, einkorn, spelt, emmer, and Khorasan are currently of particular interest for use in selected bakery products [21]. The development of new cultivars has been attempted with the aim to improve the content of all essential amino acids [17,25].

In the present work a single HILIC-MS/MS method was developed, validated and applied for the determination of 17 amino acids, in a comparative study for the determination of the TAA of 'ancient' wheat species such as emmer, spelt, and bread wheat (*Triticum aestivum*) with different extraction rates and cultivated under different practices (organic or conventional) in a fast and reliable way without the need of derivatization, preceded only by a simple hydrolysis procedure. Finally, analysis of variances and multivariate analysis were performed in order to explore potential interactions of the TAA with flour quality characteristics of samples under investigation.

2. Materials and Methods

2.1. Chemicals

Standards of amino acids (AA) namely phenylalanine (Phe), tryptophan (Trp), isoleucine (Ile), leucine (Leu), asparagine (Asn), methionine (Met), valine (Val), proline (Pro), tyrosine (Tyr), alanine (Ala), threonine (Thr), glycine (Gly), serine (Ser), glutamic acid (Glu), aspartic acid (Asp), arginine (Arg), glutamine (Gln), lysine (Lys), histidine (His), cysteine (Cys) and cystine (Cys2),were purchased from Sigma-Aldrich (Steinheim, Germany). Acetonitrile (ACN) and water (LC-MS grade) was purchased from Carlo Erba (Milan, Italy). Formic acid (LC-MS additive), TGA (thioglycolic acid; ≥98%), and ammonium formate (NH_4HCO_2) were purchased from Sigma-Aldrich (Steinheim, Germany). Analytical grade hydrochloric acid (HCl) 37% w/w was also supplied from Carlo Erba (Milan, Italy). The chromatographic column HILIC amide BEH, Acquity UPLC 1.7 m, 2 × 150 mm (Waters) were used.

2.2. Standard Solutions

Stock solutions of the compounds were prepared in 0.1 M HCl at a concentration of 10 mg mL^{-1} and stored in amber vials at −20 °C. Working standards were prepared from the stock solutions by appropriate dilution with ACN/water 95:5 (*v/v*) and stored at −20 °C.

2.3. Sample Preparation for Amino Acid Profile Analysis

Ten (10) Triticum flours were selected from the Greek market, to be tested for their AA concentration and quality characteristics. The commercial flour samples differed in their extraction rate, type of cultivation (organic or conventional), and type of wheat, as shown in Table 1. The first letter of codification corresponds to wheat type ('S' for spelt, 'B' for bread wheat and 'E' for Emmer) followed by a 2-digit number revealing the flour extraction rate and the character 'O' in the case of organically cultivated wheat.

Table 1. Studied Triticum flours with their extraction rates, type of cultivation, and type of wheat.

No.	Extraction Rate	Cultivation	Wheat Type
1	70%	Conventional	Spelt (S-70)
2	70%	Organic	Spelt (S-70-O)
3	90%	Organic	Bread wheat (B-90-O)
4	100%	Organic	Emmer (E-100-O)
5	100%	Organic	Spelt (S-100-O)
6	70%	Conventional	Emmer (E-70)
7	100%	Conventional	Spelt (S-100a)
8	100%	Conventional	Spelt (S-100b)
9	70%	Conventional	Bread wheat (B-70)
10	90%	Conventional	Bread wheat (B-90)

An amount of 10 g of flour was selected from 4 different positions of the package and mixed properly to create a homogeneous material that was then kept in a chemical dryer until analysis. The remaining original sample was kept at 2–4 °C. A modified method reported previously from Tsochatzis et al. [15] has been applied for the determination of amino acid mass fraction (g/100 g protein) in the aforementioned homogenized flour samples. In brief, for the AA profile 10.0 ± 0.5 mg of dried flour was placed in a hydrolysis tube along with 100 μL of 6 M HCl containing 4% (*v/v*) TGA as the reducing agent. The tube was flushed with N_2 gas to establish oxygen-free conditions, sealed, and heated at 110 °C for 18 h. Then, the mixture was transferred to a centrifuge tube with 0.5 mL HCl 0.1 M and 0.5 mL water and centrifuged at 4200× *g* for 5 min. The supernatant was collected and filtered through a 0.22 μm polytetrafluoroethylene (PTFE) syringe filter. An amount of 0.5 mL of the clear extract was diluted with 4.5 mL ACN/water 95:5 *v/v* and injected to the UHPLC-HILIC-ESI MS/MS system.

2.4. UHPLC-MS/MS Analysis

UHPLC–tandem mass spectrometry was based on the method described by Tsochatzis et al. [15]. In brief, the analysis was performed on an Accela TSQ Quantum TM Access MAX Triple Quadrupole Mass Spectrometer system (Thermo Scientific, San Jose, CA, USA) operating under XCalibur (Thermo Scientific, San Jose, CA, USA) Software. The mobile phase consisted of solvent A: ACN/5 mM HCOONH$_4$, pH = 3.0 adjusted with HCOOH 95:5 (*v/v*) and solvent B: ACN/5 mM HCOONH$_4$, pH = 3.0 adjusted with HCOOH, 40:60 (*v/v*). Elution was based on a linear gradient program of 13 min from 80% A:20% B to 62% A:38% B, followed by a 2 min equilibration step to the initial conditions prior to the next injection. The flow rate was 400 μL min^{-1}, and the total analysis time was 15 min.

Chromatographic separation was performed on a 2.1 mm × 150 mm ACQUITY UPLC 1.7 μm BEH HILIC amide column (Waters), equipped with an ACQUITY UPLC BEH Amide 1.7 μm Van-Guard Pre-column, maintained at 40 °C. Selected Reaction Monitoring (SRM) with Electrospray positive ionization mode (ESI +) was applied with spray voltage at 3000 V, capillary temperature: 300 °C, vaporizer temperature: 300 °C, sheath gas pressure at 40 arbitrary units (Arb), aux gas pressure at 10 Arb, ion sweep gas pressure at 2.0 Arb, ion source discharge current at 4.0 μA and collision gas pressure at 1.5 mTorr. Auto-samplers' temperature was set at 4 °C, and the injection volume set at 5 μL. Amino acid individual data regarding molecular formulas, monoisotopic masses, precursor-product ion for the aforementioned SRM, along with their respective retention times in standard solutions and after acidic hydrolysis, are given in the Supplementary material (Table S1).

2.5. Method Validation

Method linearity, precision, trueness, the limit of detection (LOD), and limit of quantification (LOQ) were calculated. The linearity of the method was firstly assessed by analyzing standard solutions mixtures at six concentration levels for all AA (0.5, 1, 10, 50, 100, 200 μg mL^{-1}), representing the working concentration range. Calibration curves were constructed by plotting the peak areas of the respective AAs followed by linear regression analysis (R^2), based on the standard addition method. LODs and LOQs were calculated according to the signal-to-noise ratio (S/N) and the slope (S), using the equations LOD = 3 SD/S and LOQ = 10 SD/S [28].

Precision and trueness were assessed in the B-90 flour sample, which was used as a reference sample. The sample was fortified at two concentration levels (20.0 and 40.0 mg/100 g) of all AAs tested. All calculations were performed using the concentration values expressed in g/100 g for each AA individually. For short-term repeatability, the fortified samples were analyzed in triplicates during the day while for intermediate precision, the aforementioned samples were analyzed in triplicates for three consecutive days. Relative standard deviation (RSD; %) and recoveries were calculated as ((amount found in the spiked sample—amount found in the sample)/amount added) × 100 [2]. All the results regarding precision and trueness are presented in Supplementary material's Table S2.

2.6. Quantitation and Matrix Effect

The calibration curves of the studied AAs, based on the linear regression coefficients (R^2) have been performed with the standard addition method. Flour samples were fortified at two concentration (5 and 10 μmol/100 g) levels of all AAs tested, followed by an analysis in triplicates. Calibration curves were constructed by linear regression analysis of the peak area (Y) versus the injected concentration (X), and they have been assessed based on the linear regression coefficients (R^2). Linear equations were established to determine the initial concentration of amino acids in the dried cereal flour samples. Evaluation of the matrix effect (ME, %) was performed by the slope comparison method as it was previously reported [2,15].

2.7. Protein Content and Flour Quality Parameters

The moisture, ash, total protein content, wet gluten, and gluten index (GI) were determined following ICC standard methods 109/1, 104/1, 105/2, and 155, respectively [29].

2.8. Data Processing and Statistical Analysis

Data were processed using the XCalibur application manager for the quantification of compounds. Regression analysis and statistics were performed using Microsoft Excel, and further statistical analysis, such as Analysis of Variance (ANOVA), followed by Tukey comparison test in all cases, has been performed with Minitab 18.0 statistical software (Minitab Inc., State College, PA, USA). The multivariate statistical analysis, combined with cluster analysis, has been performed with Simca 15 (Umetrics, Umea, Sweden).

3. Results

3.1. Acidic Hydrolysis and Antioxidant Agent

A properly performed hydrolysis is a prerequisite of a successful analysis regarding amino acid profiling in food matrices. The conditions selected were based according to the previously reported conditions by Fountoulakis et al. and the applied conditions in case of mussels from Tsochatzis et al. [15] or the ones reported by EZ Faast [30]. We selected the conditions of 110 °C for 18 h, in order to minimize (in combination with the antioxidant agent) degradation of specific amino acids, while we made a compromise in the recoveries of the more hydrophobic AA, such as valine and isoleucine and leucine [6,15,30]. In addition, the selection of this temperature was selected to minimize the potential cross-reactions of amino-acids with starch. The selection of the hydrolysis conditions was also based on the study of Tsochatzis et al. [15].

3.2. Analytical Method Development, Validation, and Optimization

The total analysis time was less than 12 min. The method exhibited good linearity in the concentration range of 0.5–200 µg mL^{-1} with a linear regression coefficient (R^2) of above 0.99 for each AA. The effects from the matrix were minimal in both cases of either standard solutions or amino acid determination after hydrolysis. A typical chromatogram of the AA analysis in flours is presented in Figure 1.

The present analytical method exhibited satisfactory sensitivity for all AA. LODs varied from 0.002 (valine, serine, leucine, isoleucine) to 0.009 g/100 g (threonine) and the LOQs from 0.007 (serine, leucine) to 0.024 g/100 g (threonine, lysine) and a minimal matrix effect was observed. The analytical figures of merit of the method are given in Table 2.

Regarding the trueness, it was assessed by the recoveries from spiked cereal bread wheat flour, after acidic hydrolysis, at two concentration levels. The resulting recoveries ranged from 85.7% (lysine) to 121.8% (leucine) in the intra-day assay and from 86.8% (lysine) to 123.3% (leucine) for the intermediate precision (Table S2). The respective precision expressed in relative standard deviation (RSD %) values ranged from 0.6% (Glutamic acid) to 13.9% (proline) and from 1.4% (serine) to 13.7% (lysine) for short-term repeatability and intermediate precision respectively. The results of the analytical method presented adequate precision and accuracy results.

3.3. Amino Acid Profile of Flour Samples

Statistical differences have been identified between the AA mass fractions (Table 3) among the studied flour samples. From Table 3, it could be concluded that phenylalanine, threonine, glycine, and histidine did not have statistical differences among the various flours. Instead, significant differences have been observed for isoleucine, serine, lysine, valine, methionine, proline, glutamic acid, and glutamine. The results were in accordance with previously reported work, regarding the study of AA content in ancient cereal grain wheat cultivars [21].

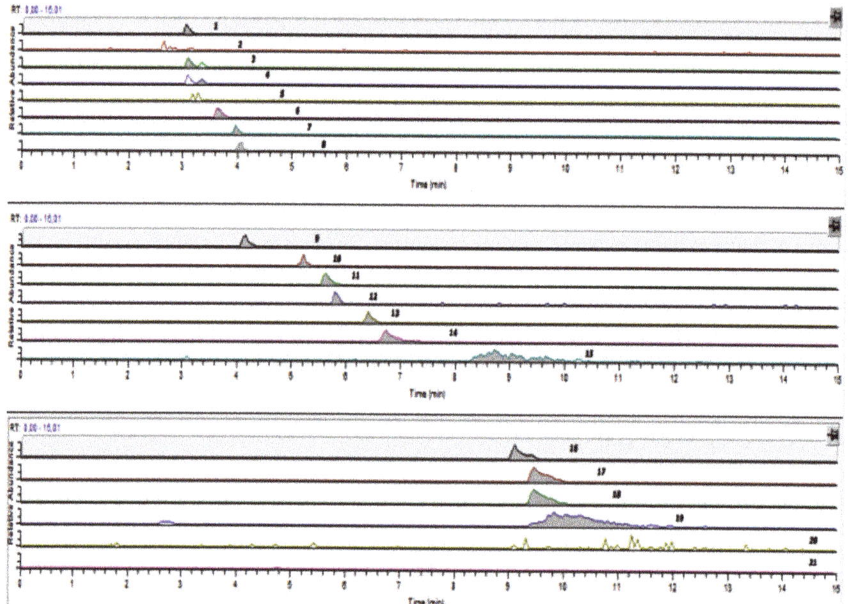

Figure 1. HILIC UHPLC–MS/MS chromatographic traces of the 17 amino acids quantified in the hydrolyzed flour sample B-90 (Elution order is 1: phenylalanine; 2: tryptophan; 3: Isoleucine; 4: leucine; 5: asparagine; 6: methionine; 7: valine; 8: proline; 9: tyrosine; 10: alanine; 11: threonine; 12: glycine; 13: serine; 14: glutamic acid; 15: aspartic acid; 16: arginine; 17: glutamine; 18: lysine; 19: histidine; 20: cystine; 21: cysteine).

Table 2. Limits of detection (LODs), limits of quantification (LOQs), linear ranges, and linear regression coefficients for the amino acids matrix match calibration in hydrolyzed cereal flour.

Amino Acid	Linear Regression (R^2)	LOD (g/100 g)	LOQ (g/100 g)	Matrix Effect (%) *
Glycine	0.993	0.005	0.010	94.6
Alanine	0.994	0.007	0.020	106.4
Serine	0.996	0.002	0.007	88.1
Proline	0.999	0.004	0.013	107.8
Valine	0.995	0.002	0.007	111.7
Threonine	0.993	0.009	0.024	108.8
Aspartic acid	0.993	0.007	0.020	102.8
Glutamic acid	0.996	0.007	0.022	92.2
Isoleucine	0.999	0.002	0.009	82.6
Leucine	0.999	0.002	0.007	88.5
Asparagine	-			-
Lysine	0.993	0.009	0.024	82.2
Methionine	0.994	0.007	0.020	86.3
Histidine	0.994	0.007	0.022	83.6
Phenylalanine	0.996	0.007	0.020	85.3
Arginine	0.997	0.004	0.016	110.1
Tyrosine	0.997	0.004	0.016	109.2
Cystine	-	-	-	-
Cysteine	-	-	-	-

* Calculated based on the slope ratio method [2].

Table 3. Average amino acids composition (g/100 g protein) of wheat flours under investigation.

Target AA	Flour Sample (g/100 g Protein; Dry Basis) *									
	1	2	3	4	5	6	7	8	9	10
	Spelta (S-70)	Spelta (S-70-O)	Bread Wheat (B-90-O)	Emmer (E-100-O)	Spelta (S-100-O)	Emmer (E-70)	Spelta (S-100a)	Spelta (S-100b)	Bread Wheat (B-70)	Bread Wheat (B-90)
Phe	3.73 [e]	3.94 [d]	4.23 [ab]	3.93 [d]	3.99 [cd]	3.52 [g]	4.13 [bc]	4.38 [a]	3.71 [ef]	3.54 [fg]
Ile	1.59 [cd]	1.75 [a]	1.74 [a]	1.55 [de]	1.57 [de]	1.50 [ef]	1.70 [ab]	1.65 [bc]	1.32 [g]	1.46 [f]
Leu	3.82 [d]	4.08 [c]	4.18 [bc]	4.03 [c]	4.09 [c]	3.76 [d]	4.28 [b]	4.49 [a]	3.76 [d]	3.46 [e]
Met	0.96 [a]	0.84 [c]	0.91 [b]	0.81 [cd]	0.89 [b]	0.81 [cd]	0.88 [b]	0.97 [a]	0.79 [d]	0.73 [e]
Tyr	1.72 [c]	1.88 [ab]	1.95 [a]	1.82 [b]	1.84 [b]	1.63 [d]	1.83 [b]	1.95 [a]	1.65 [cd]	1.64 [d]
Pro	6.18 [b]	6.87 [a]	5.44 [c]	4.44 [ef]	4.64 [de]	4.38 [f]	4.69 [d]	3.82 [g]	2.97 [i]	3.55 [h]
Val	0.96 [c]	1.10 [a]	1.05 [b]	0.87 [d]	0.89 [d]	0.81 [e]	0.95 [c]	0.82 [e]	0.69 [f]	0.73 [f]
Ala	1.91 [c]	1.94 [c]	2.09 [b]	1.68 [d]	2.05 [b]	1.88 [c]	2.38 [a]	1.12 [f]	0.96 [g]	1.52 [e]
Thr	2.29 [d]	2.59 [a]	2.51 [abc]	2.42 [c]	2.46 [bc]	2.13 [e]	2.45 [bc]	2.54 [ab]	2.11 [e]	1.86 [f]
Gly	2.67 [d]	2.07 [e]	3.48 [a]	2.02 [e]	3.00 [c]	2.13 [e]	2.72 [d]	3.44 [a]	3.47 [a]	3.15 [b]
Lys	1.57 [c]	1.43 [de]	1.46 [d]	1.28 [f]	1.36 [e]	1.25 [f]	1.43 [de]	1.42 [de]	2.38 [a]	1.89 [b]
Ser	1.27 [b]	1.43 [a]	1.39 [a]	1.28 [b]	1.16 [c]	1.25 [b]	1.29 [b]	1.12 [cd]	0.86 [e]	1.09 [d]
Arg	4.68 [e]	5.22 [bc]	5.37 [ab]	4.94 [d]	5.02 [cd]	4.60 [ef]	4.99 [d]	5.50 [a]	4.39 [f]	4.16 [g]
His	2.42 [c]	2.59 [a]	2.65 [a]	2.56 [a]	2.59 [a]	2.38 [cd]	2.58 [a]	2.54 [ab]	2.44 [bc]	2.30 [d]
Gln	1.08 [e]	1.30 [ab]	1.32 [a]	1.14 [d]	1.23 [c]	1.13 [d]	1.22 [c]	1.27 [bc]	1.09 [de]	1.09 [de]
Asp	4.36 [cd]	4.66 [b]	5.02 [a]	3.87 [f]	4.42 [c]	4.51 [bc]	4.16 [e]	4.04 [ef]	4.18 [de]	3.93 [f]
Glu	41.33 [e]	39.67 [e]	41.47 [e]	49.22 [c]	52.41 [ab]	50.10 [c]	54.28 [a]	44.83 [d]	40.90 [e]	41.25 [bc]

* Values are means of triplicate measurements. Values followed by different letters within a row are significantly different ($p < 0.05$).

3.4. Flour Quality Parameters

The results from all the studied quality characteristics of the flours are presented in Table 4. Gluten index and wet gluten have been evaluated, and ANOVA showed that the tested flours presented a significant difference with the" bread" wheat flour, also showed that flours do differ in their protein (%) content (data obtained from the Kjeldahl method), gluten index, and wet gluten. By comparing the set of ANOVA results, it could be identified that "bread" flours presented a large variety of total protein content, ranging from 12.3% (B-70) to 13.9 (B-90-O), while differences were also observed for their gluten index and wet gluten. In the case of spelta, the S-70 types, either organic or conventional, presented close results in all the studied quality characteristics, something which was also valid in case of "Emmer"-type.

Table 4. Protein content of the tested flours.

No.	Wheat Type	Protein (% dw) *	Gluten Index	Wet Gluten
1	Spelt (S-70)	13.5 [ab]	24 [a]	37.4 [d]
2	Spelt (S-70-O)	13.5 [ab]	42 [b]	39.8 [d]
3	Bread wheat (B-90-O)	12.5 [a]	94 [c]	27.5 [b]
4	Emmer (E-100-O)	13.1 [ab]	19 [a]	25.3 [b]
5	Spelt (S-100-O)	12.9 [ab]	20 [a]	26.7 [b]
6	Emmer (E-70)	14.1 [c]	18 [a]	15.1 [a]
7	Spelta (S-100a)	12.0 [a]	51 [b]	33.3 [c]
8	Spelta (S-100b)	12.0 [a]	76 [bc]	23.8 [b]
9	Bread wheat (B-70)	12.3 [a]	98 [c]	33.8 [c]
10	Bread wheat (B-90)	13.9 [c]	27 [a]	29.5 [bc]

* Values followed by different letters within a column are significantly different ($p < 0.05$).

4. Discussion

4.1. Acidic Hydrolysis

Regarding sample preparation and acidic hydrolysis, we studied the effect of the antioxidant agent as specific AA are susceptible to decomposition during acidic hydrolyses, such as tryptophan, as well as others like asparagine and glutamine to be converted to aspartic acid and glutamic acid, respectively. In this case, an antioxidant agent is needed to prevent the aforementioned decomposition or converting reactions. We selected to proceed with TGA, as it has been reported to be an effective antioxidant agent for amino acid analysis in food matrices [6,15,30]. In this case, the two levels of TGA have been tested, of 2% *v/v* [30] and 4% [15]. The results indicated that the presence of TGA at a level of 4% *v/v*, was more effective, and it was selected for the final protocol.

Even though it is a practice in acid hydrolysis to add an antioxidant agent to prevent the degradation of AAs, to our knowledge, such a protective step has not been previously applied prior to HILIC-MS analysis in case of cereal flour analysis. Moreover, no single set of conditions has been suggested so far for the effective prevention of all AA degradation in wheat flour samples.

4.2. Sample Preparation and Analytical Methods

By applying the present method, all AAs eluted prior to 11 min of run time. The selected chromatographic conditions, especially for the mobile phase, have been selected with the aim to provide as good peak shape for basic, neutral, and acidic AAs and certainly taking into consideration the effect of pH to HILIC peak shapes [15,31]. As reported previously by Tsochatzis et al., it was noticed that the retention times and the respective peak shapes of the analytes in the extracts are slightly shifted from the respective AA standard. This behavior could be due to the highly acidic conditions during hydrolysis, which affects the pH of the final injection sample and the separation of the analytes. Due to the contribution of ionic interactions in the retention mechanism, in HILIC, the charge state of the analyte affects the retention time of the compound [15,31]. To eliminate this effect, after trying different approaches, it was selected the dilution (10 times with ACN/water 95:5 *v/v* to obtain a final pH of 2.5–3) of the final solution before injection. The dilution did not present a limitation for the present method since the concentration of amino acids obtained after hydrolyzing the proteins is significantly higher than that of free amino acids present in most foods [15,32]. The Multiple Reaction Monitoring (MRM) transitions and the parameters in the tandem MS detection were selected after tuning for the optimum signal for each of the analytes.

Trueness assessed by the reported recoveries from spiked cereal flour after acidic hydrolysis. An important source of slight variations could be considered the acidic hydrolysis and especially the precision of the temperature that needs to be controlled and precisely assessed [32]. The reported results for both trueness and precision regarding the quantification of AAs are similar to the ones previously reported in other food matrices [3,5,14,15].

With the current method, only 17 amino acids could be determined following acid hydrolysis with TGA as an antioxidant agent. The amino acids tryptophan, cysteine, cystine, and asparagine were affected from the acid hydrolysis; tryptophan was unstable, asparagine and glutamine tended to convert to aspartic acid and glutamic acid, respectively, while cysteine and cystine were oxidized to cysteic acid [9,15,33].

4.3. Comparative Study of Amino Acid Profile of Flour Samples

Regarding stability and yields of amino-acids during acidic hydrolysis, serine and tyrosine are generated in low yields, methionine is sensitive to oxidation, due to acidic conditions it could be oxidized to its sulfone product, and finally, valine hydrolyzes in poor yields (longer time and temperatures are needed) [1,34,35].

In case of proline, Kapusniak et al. reported reactions of starch with α-amino acids, where proline, alanine, isoleucine, and valine most readily reacted with starch that could affect the hydrolysis [36] while, Ito et al. (2006) highlighted the significant effect of hydrolysis time on the yields of all amino acids, with changes in isoleucine, lysine and serine, while there was observed an existing binding of amino acids (glycine, alanine, and partial lysine) to starch chains [37].

In addition, partial losses of the amino acids tyrosine and serine while the other amino acids (valine, leucine, and isoleucine) are requiring longer hydrolysis time in order to obtain higher yields. Our results are in accordance with Rowan et al. [38].

It has been reported that spelt has a high concentration in methionine compared to wheat [25,39,40]. This fact was confirmed in the present study, where all tested spelt flours, presented significantly higher methionine content than emmer as well as bread wheat flours. The values of the methionine varied from 0.73–0.97 g/100 g protein (dry basis); spelt flours showed 10% higher values than the reference bread wheat flour. Aspartic acid content was significant higher in all studied spelt flours compared to whole grain wheat flour is in accordance with the results of [21,25,38].

In the case of alanine, the results indicated that all flours have significantly higher content compared to the B-90 except for E-100-O that showed similar value. The results suggest that hydrolysis has a great impact and effect on the release and analysis of the amino acids, while on the other hand, the side cultivation technique and location have a potential role in the final concentration of specific amino acids.

It is also reported in the literature that wheat proteins have low mass fractions of certain AA, especially lysine and threonine [18–21]. Contrary, in the current study, it was observed that the refined bread wheat flours and the whole grain wheat flour (reference) presented significantly higher contents of lysine (2.38 and 1.89 g/100 g protein) compared to other types of studied flours (emmer, spelt) that had significantly lower mass fractions [21]. It is also reported that wheat proteins have low mass fractions of certain AA, especially lysine and threonine [18–21].

Escarnot et al., reported that the spelt amino acid profile differs from that of bread wheat, supported by limited evidence of higher content for isoleucine, leucine, and glycine [25,40–43]. Our results are in accordance with the reported literature on higher protein content in spelt than in wheat grains under low nitrogen fertilization [25,40], although this has not been proved to be statistically significant, as it is also observed in our study. Pruska-Kedzior et al. found significantly higher protein content in spelt flour, but there should be considered that genotype and the cultivation conditions highly affect the protein content [27].

Statistical analysis (ANOVA) performed for each AA, indicated that there are significant statistical differences between all the studied flours (Table 3). Statistical analysis revealed some very interesting results about the interaction between the type of flour and the concentration of amino acids. In general, there was a significant interaction between the type of the flour studied and the amount of AA.

Glutamine and proline are the functional amino acids in dough formation [21]. For glutamine, the organic B-90 presented the higher mass fraction among the flours studied, followed by the S-70-O and conventional spelta S-100a. Especially in the case of glutamic acid (Glu) and proline, the whole grain wheat flour showed higher concentration than the B-90 (Table 3). White wheat flour and white whole grain flours showed lower proline and glutamic acid concentrations that the rest of flours studied, which is in accordance with the results of Abdel-Aal and Hucl and Escarnot et al. [21,25]. The aforementioned authors reported that the spelt is rich in proline, which is the major functional amino acid in dough formation. The data obtained showed that the organic bread wheat flour (B-90-O) showed much higher proline content than the two bread wheat flours B-70 and B-90).

4.4. Interaction with Quality Parameters

A multivariate analysis was performed to study potential interactions of the TAA with the quality characteristics of cereal flours under investigation. Thus, by developing this analytical tool that reveals the amino acid profile of flours, one could discover the amino acid profile of proteins of different *Triticum* species and use it for choosing the variety with a more balanced profile for the use in cereal product development in correlation with other quality characteristics. The performed multivariate analysis between the AA content and the flour quality characteristics showed that the studied cereal flours could be distinguished in groups, based on their origin. The score plot, along with the respective loadings is presented in Figure 2.

Results showed that the studied types of flours presented a specific AA content pattern and specific quality characteristics. From the principal component analysis (PCA) biplot and the respective score plots, it could be identified that there is a clear differentiation of the three different types of flours. Three different groups were identified as expected; the emmer grains (blue dots), the spelta (brown dots) and the bread wheat (purple dots). The right part represents the bread wheat flours, with high protein content (%), high gluten index, and lower falling number (FN). The Emmer type flours presented lower protein content and lower gluten index, while the spelt type flours are in between these two categories, presenting intermediate quality characteristics of the other two types of flours.

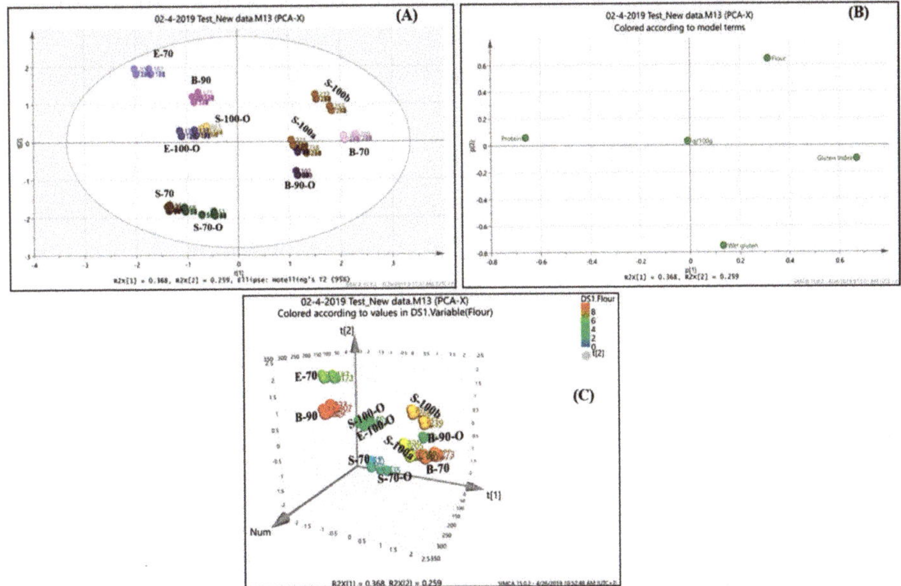

Figure 2. Principal components analysis of the studied amino acid profile and quality characteristics; (**A**) score plot and (**B**) loading plot and (**C**) 3D scatterplot (for the type studied cereal flour numbering, please see Table 1).

All the interactions and effects could be observed in the loading plot (Figure 2B). Briefly, the main effects resulted from the type of flour, and the protein content tends to play a significant factor in the clustering of the assessed grains which is resulting also to various quality characteristics and specific AA profiles, for which we have already identified statistical differences among the three types of flours. In addition, the organic cultivated grains tended to reach a higher protein content that is also reflecting to a higher content of certain AAs. On the other hand, a lower protein content reflected to a lower content of certain AA, but significant In concluding, it seems that the type of cultivation of the cereal grains affects the AA profile, as well as the quality characteristics of the flours and there is an indication of the potential effect of the cultivation (organic, conventional). Flours from organic cultivated grains seemed to be close in protein content (%) and eventually in the amino acids, but there is a clear differentiation in gluten index and wet gluten. In this case, the "spelt" flours are closer to "emmer" for the aforementioned characteristics, and all of them differentiated from the bread flour either type-70 or type-90. Spelt flours, either type-70 or 100, presented similar characteristics for either organic or conventional cultivated grains, while also "Emmer" type-70 or 100, presented also close characteristics. The biggest differentiation was noticed in the "bread" flours, where the cultivation and the extraction rate presented a significant effect for all the studied factors and the amino acid content (Table 3 and Figure 2).

5. Conclusions

A UHPLC-HILIC-tandem MS method has been developed and validated for the quantification of 17 amino acids in cereal flour samples after acid hydrolysis with HCl in the presence of a reducing agent. Tryptophan, cysteine, cystine, and asparagine were not possible to be quantified as they were degraded during hydrolysis due to harsh acidic conditions applied. The method proved to be a fast and reliable tool for acquiring information on amino acids profile from cereal flours. The developed analytical method has been applied in different flours such as spelt, dicoccum, whole grain wheat, and white wheat. Moreover, multivariate analysis showed that protein content and type of flour have the main contribution and effect, interacting with either the AA profile and with the studied quality characteristics. It has also been presented that not only with quality characteristics of the flours, but the type oz flour has significant interactions. A clear effect of an indication of the potential effect has been identified among the different flours studied.

Supplementary Materials: The following are available online at http://www.mdpi.com/2304-8158/8/10/514/s1, Table S1: Amino acids monoisotopic masses. Multiple Reaction Monitoring (MRM). Retention times (tR) and conditions in the mass spectrometer along with their respective molecular formulas, Table S2: Precision and trueness data for the analysis of AA in spiked B-90 bread wheat sample.

Author Contributions: Conceptualization—E.T., M.P. and S.K.; methodology—E.T. and S.K.; formal analysis—E.T., M.P. and S.K.; investigation—E.T.; resources—E.T., M.P. and S.K.; data curation—E.T., M.P and S.K.; writing—original draft preparation—E.T.; writing—review and editing—E.T., M.P. and S.K.; visualization—E.T.; supervision—E.T. and S.K.; project administration—E.T., M.P. and S.K.

Funding: This research received no external funding.

Conflicts of Interest: The authors declare no conflict of interest.

References

1. Belitz, H.-D.; Grosch, W.; Schieberle, P. *Food Chemistry*, 4th ed.; Springer: Berlin, Germany, 2009.
2. Jia, S.; Kang, Y.P.; Park, J.H.; Lee, J.; Kwon, S.W. Simultaneous determination of 23 amino acids and 7 biogenic amines in fermented food samples by liquid chromatography/quadrupole time-of-flight mass spectrometry. *J. Chromatogr. A* **2011**, *1218*, 9174–9182. [CrossRef] [PubMed]
3. Gökmen, V.; Serpen, A.; Mogol, B.A. Rapid determination of amino acids in foods by hydrophilic interaction liquid chromatography coupled to high-resolution mass spectrometry. *Anal. Bioanal. Chem.* **2012**, *403*, 2915–2922. [CrossRef] [PubMed]
4. Zhao, M.; Ma, Y.; Dai, L.-L.; Zhang, D.-L.; Li, J.-H.; Yuan, W.-X.; Li, Y.-L.; Zhou, H.-J. A high-performance liquid chromatographic method for simultaneous determination of 21 free amino acids in tea. *Food Anal. Methods* **2012**, *6*, 69–75. [CrossRef]
5. Zhou, G.; Pang, H.; Tang, Y.; Yao, X.; Mo, X.; Zhu, S.; Guo, S.; Qian, Y.; Su, S.; Zhang, L.; et al. Hydrophilic interaction ultra-performance liquid chromatography coupled with triple-quadrupole tandem mass spectrometry for highly rapid and sensitive analysis of underivatized amino acids in functional foods. *Amino Acids* **2013**, *44*, 1293–1305. [CrossRef] [PubMed]
6. Fountoulakis, M.; Lahm, H.-W. Hydrolysis and amino acid composition analysis of proteins. *J. Chromatogr. A* **1998**, *826*, 109–134. [CrossRef]
7. Kim, S.-L.; Lee, J.-E.; Kwon, Y.-U.; Kim, W.-H.; Jung, G.-H.; Kim, D.-W.; Lee, C.-K.; Lee, Y.-Y.; Kim, M.-J.; Hwang, T.-Y.; et al. Introduction and nutritional evaluation of germinated soy germ. *Food Chem.* **2013**, *136*, 491–500. [CrossRef] [PubMed]
8. Rombouts, I.; Lamberts, L.; Celus, I.; Lagrain, B.; Brijs, K.; Delcour, J.A. Wheat gluten amino acid composition analysis by high-performance anion-exchange chromatography with integrated pulsed amperometric detection. *J. Chromatogr. A* **2009**, *1216*, 5557–5562. [CrossRef]
9. Kelly, M.T.; Blaise, A.; Larroque, M. Rapid automated high performance liquid chromatography method for simultaneous determination of amino acids and biogenic amines in wine, fruit and honey. *J. Chromatogr. A* **2010**, *1217*, 7385–7392. [CrossRef]
10. Mayer, H.K.; Fiechter, G. Application of UHPLC for the determination of free amino acids in different cheese varieties. *Anal. Bioanal. Chem.* **2013**, *405*, 8053–8061. [CrossRef]
11. Bernal, J.; Ares, A.M.; Pól, J.; Wiedmer, S.K. Hydrophilic interaction liquid chromatography in food analysis. *J. Chromatogr. A* **2011**, *1218*, 7438–7452. [CrossRef]
12. Nováková, L.; Havlíková, L.; Vlacková, H. Hydrophilic interaction chromatography of polar and ionizable compounds by UHPLC. *Trends Anal. Chem.* **2014**, *63*, 55–64. [CrossRef]
13. Kato, M.; Kato, H.; Eyama, S.; Takatsu, A. Application of amino acid analysis using hydrophilic interaction liquid chromatography coupled with isotope dilution mass spectrometry for peptide and protein quantification. *J. Chromatogr. B* **2009**, *877*, 3059–3064. [CrossRef] [PubMed]

14. Guo, S.; Duan, J.; Qian, D.; Tang, Y.; Qian, Y.; Wu, D.; Su, D.; Shang, E. Rapid determination of amino acids in fruits of Ziziphus jujuba by hydrophilic interaction ultra-high-performance liquid chromatography coupled with triple-quadrupole mass spectrometry. *J. Agric. Food Chem.* **2013**, *61*, 2709–2719. [CrossRef] [PubMed]

15. Tsochatzis, E.D.; Begou, O.; Gika, H.G.; Karayannakidis, P.D.; Kalogiannis, S. A hydrophilic interaction chromatography-tandem mass spectrometry method for amino acid profiling in mussels. *J. Chromatogr. B* **2017**, *1047*, 197–206. [CrossRef] [PubMed]

16. Langrock, T.; Czihal, P.; Hoffmann, R. Amino acid analysis by hydrophilic interaction chromatography coupled on-line to electrospray ionization mass spectrometry. *Amino Acids* **2006**, *30*, 291–297. [CrossRef] [PubMed]

17. Papageorgiou, M.; Skendi, A. *Sustainable Recovery and Reutilisation of Cereal Processing By-Products*, 1st ed.; Elsevier: Amsterdam, The Netherlands, 2018; Chapter 1.

18. Kies, C.; Fox, H.M. Determination of the first-limiting amino acid of wheat and triticale grain for humans. *Cereal Chem.* **1970**, *47*, 615–622.

19. Fleming, S.E.; Sosulski, F.W. Nutritive value of bread fortified with concentrated plant proteins and lysine. *Cereal Chem.* **1977**, *54*, 1239–1248.

20. Abdel-Aal, E.-S.M.; Sosulski, F.W. Bleaching and fractionation of dietary fiber and protein from wheat-based stillage. *Lebensm. Wiss. Technol.* **2001**, *34*, 159–167. [CrossRef]

21. Abdel-Aal, E.-S.M.; Hucl, P. Amino Acid Composition and In Vitro Protein Digestibility of Selected Ancient Wheats and their End Products. *J. Food Comp. Anal.* **2002**, *15*, 737–747. [CrossRef]

22. Pflüger, L.A.; Martin, L.M.; Alvarez, J.B. Variation in the HMW and LMW glutenin subunits from Spanish accessions of emmer wheat (*Triticum turgidum* ssp. Dicoccum Schrank). *Theor. Appl. Genet.* **2001**, *102*, 767–772. [CrossRef]

23. Ruegger, A.; Winzerler, H.; Nosberger, J. Untersuchugen uber das keimungsverhalten von dinkel (*Triticum spelta* L.) und weizen (*Triticum aestivum* L.) unter erschwerten bedingungen. *Seed Sci. Technol.* **1990**, *18*, 311–320.

24. Kema, G.H.J.; Lange, W. Resistance in spelt wheat to yellow rust: II. Monosomic analysis of the Iranian accession 415. *Euphytica* **1992**, *63*, 219–224.

25. Escarnot, E.; Jacquemin, J.-M.; Agneessens, R.; Paquot, M. Comparative study of the content and profiles of macronutrients in spelt and wheat, a review. *Biotechnol. Agron. Soc. Environ.* **2012**, *16*, 243–256.

26. Wilson, J.D.; Bechtel, D.B.; Wilson, G.W.T.; Seib, P.A. Bread quality of spelt wheat and its starch. *Cereal Chem.* **2008**, *85*, 629–638. [CrossRef]

27. Pruska-Kedzior, A.; Kedzior, Z.; Klockiewicz-Kaminska, E. Comparison of viscoelastic properties of gluten from spelt and common wheat. *Eur. Food Res. Technol.* **2008**, *227*, 199–207. [CrossRef]

28. ICH. Guidance for Industry Q2B Validation of Analytical Procedures: Methodology. 1996. Available online: http://www.fda.gov/downloads/drugs/guidancecomplianceregulatoryinformation/guidances/ucm073384.pdf (accessed on 21 June 2018).

29. ICC-International Association for Cereal Science and Technology. 2019. Available online: https://www.icc-services.at/store/standard_methods (accessed on 31 January 2019).

30. Faast, E.Z. Phenomenex, Faast Amino Acid Analysis of Protein Hydrolysates, User's Manual. 2016. Available online: http://www.fortunesci.com/image/download2/USER%20GUIDE/Amino%20Acid%20Analysis%20of%20Protein%20Hydrolysates.pdf (accessed on 21 February 2019).

31. Gika, H.G.; Theodoridis, G.A.; Vrhovsek, U.; Mattivi, F. Quantitative profiling of polar primary metabolites using hydrophilic interaction ultra-high performance liquid chromatography-tandem mass spectrometry. *J. Chromatogr. A* **2012**, *1259*, 121–127. [CrossRef]

32. Alberto Lopes, J.; Tsochatzis, E.D.; Robouch, P.; Hoekstra, E. Influence of pre-heating of food contact polypropylene cups on its physical structure and on the migration of additives. *Food Packag. Shelf Life* **2019**, *20*, 100305. [CrossRef]

33. Zumwalt, R.W.; Gehrke, C.W. Amino Acid Analysis: A Survey of Current Techniques. In *Methods for Protein Analysis*; Cherry, J.P., Barford, R.A., Eds.; American Oil Chemists' Society (AOCS): Champaign, IL, USA, 1988; pp. 13–35.

34. Davidson, I. Hydrolysis of Samples for Amino Acid Analysis. In *Protein Sequencing Protocols (Methods in Molecular Biology)*; Smith, B.J., Ed.; Humana Press: Totowa, NJ, USA, 2003; Volume 211.

35. Otter, D. Standardised methods for amino acid analysis of food. *Br. J. Nutr.* **2012**, *108*, S230–S237. [CrossRef]

36. Kapusniak, J.; Ciesielski, W.; Koziol, J.J.; Tomasik, P. Reaction of starch with a-amino acids. *Eur. Food Res. Technol.* **1999**, *209*, 325–329. [CrossRef]

37. Ito, A.; Hattori, M.; Yoshida, T.; Watanabe, A.; Sato, R.; Takahashi, K. Regulatory effect of amino acids on the pasting behavior of porato starch is attrbutable to its binding to the starch chain. *J. Agric. Food Chem.* **2006**, *54*, 10191–10196. [CrossRef]

38. Rowan, A.M.; Moughan, P.J.; Wilson, M.N. Effect of hydrolysis time on the determination of amino acid composition of diet, ileal digesta and feces samples and on the determination of dietary amino acid digestibility coefficients. *J. Agric. Food Chem.* **1992**, *40*, 981–985. [CrossRef]

39. Ranhotra, G.S.; Gelroth, J.A.; Glaser, B.K.; Lorenz, K.J. Baking and nutritional qualities of a spelt wheat sample. *Lebensm. Wiss. Technol.* **1995**, *28*, 118–122. [CrossRef]

40. Bonafaccia, G.; Galli, V.; Francisci, R.; Mair, V.; Skrabanja, V.; Kreft, I. Characteristics of spelt wheat products and nutritional value of spelt wheat-based bread. *Food Chem.* **2000**, *68*, 437–441. [CrossRef]

41. Dvoracek, V.; Curn, V.; Moudry, J. Evaluation of amino acid content and composition in spelt wheat varieties. *Cereal Res. Commun.* **2002**, *30*, 187–193.

42. Matz, S.A. *Chemistry and Technology of Cereals as Food and Feed*; Springer Science & Business Media: New York, NY, USA, 1991; Chapter 4.

43. Matuz, J.; Bartok, T.; Morocz-Salamon, K.; Bona, L. Structure and potential allergenic character of cereal proteins. I. Protein content and amino acid composition. *Cereal Res. Commun.* **2000**, *28*, 263–270.

 *foods*

Article

Analysis of Volatile Constituents in *Platostoma palustre* (Blume) Using Headspace Solid-Phase Microextraction and Simultaneous Distillation-Extraction

Tsai-Li Kung [1], Yi-Ju Chen [2], Louis Kuoping Chao [2], Chin-Sheng Wu [2], Li-Yun Lin [3,]* and Hsin-Chun Chen [2,]*

[1]   Taoyuan District Agricultural Research and Extension Station, Council of Agriculture, Executive Yuan, Taoyuan 327, Taiwan; tlkung@tydais.gov.tw

[2]   Department of Cosmeceutics, China Medical University, Taichung 404, Taiwan; yc9429@hotmail.com (Y.-J.C.); kuoping@mail.cmu.edu.tw (L.K.C.); cswu@mail.cmu.edu.tw (C.-S.W.)

[3]   Department of Food Science and Technology, Hungkuang University, Taichung 433, Taiwan

*   Correspondence: lylin@sunrise.hk.edu.tw (L.-Y.L.); d91628004@ntu.edu.tw (H.-C.C.); Tel.: +886-4-2631-8652 (ext. 5040) (L.-Y.L.); +886-4-2205-3366 (ext. 5310) (H.-C.C.); Fax: +886-4-2236-8557 (H.-C.C.)

Received: 9 August 2019; Accepted: 10 September 2019; Published: 14 September 2019

Abstract: Hsian-tsao (*Platostoma palustre* Blume) is a traditional Taiwanese food. It is admired by many consumers, especially in summer, because of its aroma and taste. This study reports the analysis of the volatile components present in eight varieties of Hsian-tsao using headspace solid-phase microextraction (HS-SPME) and simultaneous distillation-extraction (SDE) coupled with gas chromatography (GC) and gas chromatography-mass spectrometry (GC/MS). HS-SPME is a non-heating method, and the results show relatively true values of the samples during flavor isolation. However, it is a kind of headspace analysis that has the disadvantage of a lower detection ability to relatively higher molecular weight compounds; also, the data are not quantitative, but instead are used for comparison. The SDE method uses distillation 2 h for flavor isolation; therefore, it quantitatively identifies more volatile compounds in the samples while the samples withstand heating. Both methods were used in this study to investigate information about the samples. The results showed that Nongshi No. 1 had the highest total quantity of volatile components using HS-SPME, whereas SDE indicated that Taoyuan Mesona 1301 (TYM1301) had the highest volatile concentration. Using the two extraction methods, 120 volatile components were identified. Fifty-six volatile components were identified using HS-SPME, and the main volatile compounds were α-pinene, β-pinene, and limonene. A total of 108 volatile components were identified using SDE, and the main volatile compounds were α-bisabolol, β-caryophyllene, and caryophyllene oxide. Compared with SDE, HS-SPME sampling extracted a significantly higher amount of monoterpenes and had a poorer detection of less volatile compounds, such as sesquiterpenes, terpene alcohols, and terpene oxide.

Keywords: Hsian-tsao; *Platostoma palustre* (Blume); headspace solid-phase microextraction (SPME); volatile components; simultaneous distillation-extraction (SDE)

---

1. Introduction

Hsian-tsao (*Platostoma palustre* Blume, also known as *Mesona procumbens* Hemsl. [1]), also called Liangfen Cao or black cincau, belongs to the family Lamiaceae. It is an annual plant that is mainly distributed in tropical and subtropical regions, including Taiwan, southern China, Indonesia, Vietnam, and Burma [2]. Hsian-tsao tea, herbal jelly, and sweet soup with herbal jelly are popular during the summer, and heated herbal jelly is admired by many Taiwanese, especially in winter, because of its aroma and taste. In Indonesia, janggelan (*Mesona palustris* BL) has also been made into a herbal drink and a jelly-type dessert [3]. Hsian-tsao is also used as a remedy herb in folk medicine and is supposed to be effective in treating heat-shock, hypertension, diabetes, liver diseases, and muscle and joint pains [4,5].

Hsian-tsao contains polysaccharides (gum) with a unique aroma and texture. Most research has investigated the gum of Hsian-tsao [2,6–8]; however, there are only a few studies of Hsian-tsao aroma. Wei et al. [9] identified 59 volatile compounds in *Mesona* Benth extracted using solvent extraction. They also reported that the important constituents were caryophyllene oxide, α-caryophyllene, eugenol, benzene acetaldehyde, and 2,3-butanedione. Deng et al. [10] reported the chemical constituents of essential oil from *Mesona chinensis* Benth (also known as *P. palustris* Blume [1]) using GC/MS. The major constituents were *n*-hexadecanoic acid, linoleic acid, and linolenic acid. Lu et al. [11] analyzed the volatile oil from *Mesona chinensis* Benth using GC/MS. The results indicated the main components were chavibetol, *n*-hexadecanoic acid, and α-cadinol.

Simultaneous distillation-extraction (SDE) is a traditional extraction method that was introduced by Likens and Nickerson in 1946. SDE combines the advantages of liquid–liquid extraction and steam distillation methods. It is widely used for the extraction of essential oils and volatile compounds [12,13]. In the flavor field, this technique is recognized as a superior extraction method compared to other methods, such as solvent extraction or distillation. Moreover, Gu et al. [14] indicated that SDE has excellent reproducibility and high efficiency compared with traditional extraction methods.

Headspace solid-phase microextraction (HS-SPME) is a non-destructive and non-invasive method that avoids artifact formation and solvent impurity contamination [15]. HS-SPME is a fast, simple, and solventless technique [16–18]. HS-SPME can integrate sampling, extraction, concentration and sample introduction into a single uninterrupted process, resulting in high sample throughput [19].

This study aimed to identify the volatile constituents in different varieties of Hsian-tsao and the differences in extraction methods (HS-SPME and SDE). The differences in volatile compounds caused by heating are discussed. The results from this study provide a reference for the food, horticultural, and flavor industries.

2. Materials and Methods

2.1. Plant Materials

A total of eight varieties of Hsian-tsao from throughout Taiwan were used in this study (Table 1): Nongshi No. 1 from Tongluo Township in Miaoli County; Taoyuan No. 2 from Shoufeng Township in Hualien County; Chiayi strain from Shuishang Township in Chiayi County; Taoyuan No. 1, TYM1301, and TYM1302 from Guanxi Township in Hsinchu County; and TYM1303 and TYM1304 from Shuangxi District in New Taipei City. Eight varieties of Hsian-tsao were grown at the Sinpu Branch Station (Sinpu Township in Hsinchu County) of Taoyuan District Agricultural Research and Extension Station. The identities of the plants were confirmed by Tsai-Li Kung (Chief of the Sinpu Branch Station). After shade drying, dried samples were stored at room temperature for one year before the experiment was conducted.

Table 1. The study of collections of taxa currently assigned to Hsian-tsao.

Varieties	Origin	Growing Locality
Nongshi No.1	Tongluo Township, Miaoli County	Sinpu Township, Hsinchu County
Taoyuan No.1	Guanxi Township, Hsinchu County	Sinpu Township, Hsinchu County
Taoyuan No.2	Shoufeng Township, Hualien County	Sinpu Township, Hsinchu County
Chiayi strain	Shuishang Township, Chiayi County	Sinpu Township, Hsinchu County
TYM1301	Guanxi Township, Hsinchu County	Sinpu Township, Hsinchu County
TYM1302	Guanxi Township, Hsinchu County	Sinpu Township, Hsinchu County
TYM1303	Shuangxi District, New Taipei City	Sinpu Township, Hsinchu County
TYM1304	Shuangxi District, New Taipei City	Sinpu Township, Hsinchu County

2.2. Methods

2.2.1. Optimization of the HS-SPME Procedure

The method used was modified from those of Yeh et al. [20]:

1. Comparisons of SPME fiber coatings: five different coated SPME fibers, 85 µm polyacrylate (PA), 100 µm polydimethylsiloxane (PDMS), 65 µm polydimethylsiloxane/divinylbenzene (PDMS/DVB), 75 µm carboxen/polydimethylsiloxane (CAR/PDMS), and 50/30 µm divinylbenzene/carboxen/polydimethylsiloxane (DVB/CAR/PDMS), (Supelco, Inc., Bellefonte, PA, USA) were used for the aroma extraction. Samples (Nongshi No. 1) were placed into a homogenizer (WAR7012S 7-Speed Blender 1 Qt. 120 V) (Waring commercial, Torrington, CT, USA). After being homogenized for 30 s, 1 g of homogenized samples was put into a 7 mL vial (Hole Cap Polytetrafluoroethene/Silicone Septa) (Supelco, Inc., Bellefonte, PA, USA) and sealed. The SPME method was used to extract the aroma components. The extraction temperature was 25 ± 2 °C and the extraction time was 40 min. This experiment and all other experiments in this study were performed in triplicates.

2. Comparisons of the extraction times: The above-mentioned optimal extraction fiber was used in the comparison of the extraction times. The tested extraction times were 10 min, 20 min, 30 min, 40 min, or 50 min, and the extraction temperature was maintained at 25 ± 2 °C. Sample preparation steps were the same as above.

2.2.2. Analysis of the Volatile Compounds

1. Analysis of the volatile compounds using HS-SPME extraction: a 50/30 µm divinylbenzene/carboxen/polydimethylsiloxane (DVB/CAR/PDMS) fiber (Supelco, Inc., Bellefonte, PA, USA) was used for aroma extraction. The eight different Hsian-tsao varieties were used as samples. Each sample was homogenized as described above in Section 2.2.1 (1 g was placed in a 7 mL vial (hole cap PTFE/silicone septa)). The SPME fiber was exposed to each sample for 40 min at 25 ± 2 °C; then, each sample was injected into a gas chromatograph injection unit. The injector temperature was maintained at 250 °C and the fiber was held for 10 min. The peak area of a volatile compound or total volatile compounds from the integrator was used to calculate the relative contents.

2. Analysis of volatile compounds by SDE extraction: 100 g samples of Hsian-tsao were cut with scissors into pieces approximately 1–3 cm in size and were then homogenized for 2 min with 2 L of deionized water and were placed into a 5-L round-bottom flask. The flask was attached to a simultaneous distillation-extraction apparatus and 100 °C steam was used as the heat source and passed through the sample. A 50 mL volume of solvent was prepared by mixing *n*-pentane/diethyl ether (1:1, *v/v*) into a pear-shaped flask, then placing it in a 40–45 °C water bath. This distillation circulation continued for 2 h, and the collected solvent extract was added to 200 µL of an internal standard solution of cyclohexyl acetate, and an internal standard was used to obtain the weight concentration of volatile compound in the sample; also, anhydrous sodium sulfate was used to

    remove the water. Lastly, the distillation column (40 °C, 1 h, 100 cm glass column) was used to volatilize the solvent and the condensed volatile component extracts were collected.

3.   GC analysis of the volatile compounds was conducted using a 7890A GC (Agilent Technologies, Palo Alto, CA, USA) equipped with a DB-1 (60 m × 0 .25 mm i.d. × 0.25 μm film thickness, Agilent Technologies) capillary column and a flame ionization detector. The injector and detector temperatures were maintained at 250 °C and 300 °C, respectively. The oven temperature was held at 40 °C for 1 min and then raised to 150 °C at 5 °C/min and held for 1 min, and then increased from 150 to 200 °C at 10 °C/min and held for 11 min. The carrier gas (nitrogen) flow rate was 1 mL/min. The Kovats indices were calculated for the separated components relative to a C_5–C_{25} *n*-alkanes mixture [21]. The purpose gas chromatography-flame ionization detector (GC-FID) was used both for retention indices (RI) comparison and quantitation of peak areas.

4.   GC-MS analysis of volatile compounds were identified using an Agilent 7890B GC equipped with DB-1 (60 m × 0.25 mm i.d. × 0.25 μm film thickness) fused silica capillary column coupled to an Agilent model 5977 N MSD mass spectrometer (MS). The GC conditions in the GC-MS analysis were the same as in the GC analysis. The carrier gas (helium) flow rate was 1 mL/min. The electron energy was 70 eV at 230 °C. The constituents were identified by matching their spectra with those recorded in an MS library (Wiley 7N, John Wiley & Sons, Inc. New Jersey, NJ, USA). In addition, the constituents were confirmed by comparing the Kovats indices or GC retention time data with those of authentic standards or data published in the literature. The GC and GC-MS methods used were modified from those of Yeh et al. [20].

5.   Statistical Analysis: Each sample was extracted in triplicate and the concentration of volatile compounds was determined as the mean value of three repetitions. The data were subjected to a monofactorial variance analysis with Duncan's multiple range method with a level of significance of $p < 0.05$ (SPSS Base 12.0, SPSS Inc., Chicago, IL, USA).

3. Results

3.1. Comparisons of the Optimization Conditions of HS-SPME

3.1.1. SPME Fiber Selection

The performance of five commercially available SPME fibres: 50/30-μm DVB/CAR/PDMS, 65-μm PDMS/DVB, 75-μm CAR/PDMS, 100-μm PDMS, and 85-μm PA were used to extract the volatile components of Nongshi No. 1. The 50/30-μm DVB/CAR/PDMS fiber extracted more total volatile components than the other fibers (Figure 1).

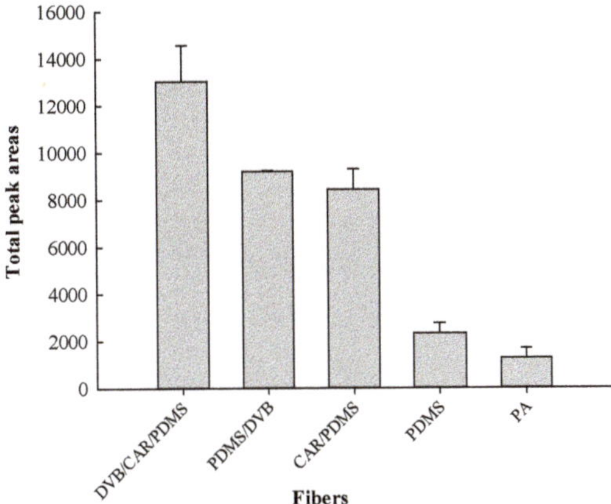

Figure 1. Comparisons of the total peak areas of total volatile compounds detected in the headspace of Nongshi No. 1 using different headspace solid-phase microextraction (HS-SPME) fibers.

Ducki et al. [22] evaluated four different types of SPME fibers (100-μm PDMS, 65-μm PDMS/DVB, 75-μm CAR/PDMS, and 50/30-μm DVB/CAR/PDMS) for the headspace analysis of volatile compounds in cocoa products. The SPME fiber coated with 50/30-μm DVB/CAR/PDMS afforded the highest extraction efficiency. Silva et al. [23] compared the performance of six fibers (PDMS, PDMS/DVB, CW/DVB, PA, CAR/PDMS, and DVB/CAR/PDMS) and found that DVB/CAR/PDMS was the most effective SPME fiber for isolating the volatile metabolites from *Mentha* × *piperita* L. fresh leaves based on the total peak areas, reproducibility, and number of extracted metabolites. Yeh et al. [20] reported the volatile components in *Phalaenopsis* Nobby's Pacific Sunset, and the optimal extraction conditions were obtained using a DVB/CAR/PDMS fiber.

The 50/30-μm DVB/CAR/PDMS was revealed to be the most suitable and was subsequently used in all further experiments.

3.1.2. HS-SPME Extraction Time

The optimal SPME fiber (50/30-μm DVB/CAR/PDMS) was used to extract Nongshi No. 1 at 25 ± 2 °C, and the extraction times from 10 to 50 min were investigated. The total peak area increased from 10–40 min and reached the peak at 40 min (Figure 2). Silva and Câmara [23] promoted the higher extraction efficiency, corresponding to the higher GC peak areas and the number of identified metabolites. This higher extraction efficiency was achieved using: DVB/CAR/PDMS coating fiber, and 40 °C and 60 min as the extraction temperature and extraction time, respectively. Zhang et al. [24] also obtained optimum extraction conditions, which were using 50/30-μm DVB/CAR/PDMS fiber for 40 min at 90 °C. According to the obtained results, 40 min was selected as the optimal extraction time.

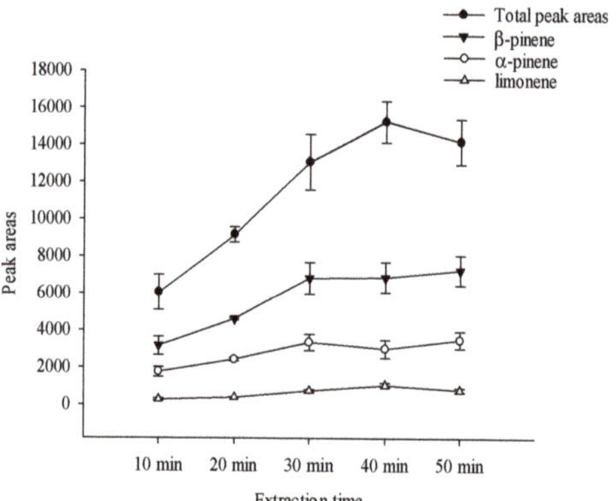

Figure 2. Comparisons of the peak areas of total volatile compounds and main components detected in the headspace of Nongshi No. 1 for different SPME extraction times at 25 °C using a DVB/CAR/PDMS fiber.

3.2. Analyses of the Volatiles of Eight Varieties of Hsian-Tsao Using HS-SPME

As shown in Figure 3, the total peak areas of volatile components was the highest in Nongshi No. 1 and lowest in TYM1304. Volatile compounds in eight varieties of Hsian-tsao were analyzed using headspace solid-phase microextraction (HS-SPME), which was coupled with GC and GC/MS. Table 2 shows a total of 56 compounds that were identified. Monoterpene compounds were the most abundant compounds in the Hsian-tsao analyzed using HS-SPME/GC. The main volatile components of Nongshi No. 1, Chiayi strain, TYM1302, and TYM1303 were β-pinene (43–50%), α-pinene (10–24%), and limonene (4–9%). β-Pinene (36–42%), α-pinene (15–17%), and β-caryophyllene (11%) were the main components from Taoyuan No. 2 and TYM1301. The main components of Taoyuan No. 1 were β-pinene (23%), limonene (21%), and α-pinene (11%). Limonene (32%), β-caryophyllene (13%), and sabinene (7%) were the main components from TYM1304. TYM1304 contained the highest content of limonene (32%), followed by Taoyuan No. 1 (21%). Limonene is a citrus note, having a pungent green and lemon-like aroma [25,26]. The peak areas of α-pinene and β-pinene were the highest in TYM1302 (24% and 50%), whereas TYM1304 was lower than the other varieties. α-Pinene was described as having a fruity, piney, and turpentine-like aroma [27,28], and β-pinene was described as having a dry-woody, pine-like, and citrus aroma [29,30]. β-Caryophyllene was described as having a dry-woody, pine-like, and spicy aroma, and TYM1304 contained the highest content (13%), followed by TYM1301, and Taoyuan No. 2 (11%).

Table 2. Comparisons of volatile compounds from eight varieties of Hsian-tsao extracted using HS-SPME.

Compound [a]	RI [b]	Relative Content (%) [c]							
		Nongshi No. 1	Taoyuan No. 1	Taoyuan No. 2	Chiayi Strain	TYM1301	TYM1302	TYM1303	TYM1304
Aliphatic alcohol									
1-Octen-3-ol [f]	962	0.29 ± 0.11	0.69 ± 0.24	1.16 ± 0.19	0.17 ± 0.03	2.25 ± 0.40	1.39 ± 0.28	1.61 ± 0.22	3.01 ± 0.82
Aliphatic aldehydes									
Hexanal [f]	777	3.63 ± 0.52	1.85 ± 0.68	0.29 ± 0.10	0.39 ± 0.15	1.19 ± 0.26	1.15 ± 0.22	1.81 ± 0.24	1.74 ± 0.19
(E)-2-heptenal [f]	931	tr [d]	1.06 ± 0.20	tr	- [e]	tr	tr	tr	tr
Decanal [f]	1181	-	-	-	tr	-	-	-	-
Aliphatic ketones									
1-Octen-3-one [f]	956	tr	0.14 ± 0.02	0.12 ± 0.02	tr	0.29 ± 0.03	tr	0.24 ± 0.02	0.47 ± 0.07
3-Octanone [f]	965	-	-	-	tr	-	-	tr	tr
Aliphatic ester									
Linalyl isobutyrate	1387	0.32 ± 0.05	-	-	-	-	-	-	-
Aromatic alcohol									
Eugenol [g]	1328	-	-	-	0.08 ± 0.02	-	-	-	-
Aromatic aldehyde									
Benzaldehyde [f]	933	tr	-	-	tr	-	-	-	-
Aromatic hydrocarbon									
p-Cymene	1004	-	0.68 ± 0.05	-	tr	tr	tr	tr	tr
Terpene alcohol									
α-Bisabolol	1680	0.11 ± 0.05	0.22 ± 0.09	0.09 ± 0.03	0.13 ± 0.03	0.36 ± 0.04	0.08 ± 0.01	0.08 ± 0.01	-
Monoterpenes									
α-Thujene	927	1.02 ± 0.11	0.12 ± 0.02	1.28 ± 0.17	0.86 ± 0.21	1.16 ± 0.11	1.10 ± 0.05	1.10 ± 0.04	3.19 ± 0.24
α-Pinene [f]	937	19.73 ± 1.72	11.14 ± 1.10	16.65 ± 1.26	22.67 ± 0.45	14.76 ± 0.68	24.20 ± 0.40	20.20 ± 0.69	4.72 ± 0.19
Camphene	949	0.79 ± 0.04	0.46 ± 0.06	0.63 ± 0.02	1.22 ± 0.09	0.70 ± 0.02	0.93 ± 0.02	0.82 ± 0.05	-
sabinene	967	2.75 ± 0.25	3.60 ± 0.15	3.56 ± 0.46	1.46 ± 0.41	2.87 ± 0.40	2.86 ± 0.10	3.07 ± 0.21	7.37 ± 0.62
β-Pinene [f]	972	44.99 ± 2.09	23.19 ± 2.17	42.28 ± 1.31	46.60 ± 1.89	36.18 ± 0.83	49.51 ± 1.45	43.24 ± 0.19	2.15 ± 0.36
β-Myrcene	983	0.97 ± 0.02	2.26 ± 0.16	0.41 ± 0.02	0.28 ± 0.10	0.64 ± 0.09	0.37 ± 0.01	0.82 ± 0.07	2.40 ± 0.09
δ-3-Carene	998	-	0.85 ± 0.04	-	0.43 ± 0.04	-	-	0.46 ± 0.06	1.92 ± 0.09
α-Terpinene	1007	0.52 ± 0.05	0.05 ± 0.00	0.61 ± 0.07	0.65 ± 0.16	0.63 ± 0.06	0.56 ± 0.03	0.60 ± 0.02	1.64 ± 0.09
Limonene [f]	1015	7.06 ± 1.17	20.98 ± 1.01	6.45 ± 0.31	4.66 ± 1.96	7.47 ± 1.68	4.29 ± 0.08	9.05 ± 1.05	32.08 ± 2.81
γ-Terpinene	1045	-	0.55 ± 0.09	0.60 ± 0.07	0.38 ± 0.12	-	0.52 ± 0.04	-	1.63 ± 0.09
α-Terpinolene	1081	-	0.39 ± 0.05	-	0.26 ± 0.04	0.46 ± 0.01	0.32 ± 0.01	-	1.14 ± 0.07
Sesquiterpenes									
α-Cubebene	1353	-	0.16 ± 0.01	0.16 ± 0.05	0.07 ± 0.02	-	-	-	-
α-Ylangene	1378	tr	tr	tr	0.41 ± 0.23	tr	tr	tr	tr
α-Copaene	1382	0.20 ± 0.03	1.45 ± 0.17	0.24 ± 0.05	0.25 ± 0.14	0.62 ± 0.09	0.32 ± 0.02	0.09 ± 0.01	0.24 ± 0.01
β-Elemene	1392	1.13 ± 0.19	3.26 ± 0.12	3.05 ± 0.35	1.27 ± 0.17	2.60 ± 0.30	0.58 ± 0.06	1.04 ± 0.14	4.38 ± 0.28
β-Bourbonene	1393	-	-	tr	0.08 ± 0.00	-	tr	-	-
α-Cedrene	1401	-	0.16 ± 0.01	-	-	-	-	-	-
cis-α-Bergamotene	1414	-	0.49 ± 0.07	-	-	-	-	-	-
α-Gurgujene	1418	-	-	-	0.07 ± 0.00	-	-	-	-
β-Caryophyllene [f,g,h]	1427	2.68 ± 0.40	8.49 ± 1.16	10.71 ± 1.56	2.74 ± 0.74	10.81 ± 1.03	2.76 ± 0.52	2.61 ± 0.25	12.53 ± 1.51
Aromadendrene	1433	-	-	-	tr	-	-	-	-
α-Bergamotene [h]	1436	0.77 ± 0.10	1.31 ± 0.25	-	0.55 ± 0.08	0.90 ± 0.01	-	-	0.71 ± 0.04
cis-Thujopsene	1437	-	0.15 ± 0.02	-	0.09 ± 0.00	-	-	-	-
β-Gurjunene	1442	-	-	0.37 ± 0.06	-	-	-	-	-
cis-β-Farnesene [h]	1445	-	0.69 ± 0.05	-	0.39 ± 0.11	0.59 ± 0.08	-	-	-
trans-β-Farnesene	1452	-	0.20 ± 0.02	-	-	0.18 ± 0.01	-	0.05 ± 0.01	-
α-Caryophyllene [f,h]	1461	0.87 ± 0.10	1.33 ± 0.18	1.51 ± 0.20	0.49 ± 0.10	2.16 ± 0.31	0.50 ± 0.08	0.59 ± 0.25	1.84 ± 0.33

Table 2. *Cont.*

Compound [a]	RI [b]	Relative Content (%) [c]							
		Nongshi No. 1	Taoyuan No. 1	Taoyuan No. 2	Chiayi Strain	TYM1301	TYM1302	TYM1303	TYM1304
α-Muurolene	1479	-	0.36 ± 0.13	-	0.16 ± 0.02	-	-	0.10 ± 0.02	-
γ-Muurolene	1479	-	-	-	-	-	-	-	0.17 ± 0.03
Valencene	1486	0.12 ± 0.01	-	-	-	-	-	-	-
Germacrene D	1487	-	-	0.18 ± 0.00	0.11 ± 0.02	-	-	-	-
β-Selinene	1494	0.35 ± 0.06	0.65 ± 0.17	0.56 ± 0.23	2.72 ± 0.33	1.01 ± 0.11	-	0.38 ± 0.17	1.43 ± 0.12
α-Selinene	1501	-	-	-	1.50 ± 0.17	-	-	-	1.17 ± 0.09
β-Bisabolene	1505	0.42 ± 0.10	0.89 ± 0.09	-	tr	1.20 ± 0.16	-	0.34 ± 0.02	-
α-Chamigrene	1514	-	-	-	0.07 ± 0.01	-	-	-	-
α-Amorphene	1518	-	-	-	0.07 ± 0.02	-	-	-	-
δ-Cadinene	1523	-	-	0.15 ± 0.04	0.08 ± 0.02	-	-	-	-
trans-γ-Bisabolene	1526	-	0.16 ± 0.00	0.06 ± 0.00	-	-	-	-	-
trans-α-Bisabolene	1537	0.18 ± 0.04	0.31 ± 0.05	0.10 ± 0.04	0.43 ± 0.07	-	-	-	-
Terpene oxide									
Caryophyllene oxide [f,g]	1585	-	0.04 ± 0.01	-	-	-	-	-	-
Hydrocarbons									
2-Methyl-octane	869	-	-	0.53 ± 0.05	-	-	-	-	-
Undecane	1098	0.51 ± 0.07	0.44 ± 0.11	0.27 ± 0.02	0.12 ± 0.01	0.24 ± 0.02	-	0.25 ± 0.00	-
Dodecane	1197	0.35 ± 0.07	0.19 ± 0.04	-	tr	0.23 ± 0.00	-	-	0.06 ± 0.01
Tridecane	1294	-	0.12 ± 0.06	0.08 ± 0.01	tr	0.09 ± 0.01	-	0.08 ± 0.01	-
Furan									
2-Pentylfuran	978	tr	tr	tr	tr	tr	tr	tr	tr

[a] Identification of components based on the GC/MS library (Wiley 7N). [b] Retention indices, using paraffin (C_5–C_{25}) as references. [c] Relative percentages from GC-FID, values are means ± standard deviation (SD) of triplicates. [d] Trace. [e] Undetectable. [f] Published in the literature (Wei et al. [9]). [g] Published in the literature (Lu et al. [11]). [h] Published in the literature (Deng et al. [10]).

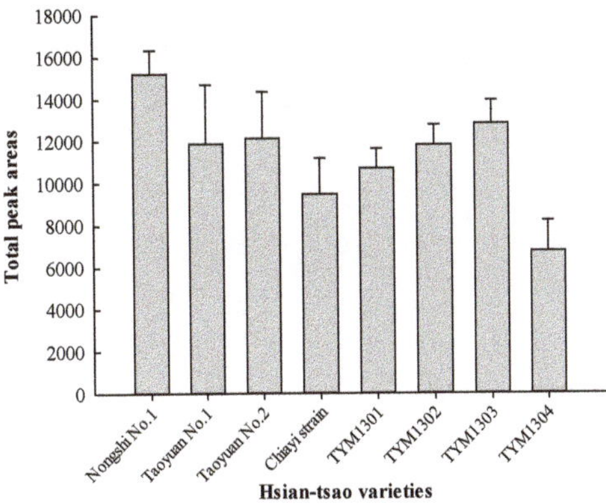

Figure 3. Comparisons of the total peak areas of total volatile compounds of eight varieties of Hsian-tsao (1 g) extracted at 25 °C for 40 min using a DVB/CAR/PDMS fiber. The peak area of a volatile compound or total volatile compounds from the integrator was used to calculate the relative contents using gas chromatography-flame ionization detector (GC-FID). The data corresponds to the mean ± standard deviation (SD) of triplicates.

The eight varieties of Hsian-tsao shared 15 volatile components; the differences in percentage were: 1-octen-3-ol (trace–3%), hexanal (trace–4%), 1-octen-3-one (trace–1%), α-thujene (trace–3%), α-pinene (5–24%), sabinene (2–7%), β-pinene (2–50%), β-myrcene (trace–2%), α-terpinene (trace–2%), limonene (4–32%), α-ylangene (trace), α-copaene (trace–2%), β-elemene (1–4%), β-caryophyllene (3–13%), and α-caryophyllene (1–2%). Among them, hexanal was described as having a green and cut-grass aroma [31], and was responsible for green, apple, and green fruit perceptions [32]. Nongshi No. 1 contained the highest content of hexanal (4%).

3.3. Analysis of the Volatiles of Eight Varieties (Clones) of Hsian-Tsao Using SDE

As shown in Table 3, the volatile components content peaked in TYM1301 and was the lowest in TYM1303. Table 4 shows the results of the SDE analysis: 108 components were identified, including 11 aliphatic alcohols, 14 aliphatic aldehydes, 9 aliphatic ketones, 1 aliphatic ester, 3 aromatic alcohols, 2 aromatic aldehydes, 1 aromatic ketones, 2 aromatic esters, 4 aromatic hydrocarbons, 8 terpene alcohols, 2 terpene aldehydes, 2 terpene ketones, 10 monoterpenes, 22 sesquiterpenes, 1 terpene oxide, 7 hydrocarbons, 3 straight-chain acids, 3 furans, 2 methoxy-phenolic compounds, and 1 nitrogen-containing compound. Sesquiterpene compounds, terpene alcohols, and terpene oxide were the main compounds in the Hsian-tsao analyzed using SDE. The major components of Hsain-tsao (Nongshi No. 1, Taoyuan No. 1, and TYM1301) were α-bisabolol (59–144 mg/kg), caryophyllene oxide (9–28 mg/kg), and β-caryophyllene (21–54 mg/kg). β-Caryophyllene (21–56 mg/kg) and caryophyllene oxide (32–50 mg/kg) were the main compounds of Taoyuan No. 2 and TYM1302. α-Bisabolol (116 mg/kg), β-bisabolene (24 mg/kg), β-selinene (23 mg/kg), and β-caryophyllene (19 mg/kg) were the main components of the Chiayi strain. α-Bisabolol (144 mg/kg), α-bisabolol (43 mg/kg), and caryophyllene oxide (15 mg/kg) were the main components of TYM1303. β-Caryophyllene (53 mg/kg), β-elemene (35 mg/kg), α-selinene (15 mg/kg), and α-caryophyllene (13 mg/kg) were the main compounds in TYM1304. Nongshi No. 1, Taoyuan No. 1, Chiayi strain, TYM1301, and TYM1303 contained a higher content of α-bisabolol; Taoyuan No. 2 and TYM1304 contained a higher content of β-caryophyllene; and TYM1302 contained a higher content of caryophyllene oxide. Wei et al. [9] analyzed the volatile components of *Mesona* Benth, where they reported that the main components were caryophyllene oxide and caryophyllene, similar with these experimental results. β-Caryophyllene had a woody aroma, and Taoyuan No. 2 had the highest concentration, followed by TYM1301 and TYM1304 (53–56 mg/kg).

Table 3. Comparisons of the total volatile compounds in eight varieties of Hsian-tsao extracted using SDE.

Varieties	Concentration (mg/kg) [a]
Nongshi No. 1	230.71 ± 89.56 [b]
Taoyuan No. 1	206.90 ± 50.00 [bc]
Taoyuan No. 2	194.20 ± 50.37 [bc]
Chiayi Strain	264.61 ± 22.87 [ab]
TYM1301	329.82 ± 82.32 [a]
TYM1302	205.43 ± 69.26 [bc]
TYM1303	124.67 ± 79.29 [c]
TYM1304	207.50 ± 53.42 [bc]

[a] The 100 g samples of Hsian-tsao were extracted using SDE for 2 h, quantification using cyclohexyl acetate as an internal standard. The data correspond to the mean ± SD of triplicates. Values having different superscripts were significantly ($p < 0.05$) different.

Table 4. Comparisons of volatile compounds from eight varieties of Hsian-tsao extracted using SDE.

Compound [a]	RI [b]	Concentration (mg/kg) [c]							
		Nongshi No. 1	Taoyuan No. 1	Taoyuan No. 2	Chiayi Strain	TYM1301	TYM1302	TYM1303	TYM1304
Aliphatic alcohols									
Isobutanol	643	tr [d]	- [e]	-	-	-	-	-	-
1-Penten-3-ol	693	tr	tr	tr	tr	tr	tr	tr	tr
1-Pentanol [f]	739	-	tr	-	-	-	-	-	-
Isoamyl alcohol [f]	767	0.77 ± 0.89	0.46 ± 0.29	0.16 ± 0.09	0.47 ± 0.29	1.28 ± 0.12	3.10 ± 0.69	1.31 ± 0.82	0.87 ± 0.20
3-Hexanol	791	-	-	0.04 ± 0.01	-	-	-	-	-
(Z)-hex-3-en-1-ol	845	1.39 ± 0.53	0.59 ± 0.09	0.77 ± 0.26	0.61 ± 0.03	1.12 ± 0.01	1.29 ± 0.13	0.53 ± 0.22	0.77 ± 0.09
(E)-2-hexen-1-ol	855	0.02 ± 0.00	tr	0.02 ± 0.03	0.01 ± 0.02	-	0.20 ± 0.11	0.05 ± 0.04	-
Hexanol [f]	856	0.34 ± 0.26	0.22 ± 0.05	0.36 ± 0.08	0.41 ± 0.54	0.19 ± 0.03	0.45 ± 0.52	0.23 ± 0.03	0.17 ± 0.01
1-Octen-3-ol [f]	962	0.44 ± 0.25	0.32 ± 0.00	0.31 ± 0.07	0.50 ± 0.12	1.14 ± 0.15	0.69 ± 0.02	0.23 ± 0.15	1.05 ± 0.06
3-Octanol [f]	979	0.07 ± 0.03	tr	0.06 ± 0.03	0.07 ± 0.01	-	0.12 ± 0.10	0.14 ± 0.06	0.16 ± 0.02
Aliphatic aldehydes									
Pentanal [f]	700	tr	tr	tr	tr	tr	tr	tr	tr
Hexanal [f]	777	0.69 ± 0.59	tr	0.03 ± 0.03	0.14 ± 0.04	0.22 ± 0.14	5.07 ± 4.73	0.37 ± 0.40	0.07 ± 0.03
(E)-2-hexenal [f]	832	2.63 ± 1.36	0.48 ± 0.08	0.19 ± 0.06	0.22 ± 0.02	0.86 ± 0.01	0.69 ± 0.11	0.54 ± 0.27	0.96 ± 0.08
(Z)-4-heptenal	877	0.02 ± 0.01	tr	0.04 ± 0.01	0.04 ± 0.00	0.01 ± 0.01	0.16 ± 0.06	tr	0.11 ± 0.08
(E,E)-2,4-hexadienal	879	0.05 ± 0.02	tr	tr	tr	0.05 ± 0.02	tr	tr	0.04 ± 0.01
(E)-2-heptenal	931	tr	tr	tr	0.03 ± 0.00	0.01 ± 0.01	0.05 ± 0.00	tr	tr
(E,E)-2,4-heptadienal [f]	979	tr	-	-	tr	0.14 ± 0.02	tr	-	-
(E)-2-octenal [f]	1024	0.39 ± 0.09	0.45 ± 0.08	0.43 ± 0.08	0.56 ± 0.07	0.37 ± 0.04	0.72 ± 0.04	0.42 ± 0.21	0.50 ± 0.09
Nonanal [f]	1078	3.99 ± 1.18	0.11 ± 0.00	0.18 ± 0.00	0.16 ± 0.02	1.07 ± 0.17	0.93 ± 0.03	tr	2.20 ± 0.27
(E)-2-nonenal [f]	1130	0.34 ± 0.22	0.23 ± 0.01	0.06 ± 0.11	0.65 ± 0.08	0.51 ± 0.03	0.82 ± 0.05	0.45 ± 0.17	0.69 ± 0.12
Safranal	1175	0.80 ± 0.22	tr	0.73 ± 0.20	-	0.54 ± 0.03	-	-	0.76 ± 0.17
Decanal [f]	1181	-	0.34 ± 0.30	0.32 ± 0.04	0.21 ± 0.03	0.12 ± 0.03	0.34 ± 0.01	0.23 ± 0.07	0.16 ± 0.02
(E)-2-decenal	1241	0.40 ± 0.21	0.19 ± 0.03	0.32 ± 0.07	-	0.12 ± 0.05	0.64 ± 0.01	0.33 ± 0.05	0.17 ± 0.08
(E,E)-2,4-decadienal [f]	1285	0.20 ± 0.16	0.56 ± 0.58	1.17 ± 0.49	1.07 ± 0.14	0.42 ± 0.08	1.48 ± 0.04	0.85 ± 0.52	0.89 ± 0.25

Table 4. *Cont.*

Compound [a]	RI [b]	Concentration (mg/kg) [c]							
		Nongshi No. 1	Taoyuan No. 1	Taoyuan No. 2	Chiayi Strain	TYM1301	TYM1302	TYM1303	TYM1304
Aliphatic ketones									
3-Hexen-2-one	820	-	-	0.07 ± 0.04	-	-	-	-	-
2-Heptanone [f]	872	0.18 ± 0.04	tr	0.10 ± 0.03	0.26 ± 0.02	0.17 ± 0.08	0.36 ± 0.05	0.22 ± 0.13	0.17 ± 0.09
1-Octen-3-one [f]	956	4.79 ± 1.63	1.97 ± 0.03	1.89 ± 0.76	2.19 ± 0.31	6.47 ± 0.43	3.49 ± 0.14	2.66 ± 0.93	8.29 ± 0.94
3-Octanone	965	2.43 ± 2.56	tr	0.25 ± 0.05	2.97 ± 1.25	tr	5.51 ± 1.15	0.45 ± 0.31	tr
3,5-Octadien-2-one	1066	0.16 ± 0.02	0.07 ± 0.00	0.10 ± 0.00	0.09 ± 0.01	0.08 ± 0.03	0.30 ± 0.12	0.22 ± 0.13	0.11 ± 0.04
2-Nonanone	1070	0.28 ± 0.08	-	-	0.37 ± 0.05	-	0.51 ± 0.05	0.29 ± 0.13	-
6-Methyl-3,5-heptadien-2-one [f]	1074	1.26 ± 0.17	1.08 ± 0.51	1.04 ± 0.39	2.07 ± 0.30	0.73 ± 0.02	1.30 ± 0.07	1.63 ± 0.43	0.93 ± 0.14
3-Nonen-2-one	1110	0.18 ± 0.06	0.13 ± 0.01	0.15 ± 0.02	-	-	0.26 ± 0.04	0.17 ± 0.07	-
β-Damascenone [f,g]	1365	1.17 ± 0.45	0.70 ± 0.02	1.50 ± 0.33	0.97 ± 0.09	0.63 ± 0.55	1.87 ± 0.02	1.03 ± 0.37	0.51 ± 0.16
Aliphatic ester									
Ethyl acetate	631	-	-	tr	tr	tr	tr	tr	tr
Aromatic alcohols									
Benzyl alcohol [f]	999	0.18 ± 0.06	tr	0.10 ± 0.02	0.18 ± 0.01	0.12 ± 0.02	0.57 ± 0.05	0.35 ± 0.23	0.11 ± 0.06
1-Octanol	1048	0.18 ± 0.05	tr	0.10 ± 0.01	0.13 ± 0.01	-	0.36 ± 0.12	-	-
Eugenol [g]	1328	2.16 ± 0.96	1.04 ± 0.28	0.62 ± 0.44	2.17 ± 0.25	1.27 ± 0.29	0.73 ± 0.02	0.35 ± 0.18	1.00 ± 0.35
Methyleugenol	1369	0.23 ± 0.19	0.15 ± 0.06	2.43 ± 0.49	0.44 ± 0.07	-	-	-	-
Aromatic aldehydes									
Benzaldehyde [f]	933	tr	tr	tr	tr	tr	tr	tr	tr
Benzeneacetaldehyde	1002	tr	0.11 ± 0.01	tr	tr	0.20 ± 0.01	0.80 ± 0.33	tr	0.28 ± 0.04
Aromatic ketone									
Acetophenone	1033	0.46 ± 0.07	0.29 ± 0.03	0.38 ± 0.03	0.31 ± 0.04	0.03 ± 0.02	0.62 ± 0.05	0.26 ± 0.12	0.05 ± 0.02
Aromatic esters									
Hexyl acetate	991	tr	-	-	-	-	-	-	-
Methyl salicylate	1170	1.29 ± 0.49	0.82 ± 0.11	1.78 ± 0.25	-	0.56 ± 0.08	-	1.70 ± 0.48	0.42 ± 0.09
Aromatic hydrocarbons									
p-Cymene	1004	0.30 ± 0.08	0.29 ± 0.01	0.27 ± 0.03	0.31 ± 0.07		-	0.34 ± 0.25	0.24 ± 0.05
Aromatic hydrocarbons									
β-Methylnaphthalene	1275	0.30 ± 0.20	-	-	0.39 ± 0.05	-	-	-	0.25 ± 0.14
α-Ionene [g]	1397	-	0.45 ± 0.11	-	-	-	-	-	-
Cuparene	1514	-	-	-	-	-	-	2.14 ± 2.60	10.00 ± 5.81
Terpene alcohols									
Linalool [f]	1082	0.20 ± 0.07	-	0.21 ± 0.00	0.18 ± 0.06	0.15 ± 0.03	0.34 ± 0.04	0.21 ± 0.11	0.13 ± 0.05
Borneol [g]	1149	0.63 ± 0.24	0.26 ± 0.12	0.35 ± 0.09	-	-	-	0.35 ± 0.29	-
Menthol	1157	-	-	-	-	-	-	0.19 ± 0.07	-
4-Terpineol [f]	1163	0.48 ± 0.23	0.17 ± 0.01	0.26 ± 0.03	0.58 ± 0.09	0.08 ± 0.03	0.83 ± 0.17	0.28 ± 0.17	0.13 ± 0.04
α-Terpineol [f]	1173	0.27 ± 0.17	0.27 ± 0.16	tr	1.58 ± 0.28	0.17 ± 0.04	1.45 ± 0.22	tr	0.17 ± 0.05
Myrtenol	1184	0.18 ± 0.13	-	0.11 ± 0.09	0.19 ± 0.03	tr	1.14 ± 0.04	0.64 ± 0.27	-
Gossonorol	1617	0.99 ± 0.77	2.03 ± 0.11	-	-	1.51 ± 0.46	-	1.81 ± 0.83	1.28 ± 0.51
α-Bisabolol	1680	58.69 ± 18.60	72.25 ± 12.19	4.23 ± 0.16	116.29 ± 3.79	144.20 ± 32.40	2.78 ± 0.10	42.90 ± 15.95	0.67 ± 0.29
Terpene aldehydes									
β-Cyclocitral	1198	0.83 ± 0.42	-	1.41 ± 0.13	0.84 ± 0.15	0.91 ± 0.07	2.60 ± 0.06	0.65 ± 0.32	1.11 ± 0.22
β-Citral [f]	1217	-	-	-	-	-	-	-	0.07 ± 0.06
Terpene ketones									
Camphor	1128	-	0.43 ± 0.07	0.65 ± 0.26	-	-	-	-	-
β-Ionene [g]	1469	4.84 ± 3.53	-	3.10 ± 0.55	3.29 ± 0.28	2.46 ± 0.46	8.28 ± 0.48	1.30 ± 1.06	4.09 ± 1.17

Table 4. *Cont.*

Compound [a]	RI [b]	Concentration (mg/kg) [c]							
		Nongshi No. 1	Taoyuan No. 1	Taoyuan No. 2	Chiayi Strain	TYM1301	TYM1302	TYM1303	TYM1304
Monoterpenes									
α-Thujene	927	0.63 ± 0.83	0.47 ± 0.24	0.34 ± 0.32	0.80 ± 0.65	4.45 ± 0.54	1.28 ± 0.52	1.00 ± 1.46	0.62 ± 0.09
α-Pinene [f]	937	0.02 ± 0.02	tr	tr	0.03 ± 0.00	0.05 ± 0.05	0.04 ± 0.00	tr	0.03 ± 0.01
Camphene	949	0.87 ± 0.32	0.96 ± 0.02	0.67 ± 0.21	1.18 ± 0.04	4.16 ± 0.30	2.48 ± 0.17	1.18 ± 0.74	7.17 ± 0.46
Sabinene	967	-	1.56 ± 0.52	1.83 ± 1.43	-	12.57 ± 1.41	-	3.38 ± 4.04	1.41 ± 0.23
β-Pinene [f]	972	2.39 ± 0.66	0.73 ± 0.26	1.12 ± 0.26	0.97 ± 0.20	0.59 ± 0.05	1.66 ± 0.10	0.95 ± 0.53	-
β-Myrcene	983	-	tr	-	-	-	-	-	0.06 ± 0.02
α-Phellandrene	998	-	0.13 ± 0.00	0.21 ± 0.01	-	-	0.22 ± 0.20	-	-
Limonene [f]	1015	0.70 ± 0.23	1.43 ± 0.13	0.63 ± 0.12	0.90 ± 0.12	1.41 ± 1.14	0.57 ± 0.03	0.59 ± 0.35	2.43 ± 2.05
γ-Terpinene	1045	0.28 ± 0.07	0.23 ± 0.01	0.27 ± 0.11	0.29 ± 0.03	0.23 ± 0.03	0.37 ± 0.00	-	0.16 ± 0.02
α-Terpinolene	1081	-	0.11 ± 0.01	-	-	-	-	-	-
Sesquiterpenes									
α-Cubebene	1353	0.28 ± 0.22	0.26 ± 0.09	0.51 ± 0.13	0.48 ± 0.01	-	1.10 ± 0.27	0.30 ± 0.22	-
α-Ylangene	1378	0.82 ± 0.39	1.80 ± 0.13	0.80 ± 0.07	1.17 ± 0.11	1.04 ± 0.60	2.66 ± 0.14	0.31 ± 0.20	0.99 ± 0.31
α-Copaene	1382	0.47 ± 0.33	0.69 ± 0.20	1.60 ± 0.20	1.04 ± 0.09	-	1.65 ± 0.00	0.75 ± 0.41	1.76 ± 1.77
β-Elemene	1392	8.07 ± 3.08	7.73 ± 0.54	16.89 ± 4.16	-	11.94 ± 2.44	-	4.48 ± 2.02	35.25 ± 8.75
β-Bourbonene	1393	-	-	-	6.78 ± 0.57	-	6.40 ± 0.07	-	-
β-Caryophyllene [f,g,h]	1427	24.11 ± 9.73	20.57 ± 3.12	55.71 ± 16.29	18.99 ± 2.29	53.59 ± 10.44	20.68 ± 18.88	4.13 ± 1.55	52.80 ± 10.04
α-Bergamotene [h]	1436	7.64 ± 1.99	3.51 ± 0.16	-	5.36 ± 0.30	5.90 ± 1.07	-	-	5.91 ± 1.86
cis-Thujopsene	1437	-	-	-	-	-	-	-	-
β-Gurjunene	1442	tr	-	-	-	-	-	1.32 ± 1.01	-
cis-β-Farneseneh	1445	2.57 ± 0.89	1.86 ± 0.51	-	-	0.79 ± 0.04	-	-	5.90 ± 1.83
α-Caryophyllene [f,h]	1461	6.73 ± 3.56	5.50 ± 0.77	3.85 ± 3.46	5.68 ± 0.12	11.83 ± 2.62	8.57 ± 2.49	2.59 ± 1.53	12.51 ± 2.53
γ-Muurolene	1479	-	2.55 ± 0.22	-	-	-	-	-	-
Germacrene D	1487	4.77 ± 3.40	5.12 ± 0.77	7.63 ± 1.31	6.03 ± 0.26	5.73 ± 1.74	-	-	6.40 ± 1.65
β-Selinene	1494	11.48 ± 2.20	5.92 ± 2.25	5.20 ± 1.26	22.45 ± 2.48	8.32 ± 6.00	7.31 ± 0.89	5.42 ± 2.70	tr
α-Selinene	1501	5.68 ± 0.88	3.96 ± 2.85	10.22 ± 1.00	tr	12.66 ± 8.28	8.14 ± 1.48	5.91 ± 2.72	14.45 ± 3.13
β-Bisabolene	1505	10.74 ± 2.11	8.08 ± 3.25	-	23.72 ± 0.53	8.76 ± 3.09	6.39 ± 2.50	-	-
δ-Guaiene	1510	-	-	-	-	-	-	-	3.93 ± 0.92
α-Chamigrene	1514	2.14 ± 1.82	-	-	4.60 ± 0.02	-	-	2.37 ± 3.11	-
γ-Cadinene	1517	11.17 ± 3.59	3.98 ± 3.80	7.72 ± 4.82	6.38 ± 1.42	-	13.13 ± 1.01	-	-
δ-Cadinene	1523	-	8.54 ± 9.04	4.23 ± 1.65	6.14 ± 0.39	-	8.52 ± 3.31	4.40 ± 6.68	2.15 ± 1.45
trans-γ-Bisabolene	1526	5.81 ± 7.90	3.43 ± 3.46	-	4.47 ± 4.18	0.56 ± 0.17	-	-	2.68 ± 1.83
trans-α-Bisabolene	1537	3.38 ± 2.03	3.49 ± 0.37	-	3.97 ± 0.22	6.90 ± 1.60	-	2.99 ± 3.22	-
Terpene oxide									
Caryophyllene oxide [f,g]	1585	24.16 ± 5.73	27.53 ± 1.85	32.15 ± 3.37	2.13 ± 0.21	8.67 ± 4.24	49.59 ± 20.74	15.07 ± 9.85	9.51 ± 1.81
Hydrocarbons									
Heptane	721	tr	-	-	-	-	-	-	-
Octane	805	-	-	-	0.03 ± 0.00	-	0.05 ± 0.05	-	-
Nonane	898	-	tr	0.06 ± 0.01	0.07 ± 0.01	0.04 ± 0.00	0.11 ± 0.02	-	0.02 ± 0.01
Undecane	1098	0.21 ± 0.06	0.21 ± 0.07	0.13 ± 0.03	-	0.06 ± 0.03	-	-	-
Dodecane	1197	-	-	-	-	0.24 ± 0.02	-	-	-
Tridecane	1294	-	-	-	-	0.13 ± 0.07	-	-	-
Pentadecane	1498	-	-	-	-	-	12.41 ± 6.36	-	-
Straight-chain acids									
Nonanoic acid	1253	tr	-	0.22 ± 0.04	-	-	-	0.22 ± 0.17	-
Decanoic acid	1344	-	-	5.32 ± 4.16	-	-	1.25 ± 0.06	tr	-

Table 4. *Cont.*

Compound [a]	RI [b]	Concentration (mg/kg) [c]							
		Nongshi No. 1	Taoyuan No. 1	Taoyuan No. 2	Chiayi Strain	TYM1301	TYM1302	TYM1303	TYM1304
Dodecanoic acid	1542	tr	-	tr	-	-	-	1.22 ± 1.14	-
Furans									
2-Ethylfuran	712	tr	tr	-	-	tr	tr	tr	tr
Furfural	795	0.02 ± 0.00	tr	tr	-	tr	-	0.04 ± 0.04	tr
2-Pentylfuran	978	0.15 ± 0.04	-	0.02 ± 0.02	-	0.90 ± 0.17	-	-	1.28 ± 0.46
Methoxy-phenolic compounds									-
2-Methoxy-phenol	1058	-	0.09 ± 0.00	0.14 ± 0.02	-	0.04 ± 0.02	-	-	
2-Methoxy-4-vinylphenol	1279	tr	tr	0.19 ± 0.02	-	0.18 ± 0.05	-	-	0.35 ± 0.16
Nitrogen-containing compound									
Indole	1260	0.22 ± 0.16	0.23 ± 0.01	0.25 ± 0.03	-	0.17 ± 0.11	-	0.23 ± 0.16	0.15 ± 0.10

[a] Identification of components based on the GC/MS library (Wiley 7N); [b] Retention indices, using paraffin (C_5–C_{25}) as references. [c] Relative percentages from GC-FID, values are means ± SD of triplicates; [d] Trace; [e] Undetectable. [f] Published in the literature (Wei et al. [9]); [g] Published in the literature (Lu et al. [11]); [h] Published in the literature (Deng et al. [10]).

3.4. Comparisons of the Differences between HS-SPME and SDE

Similar to Table 2, Table 4 shows the eight different Hsian-tsao varieties, along with the 120 components identified using HS-SPME and SDE, of which, 44 were found using both extraction methods, 12 (mainly α-terpinene, δ-3-carene, and *cis*-α-bergamotene) were identified using HS-SPME but not detected using SDE, and 64 (mainly nonanal, 6-methyl-3,5-heptadien-2-one, and gossonorol) were identified using SDE but not detected using HS-SPME.

Table 5 and Figure 4 show that the monoterpene relative content was higher than that of sesquiterpene. Table 6 and Figure 5 show that the SDE samples had a high content of sesquiterpenes, terpene oxide, and terpene alcohols, but a lower content of monoterpenes than the SPME samples. Tersanisni and Berry [33] reported that certain hydrocarbon compounds, such as linalool and α-terpineol, as well as their hydrocarbon interactions, can be interrupted by heat stress, resulting in the induction of volatilization. We detected α-terpineol using SDE but by using HS-SPME. However, both methods identified terpene hydrocarbons as the major components. HS-SPME extracted more terpene hydrocarbons, and the majority was highly volatile monoterpenes with a low molecular weight. SDE extracted mainly sesquiterpenes with higher molecular weights. SDE also identified components that HS-SPME was unable to identify, such as straight-chain acids, aromatic ketones, aromatic esters, terpene aldehydes, terpene ketones, methoxy phenols, and nitrogen-containing compounds. Montserrat et al. [34] analyzed the volatile composition of white salsify (*Tragopogon porrifolius* L.) and found that SDE used high temperature and a long extraction time, and large quantities of volatile components were lost during the extraction process. Therefore, the SDE method may increase the low volatile compounds with a high molecular weight, such as sesquiterpenes and straight-chain acids. HS-SPME used shorter extraction times, so it was able to extract highly volatile monoterpenes with lower molecular weights. As such, HS-SPME is more appropriate for quality control. This study found that although HS-SPME was more rapid and SDE had a higher temperature and longer extraction time, SDE was able to extract more Hsian-tsao compounds; therefore, both methods can be used to complement each other. Yang et al. [35] compared HS-SPME with traditional methods in the analysis of *Melia azedarach* and reported that the HS-SPME method is a powerful analytic tool and is complementary to traditional methods for the determination of the volatile compounds in herbs. Comparing both techniques, HS-SPME samples were smaller (1 g) and did not require heating, the data was accurate, and involved less chemical reactions and changes, but the yield of larger molecules were lower, and the identified components were fewer, while SDE needed the use of 100 g of plant material and heating (2 h). The popularity of this method comes from the fact that volatiles with medium to high boiling points are recovered well. The aroma profile can be greatly altered via the formation of artifacts due to heating the sample during isolation. However, Hsian-tsao food needs to be processed using heat; therefore, by combining the HS-SPME and SDE methods of volatile compounds isolation, each isolation technique provides a part of the overall Hsian-tsao profile.

Table 5. Percentages of extracted chemical groups of Hsian-tsao analyzed using HS-SPME.

	Nongshi No. 1	Taoyuan No.1	Taoyuan No. 2	Chiayi Strain	TYM1301	TYM1302	TYM1303	TYM1304
Aliphatic alcohol	0.29	0.69	1.16	0.17	2.25	1.39	1.61	3.01
Aliphatic aldehydes	3.63	2.91	0.29	0.39	1.19	1.15	1.81	1.74
Aliphatic ketones	tr	0.14	0.12	tr	0.29	tr	0.24	0.47
Aliphatic ester	0.32	-	-	-	-	-	-	-
Aromatic aldehyde	tr	-	-	tr	-	-	-	-
Aromatic alcohol	-	-	-	0.08	-	-	-	-
Aromatic hydrocarbon	-	0.68	-	tr	tr	tr	tr	tr
Terpene alcohol	0.11	0.22	0.09	0.13	0.36	0.08	0.08	-
Monoterpenes	77.83	63.59	72.47	79.47	64.87	84.63	79.36	58.24
Sesquiterpenes	6.72	20.06	16.93	11.28	20.5	4.16	5.2	22.47
Terpene oxide	-	0.04	-	-	-	-	-	-
Hydrocarbons	0.86	0.75	0.88	0.12	0.56	-	0.33	0.06
Furan	tr	tr	tr	tr	tr	tr	tr	tr

All the definitions of the symbols used in Table 2 mean values were also used in Table 5.

Table 6. Concentrations of chemical groups of overall extracted Hsian-tsao analyzed using SDE.

	Nongshi No. 1	Taoyuan No. 1	Taoyuan No. 2	Chiayi Strain	TYM1301	TYM1302	TYM1303	TYM1304
Aliphatic alcohols	3.03	1.59	1.72	2.07	3.73	5.85	2.49	3.02
Aliphatic aldehydes	9.51	2.36	3.47	3.08	4.44	10.90	3.19	6.55
Aliphatic ketones	10.45	3.95	5.10	8.92	8.08	13.60	6.67	10.01
Aliphatic ester	-	-	tr	tr	tr	tr	tr	tr
Aromatic alcohols	2.75	1.19	3.25	2.92	1.39	1.66	0.7	1.11
Aromatic aldehydes	tr	0.11	tr	tr	0.2	0.8	tr	0.28
Aromatic ketone	0.46	0.29	0.38	0.31	0.03	0.62	0.26	0.05
Aromatic esters	1.29	0.82	1.78	-	0.56	-	1.7	0.42
Aromatic hydrocarbons	0.60	0.74	0.27	0.7	-	-	2.48	10.49
Terpene alcohols	61.44	74.98	5.16	118.82	146.11	6.54	46.38	2.38
Terpene aldehydes	0.83	-	1.41	0.84	0.91	2.6	0.65	1.18
Terpene ketones	4.84	0.43	3.75	3.29	2.46	8.28	1.3	4.09
Monoterpenes	4.89	5.62	5.07	4.17	23.46	6.62	7.1	11.88
Sesquiterpenes	105.86	86.99	124.36	117.26	128.02	84.55	34.97	144.73
Terpene oxide	24.16	27.53	32.15	2.13	8.67	49.59	15.07	9.51
Hydrocarbons	0.21	0.21	0.19	0.10	0.47	12.57	0	0.02
Straight-chain acids	tr	-	5.54	-	-	1.25	1.44	-
Furans	0.17	tr	0.02	-	0.9	-	0.04	1.28
Methoxy-phenolic compounds	tr	0.09	0.33	-	0.22	-	-	0.35
Nitrogen-containing compound	0.22	-	0.25	-	0.17	-	0.23	0.15

Notes: All the definitions of the symbols used in Table 3 mean values were also used in Table 4.

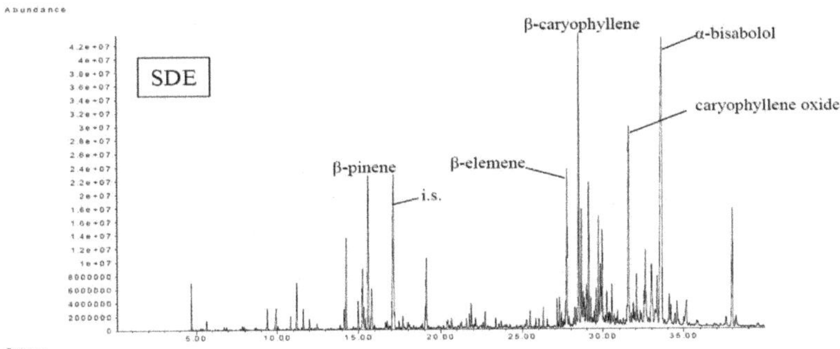

Figure 4. Total ion chromatogram of volatile components of Nongshi No. 1 determined using HS-SPME.

Figure 5. Total ion chromatogram of volatile components of Nongshi No. 1 determined using SDE.

4. Conclusions

This study determined the volatile components present in eight varieties of Hsian-tsao using HS-SPME and SDE methods. A total of 120 volatile components were identified, of which, 56 were verified using HS-SPME and 108 using SDE. HS-SPME extracted more monoterpenes; however, SDE extracted more sesquiterpenes and terpene alcohols, and a terpene oxide, such as β-caryophyllene, α-bisabolol, and caryophyllene oxide. SDE was able to detect more components, but HS-SPME analysis was more convenient. In the future, the two extraction methods can be used in a complementary manner for Hsian-tsao analysis and research.

Author Contributions: Conceptualization, T.-L.K., L.-Y.L., and H.-C.C.; methodology, Y.-J.C. and T.-L.K.; validation, L.K.C., C.-S.W., and H.-C.C.; formal analysis, Y.-J.C. and L.K.C.; investigation, T.-L.K. and L.-Y.L.; writing—original draft preparation, Y.-J.C. and H.-C.C.; writing—review and editing, L.-Y.L. and H.-C.C.

Funding: This work was supported by a research grant from the Council of Agriculture, Executive Yuan (Taiwan) (108AS-7.2.5-FD-Z1 (2), Ministry of Science and Technology (Taiwan) (107-2221-E-039-008-) and Ministry of Education (Taiwan) (1038142*).

Acknowledgments: Financial support from from the Council of Agriculture, Executive Yuan (Taiwan) (108AS-7.2.5-FD-Z1 (2), Ministry of Science and Technology (Taiwan) (107-2221-E-039-008-) and Ministry of Education (Taiwan) (1038142*) are gratefully acknowledged.

Conflicts of Interest: The authors declare no conflict of interest.

References

1. Paton, A.J. Classification and species of *Platostoma* and its relationship with *Haumaniastrum* (Labiatae). *Kew Bull.* **1997**, *52*, 257–292. [CrossRef]
2. Feng, T.; Ye, R.; Zhuang, H.; Fang, Z.; Chen, H. Thermal behavior and gelling interactions of *Mesona Blumes* gum and rice starch mixture. *Carbohydr. Polym.* **2012**, *90*, 667–674. [CrossRef]
3. Widyaningsih, T.D. Cytotoxic effect of water, ethanol and ethyl acetate extract of black cincau (*Mesona Palustris* BL) against HeLa cell culture. *APCBEE Procedia* **2012**, *2*, 110–114. [CrossRef]
4. Yen, G.-C.; Hung, C.-Y. Effects of alkaline and heat treatment on antioxidative activity and total phenolics of extracts from Hsian-tsao (*Mesona procumbens* Hemsl.). *Food Res. Int.* **2000**, *33*, 487–492. [CrossRef]
5. Hung, C.-Y.; Yen, G.-C. Antioxidant activity of phenolic compounds isolated from *Mesona procumbens* Hemsl. *J. Agric. Food Chem.* **2002**, *50*, 2993–2997. [CrossRef]
6. Lai, L.-S.; Liao, C.-L. Dynamic rheology of structural development in starch/decolourised hsian-tsao leaf gum composite systems. *J. Sci. Food Agric.* **2002**, *82*, 1200–1207. [CrossRef]
7. Lai, L.-S.; Liu, Y.-L.; Lin, P.-H. Rheological/textural properties of starch and crude hsian-tsao leaf gum mixed systems. *J. Sci. Food Agric.* **2003**, *83*, 1051–1058. [CrossRef]
8. Zhuang, H.; Feng, T.; Xie, Z.; Toure, A.; Xu, X.; Jin, Z.; Su, Q. Effect of *Mesona* Blumes gum on physicochemical and sensory characteristics of rice extrudates. *Int. J. Food Sci. Technol.* **2010**, *45*, 2415–2424. [CrossRef]
9. Wei, J.; Zheng, E.-L.; Cai, X.-K.; Ji, X.-D.; Xu, C.-H. Preparation of water-soluble extracts from *Mesona Benth* and analysis of the volatile aroma components by GC-MS. *Food Sci. Technol.* **2014**, *39*, 190–192.
10. Deng, C.; Li, R. Analysis of the chemical constituents of the essential oils from *Mesona chinensis* benth by gas chromatography-mass spectrometry. *China Modern Med.* **2012**, *19*, 68–69.
11. Lu, S.-P.; Tian, Y.-K. Analysis of the chemical constituents of the volatile oil from *Mesona chinensis* Benth in Guangdong Yangjiang. *Strait. Pharm. J.* **2014**, *26*, 33–36.
12. Orav, A.; Stulova, I.; Kailas, T.; Müürisepp, M. Effect of storage on the essential oil composition of *Piper nigrum* L. fruits of different ripening states. *J. Agric. Food Chem.* **2004**, *52*, 2582–2586. [CrossRef] [PubMed]
13. Blanch, G.P.; Reglero, G.; Herraiz, M. Rapid extraction of wine aroma compounds using a new simultaneous distillation-solvent extraction device. *Food Chem.* **1996**, *56*, 439–444. [CrossRef]
14. Gu, X.; Zhang, Z.; Wan, X.; Ning, J.; Yao, C.; Shao, W. Simultaneous distillation extraction of some volatile flavor components from Pu-erh tea samples comparison with steam distillation-liquid/liquid extraction and soxhlet extraction. *Int. J. Anal. Chem.* **2009**, *276713*. [CrossRef] [PubMed]
15. Heath, H.B.; Reineccius, G. *Flavour Chemistry and Technology*; Van Nostrand Reinhold Co.: New York, NY, USA, 1986.
16. Holt, R.U. Mechanisms effecting analysis of volatile flavour components by solid-phase microextraction and gas chromatography. *J. Chromatogr. A* **2001**, *937*, 107–114. [CrossRef]
17. Kataoka, H.; Lord, H.L.; Pawliszyn, J. Applications of solid-phase microextraction in food analysis. *J. Chromatogr. A* **2000**, *880*, 35–62. [CrossRef]

18. Zhang, Z.; Yang, M.J.; Pawliszyn, J. Solid-phase microextraction. A solvent-Free Alternative for Sample Preparation. *Anal. Chem.* **1994**, *66*, 844A–853A. [CrossRef]

19. Adam, M.; Juklová, M.; Bajer, T.; Eisner, A.; Ventura, K. Comparison of three different solid-phase microextraction fibres for analysis of essential oils in yacon (*Smallanthus sonchifolius*) leaves. *J. Chromatogr. A* **2005**, *1084*, 2–6. [CrossRef] [PubMed]

20. Yeh, C.-H.; Tsai, W.-Y.; Chiang, H.-M.; Wu, C.-S.; Lee, Y.-I.; Lin, L.-Y.; Chen, H.-C. Headspace solid-phase microextraction analysis of volatile components in *Phalaenopsis* Nobby's Pacific Sunset. *Molecules* **2014**, *19*, 14080–14093. [CrossRef] [PubMed]

21. Schomburg, G.; Dielmann, G. Identification by means of retention parameters. *J. Chromatogr. Sci.* **1973**, *11*, 151–159. [CrossRef]

22. Ducki, S.; Miralles-Garcia, J.; Zumbe, A.; Tornero, A.; Storey, D.M. Evaluation of solid-phase micro-extraction coupled to gas chromatography-mass spectrometry for the headspace analysis of volatile compounds in cocoa products. *Talanta* **2008**, *74*, 1166–1174. [CrossRef] [PubMed]

23. Silva, C.L.; Câmara, J.S. Profiling of volatiles in the leaves of Lamiaceae species based on headspace solid phase microextraction and mass spectrometry. *Food Res. Int.* **2013**, *51*, 378–387. [CrossRef]

24. Zhang, C.; Qi, M.; Shao, Q.; Zhou, S.; Fu, R. Analysis of the volatile compounds in Ligusticum chuanxiong Hort. using HS-SPME-GC-MS. *J. Pharm. Biomed Anal.* **2007**, *44*, 464–470. [CrossRef] [PubMed]

25. Minh Tu, N.T.; Onishi, Y.; Choi, H.S.; Kondo, Y. Characteristic Odor Components of Citrus sphaerocarpa Tanaka (Kabosu) Cold-Pressed Peel Oil. *J. Agric. Food Chem.* **2002**, *50*, 2908–2913. [CrossRef] [PubMed]

26. Guillot, S.; Peytavi, L.; Bureau, S.; Boulanger, R.; Lepoutre, J.; Crouzet, J.; Schorrgalindo, S. Aroma characterization of various apricot varieties using headspace-solid phase microextraction combined with gas chromatography—mass spectrometry and gas chromatography-olfactometry. *Food Chem.* **2006**, *96*, 147–155. [CrossRef]

27. Högnadóttir, Á.; Rouseff, R.L. Identification of aroma active compounds in orange essence oil using gas chromatography-olfactometry and gas chromatography-mass spectrometry. *J. Chromatogr. A.* **2003**, *998*, 201–211. [CrossRef]

28. Ravi, R.; Prakash, M.; Bhat, K.K. Aroma characterization of coriander (*Coriandrum sativum* L.) oil samples. *Eur. Food Res. Technol.* **2007**, *225*, 367–374. [CrossRef]

29. Rega, B.; Fournier, N.; Guichard, E. Solid phase microextraction (SPME) of orange juice flavor: Odor representativeness by direct gas chromatography olfactometry (D-GC-O). *J. Agric. Food Chem.* **2003**, *51*, 7092–7099. [CrossRef]

30. Jirovetz, L.; Buchbauer, G.; Ngassoum, M.B.; Geissler, M. Aroma compound analysis of *Piper nigrum* and *Piper guineense* essential oils from cameroon using solid-phase microextraction-gas chromatography, solid-phase microextraction-gas chromatography-mass spectrometry and olfactometry. *J. Chromatogr. A.* **2002**, *976*, 265–275. [CrossRef]

31. Friedrich, J.E.; Acree, T.E. Gas chromatography olfactometry (GC/O) of dairy products. *Int. Dairy J.* **1998**, *8*, 235–241. [CrossRef]

32. Morales, M.T.; Aparicio, R. Effect of extraction conditions on sensory quality of virgin olive oil. *J. Am. Oil Chem. Soc.* **1999**, *76*, 295–300. [CrossRef]

33. Tersanisni, P.; Berry, R.G. Chemical changes in flavor components. In *Chemical Changes in Food During Processing*; Richardson, T., Finely, J.W., Eds.; Springer: Berlin, Germany, 1985; pp. 327–346.

34. Montserrat, R.-A.; Vargas, L.; Vichi, S.; Guadayol, J.M.; López-Tamames, E.; Buxaderas, S. Characterisation of volatile composition of white salsify (*Tragopogon porrifolius* L.) by headspace solid-phase microextraction (HS-SPME) and simultaneous distillation–extraction (SDE) coupled to GC–MS. *Food Chem.* **2011**, *129*, 557–564.

35. Yang, Y.; Xiao, Y.; Liu, B.; Fang, X.; Yang, W.; Xu, J. Comparison of headspace solid-phase microextraction with conventional extraction for the analysis of the volatile components in *Melia azedarach*. *Talanta* **2011**, *86*, 356–361. [CrossRef] [PubMed]

Review

Analytical and Sample Preparation Techniques for the Determination of Food Colorants in Food Matrices

Konstantina Ntrallou [1], Helen Gika [2,3] and Emmanouil Tsochatzis [1,3,*]

[1] Department of Chemical Engineering, Aristotle University of Thessaloniki, 54124 Thessaloniki, Greece; kondrallou@gmail.com

[2] Laboratory of Forensic Medicine & Toxicology, Department of Medicine, Aristotle University of Thessaloniki, 54124 Thessaloniki, Greece; gkikae@auth.gr

[3] BIOMIC AUTH Center for Interdisciplinary Research of the Aristotle University of Thessaloniki, Innovation Area of Thessaloniki, 57001 Thermi, Greece

* Correspondence: tsochatzism@gmail.com; Tel.: +30-6977-441091

Received: 28 November 2019; Accepted: 3 January 2020; Published: 7 January 2020

Abstract: Color additives are widely used by the food industry to enhance the appearance, as well as the nutritional properties of a food product. However, some of these substances may pose a potential risk to human health, especially if they are consumed excessively and are regulated, giving great importance to their determination. Several matrix-dependent methods have been developed and applied to determine food colorants, by employing different analytical techniques along with appropriate sample preparation protocols. Major techniques applied for their determination are chromatography with spectophotometricdetectors and spectrophotometry, while sample preparation procedures greatly depend on the food matrix. In this review these methods are presented, covering the advancements of existing methodologies applied over the last decade.

Keywords: food colorants (synthetic, natural); food matrices; instrumental analysis; sample preparation

---

1. Introduction

Codex Alimentarius gives a definition for food additives as "any substance that its intentional addition of which to a food aiming for a technological (including organoleptic) purpose in the manufacture, processing, preparation treatment, packing, packaging, transport or holding of such food results, or may be reasonably expected to result, in it or its by-products becoming a component of the food or otherwise affecting the characteristics of such foods" [1,2]. Carocho et al. highlighted that the definition given by the Codex Alimentarius does not include the term contaminants or substances added to food for maintaining or improving nutritional qualities [2].

In food technology, food colorants, of several types, are chemical substances that are added to food matrices, to enhance or sustain the sensory characteristics of the food product, which may be affected or lost during processing or storage, and in order to retain the desired color appearance [3–5]. These are classified based on several criteria: firstly, based on their origin in nature, nature-identical or, if synthetic, whether they are organic or inorganic. Another classification could be based on their solubility (e.g., soluble or insoluble) or covering ability (e.g., transparent or opaque), though an overlap may exist among one or more of these classifications. The most common and widely used classification is based on the distinction between soluble and insoluble color additives (colorants or pigments), which can be further categorized as natural or synthetic [4].

In addition, as described by Martins et al., there were several food additives that had been used extensively in the past but are no longer allowed, due to existing evidence of their side effects, toxicity in the medium- and long-term, as well as a high frequency of potential health incidents [6]. It is also important to note that, apart from synthetic food colorants, certain commercial additives of plant or animal origin have also been suspended [3,6–8].

It is clear that the analysis of trace amounts of food colorants is essential with the proper analytical techniques applied, with high specificity and selectivity. Ni et al. has reported that there is increasing interest in the monitoring of the concentration of synthetic food colorants in various products [9].

The analytical methods and sample preparation protocols presented hereafter cover the main techniques that have been applied over the last decade (2008 onwards).

2. Natural Food Colorants

Natural additives have been used since ancient times. In certain cases, they were used for the preservation of foodstuffs. Nowadays, most consumers seem to be in favor of the use of the natural, as opposed to the synthetic ones, which are considered by the food industry to be more efficient. In the meantime, there is also considerable interest in the overall reduction of food colorants to food products [4,5,10]. The classification of naturally derived colorants can become very complex because of the wide variety of innate properties of the coloring substances. They can be derived from a variety of sources in nature, and therefore, natural colorants also exhibit a wide variety of chemical compositions that affect properties, solubilities, and stabilities differently, and they can have different sources as plant-origin or animal-origin [10].

As reported by Carocho et al., there are benefits linked with the use of natural additives over their respective synthetic ones, which in certain cases present a greater potency over the synthetic ones. The latter in most cases present a single effect on the foodstuff in question. Nevertheless, natural additives are often produced using different methods, i.e., extraction from plants or produced by microorganisms, although there is a tendency to consider them safer than their respective synthetic additive. In general, toxicity is a factor that must be thoroughly assessed and evaluated, to ensure health and safety [2,5,10].

Synthetic colorants have a large span of application and are proportionally lower in cost, than their respective natural substances. However, natural colorants are gradually replacing the synthetic ones as they tend to be considered safer, while presenting higher color specificity, no side effects or related toxicity, and conferring health improving effects and functional benefits to the food itself [6,11,12]. A good example for this beneficial effect is the class of yeast-derived natural pigments (e.g., monascin; a yellow natural pigment). These present certain features, apart from food coloring, such as biological activity, reported potential anti-cancer, anti-inflammatory, anti-diabetic, and anti-cholesterolemic effects [6,13,14].

As reported by Martins et al., numerous references highlighted the effective and/or selective use of food colorants. Therefore, for the approved food colorants with an "E" code, individual Acceptable Daily Intakes (ADI) have been approved and established, expressed mostly as mass fractions (i.e., mg/kg per body weight (b.w), which can be used for specific purposes (i.e., colorants) in specific food products (i.e., biscuits, chocolates, cheeses etc.) [6].

Commonly, naturally occurring food colorants can be allocated in different sub-categories, namely anthocyanins, carotenoids, beet colorants, and phenolic compounds. In addition, annatto, carminic acid, and some curcuminoids have been studied, particularly curcumin. Finally, other colorants remain to be assessed and evaluated in order to be authorized with an "E" code.

Anthocyanins are a widely studied natural food colorants group, mainly obtained from flowers, fruits, leaves, and even whole plants with a color range that goes from red to purple and blue. Commercial anthocyanins, such as cyanidin 3-glucoside, pelargonidin 3-glucoside, and peonidin 3-glucoside have been used effectively [2,4,6].

Carotenoids are another cluster of naturally derived colorants with a renowned technological effect. They present coloring attributes along with certain bioactive as well as antioxidant properties and are being used extensively in the food industry as natural preservatives [4,6,7,10,15] apart from food colorants [7]. Their main source is extracts from plant roots, flowers, and leaves, as well as from algae, yeasts, and aquatic animals. This category mainly includes Lutein, astaxanthin, and lycopene [2,6], the most widely used carotenoids used with others such as crocin and crocetin, still under investigation [4–6].

Red-purple colorants derived from beets and beetroot (Beta vulgaris L.) root is the principal source of these natural colorants but also fruit of *Hylocereus polyrhizus* (Weber) Britton and Rose, *Opuntia ficus-indica* [L.] Miller, *Opuntia stricta* (Haw). Haw and *Rivina humilis* L. are also rich in these colorant substances, namely, the betacyanins and betalains, which are the most frequently studied and already authorized (E162). They are being used in various food products such as burgers, desserts, ice creams, jams, jellies, soups, sauces, sweets, drinks, dairy products, and yogurts [2,4–6].

Other natural food colorants are considered the phenolic compounds, where flavanones, flavones (4′,5,7-trihydroxyflavones), and flavonols (fisetin, myricetin, myricitin, quercetin, and rutin) have been studied. As reported by Martins et al., currently only the commercially available products are being used (i.e., myricetin and myricitrin from *Myrica cerifera* L. roots). Phenolic compounds do not yet have an approved "E" code nor an ADI value [6] with many still being studied and examined since their safety, stability, and spectrum of activity still remain unclear [6,16].

Another category of natural food colorant is the curcuminoids with the most widely known and used colorant in this group being curcumin (E100), usually isolated from Curcuma longa L. rhizomes.

Other natural used colorants are the annatto (E160b) group, as well as bixin and norbixin which are extracts from *Bixa orellana* L. seeds [2,4–6]. In addition, carminic acid (E120) with a yellow to red-orange food color is already largely used, either naturally occurring or of synthetic origin with an ADI of 5 mg/kg b.w [6] or crocin. Nevertheless, there are other food colorants under investigation, including c-phycocyanin (blue pigment isolated from Arthrospira platensis) and c-phycoerythrin (red-orange pigment from blue-green algae). Other naturally occurring pigments, which are commercially available, are being studied, such as geniposide, monascorubrin, and purple corn color [4–6].

3. Synthetic Food Colorants

Based on increasing demand, mainly from the consumer, for products that are more visually attractive, several synthetic food colorants have been developed for use in food production, to increase certain quality and organoleptic characteristics. However, it is reported that over time, most of the synthetic food colorants were excluded due to repeated side effects as well as to their short- and/or long-term toxicity and eventually to potential carcinogenic effects [3,6,11].

Thus, a change in consumer expectations has been reported, which is largely in favor of the natural colorants [6,17].

Apart from this, also from a regulatory point of view, there is increasing attention and interest related to the risk assessment of these colorants used in food products (i.e., azo-dyes). In case of the azo-dyes, a limiting factor for their use is their potential carcinogenicity, which occurs after their reduction to carcinogenic metabolites into the intestine [3,18,19]. These metabolites are produced in the human body, though their toxic effect depends on the ingested amount of the target substance/colorant [3,18,20]. However, it is reported that regular evaluation and assessment of potential toxicity of food colorants by regulatory authorities is necessary [3,18,21].

4. Toxicological Aspects and Regulatory Framework

Based on various scientific findings, several toxicity effects, have been reported including behavioral effects on children, effects on the respiratory system, connection with allergies, development of attention deficit hyperactivity disorder (ADHD) in children, or neuro-developmental effects at the No-Adverse Effect Limit levels [3,18,21]. In any case, further investigation to assess the potential associated risks of these compounds is needed [3–9,11,14,18].

Several groups have indicated the toxic effect of some of groups of these substances. As an example, Mpountoukas et al. have tested the food colorants amaranth, erythrosine, and tartrazine by in vitro experiments, and they concluded there was an in vitro toxic effect on human lymphocytes as they bound to DNA [22]. Many other studies have shown the chemical property of synthetic colorants, namely, Tartrazine [23], azorubine [17,24,25], Allura Red [17,26,27] Sunset Yellow, Quinoline Yellow [17], and Patent Blue [28], to bind to human serum albumin (HSA). Masone and Chanforan compared binding affinities of artificial colorants to human serum albumin (HSA), exhibiting more affinity to HSA than to their natural equivalents' colorants and interacting with its functions. The results supported the hypothesis of their potential risk to human health [17]. Finally, there are dyes, which are rather inexpensive, and which have been used in the food industry, such as Sudan I–IV, which are classified as both a toxic and carcinogenic [24–31]. In Figure 1, basic structures of colorants used in the food industry some of them with toxicological concern (Sudan I–IV) are presented.

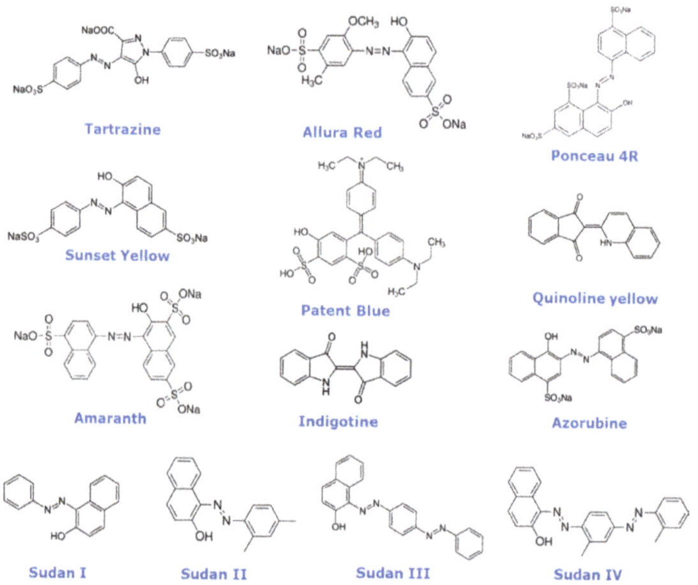

Figure 1. Chemical structures of selected regulated food colorants.

The main regulatory authorities, EFSA in Europe and the US Food and Drug Administration (FDA) in the United States, are responsible for the evaluation and assessment of food products to enhance and promote health safety [2,4,5]. The European Union, set a re-evaluation program of food additives, including food colorants, to be performed by EFSA by 2020, based on the EU Regulation 257/2010. This re-evaluation program was set in order to assess the safety of all authorized food additives in the European Union before 20 January 2009 [32].

The regulatory framework in Europe, in brief, contains the authorization procedure in Regulation (EU) No. 1331/2008, the rules on food additives with a list of approved color additives and their conditions of use in Regulations (EU) No 1333/2008 and 1129/2011, the specifications for food additives in Regulation (EU) 231/2012, and finally for labelling in Regulations (EU) No. 1169/2011 and 1333/2008. Respectively, in the United States, the color additives are included in Title 21 CFR Part 70, listing food additives (exempt from certification, including specifications and conditions of use) in Title 21 CFR Part 73, and certification of donor additives in Title 21 CFR Part 80 [4,5,10,33].

However, despite the existence of different regulatory frameworks, the overall approach follows similar steps, which are based on well-established risk assessment procedures [33].

Authorization for the use of food colorants in the production of food products is subject to a number of toxicity tests, in order to define and evaluate acute, sub-chronic and chronic toxicity, hepatotoxicity, carcinogenicity, mutagenicity, teratotoxicity, genotoxicity, reproductive toxicity, accumulation in the body, bioenergy effects, and immunotoxicity [3–9,11,14,18].

5. Analytical Methodologies for the Determination of Food Colorants

5.1. Analytical Techniques in the Use of Natural Food Colorants Determinations

The available bibliography concerning the methods of analysis for the natural colorants is limited, compared to that for the synthetic ones, and it is exclusively oriented to their determination in the different naturally deriving products.

All the relative information concerning analytical methods for natural colorants, including tested matrices, analytical technology, type of detection and settings, analytical columns if used, elution parameters, mobile phases, injection volumes, and analytical figures of merit (LOD, LOQ), have been reviewed and are summarized in Table 1.

It can be concluded from Table 1 that evaluation of methods' performance criteria was not within the aims of the above-mentioned reports, as they were focusing in activity, bioavailability, processing impact, and adulteration. Thus, no analytical figures of merit are reported in these papers.

From Table 1 and Figure 2 it could be perceived that the predominant technique is HPLC combined with spectrophotometric (UV-Vis) or Diode Array (DAD) detectors, followed by HPLC by MS/MS. Spectrophotometric UV-Vis methods seem also to be preferred by the researchers in this field as they show low instrument cost and do not involve expert skill. However, it should be considered that the individual features of the spectra obtained for single colors are highly dependent on the pH-adjustment of the solution or the mobile phase, using proper acid or alkali. The pH adjustment certainly affects maximum absorption wavelength, where shifts and intensities based on the different pH can be observed. Although sample preparation is much less demanding in comparison to the LC methods, these techniques present a significant disadvantage, which is the lack of ability to analyze simultaneously a bigger number of food colorants.

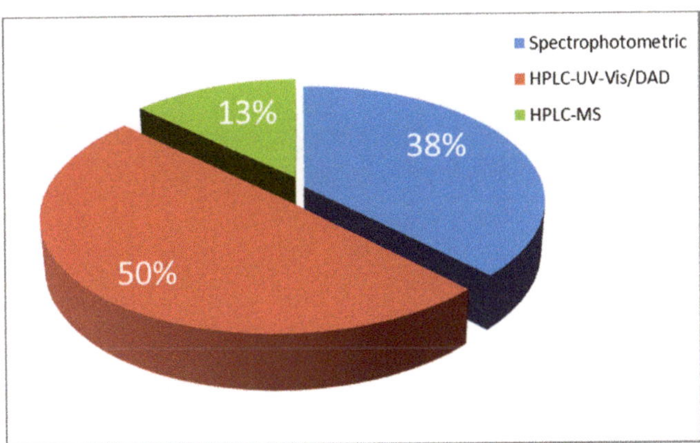

Figure 2. Distribution of techniques used for the analysis of natural food colorants.

Table 1. Methods for the analysis of natural food colorants in various food products.

Food Colorant	Food Matrix	Analytical Technique	Detection	Detection Settings (i.e., λ, Ionisation)	Column	Elution	Mobile Phase	Inj. Volume	Figures of Merit (LOD, LOQ, Linear Range)	Ref.
3-Deoxy-anthocyanidins	*Sorghum bicolor* (L.) Moench seeds	High Pressure Liquid Chromato-graphy (HPLC)	Diode Array Detection (DAD)	485 nm	Luna C18 column (150 × 4.6 mm, 5 mm)	Gradient	4% HCOOH in H_2O (*v/v*) (Solvent A) and acetonitrile (Solvent B)	20 µL	n/a	[34]
Anthocyanin-derived extracts	*Acacia decurrens* Willd. Bark	Spectrophotometric analysis	UV-Vis	400–800 nm	n/a	n/a	n/a		n/a	[35]
	-*Tulipa gesneriana* L.	Spectrophotometric analysis	UV-Vis	765 nm	n/a	n/a	n/a	n/a	n/a	[36]
Cyanidin 3-glucoside	-*Pistacia lentiscus* L. fruits; -*Santalum album* L. fruits	HPLC	DAD	520 nm, 440 nm, 310 nm and 280 nm	SS Wakosil C18 (150 × 4.6 mm, 5 µm)	Gradient	0.1% trifluoroacetic acid (TFA) in H_2O (solvent A) and 0.1% TFA in acetonitrile (Solvent B)	20 µL	n/a	[37]
Cyanidin 3-glucoside	-*Pistacia lentiscus* L. fruits; -*Santalum album* L. fruits	HPLC	ESI-MS		SS Wakosil C18 (150 × 4.6 mm, 5 µm)	Gradient	0.1% TFA in H_2O (solvent A) and 0.1% TFA in acetonitrile (Solvent B	20 µL	n/a	[37]
Betacyanins	-*Hylocereus polyrhizus*	HPLC	MS	ESI (+)	AQUA C18-reversed phase column, 5 µm	Gradient	(A) 2% (*v/v*) CH_3COOH in H_2O and (B)0.5% CH_3COOH in H_2O/acetonitrile (50/50, *v/v*)		n/a	[38]
Betalains	*Beta vulgaris* L. roots	HPLC	UV-Vis	538 nm; 480 nm	Lichrocart 250 × 4 RP-18 (5 µm)	Gradient	H_2O (A) and acetonitrile (B).	20 µL	n/a	[39]
Betalains	*Opuntia ficus-indica* [L.]	HPLC	UV-Vis	245 nm	Luna C18(2) column (250 × 4.6 mm, 5 µm)	Isocratic	20 mM KH_2PO_4/ Acetonitrile 95:5 *v/v*	20 µL	n/a	[40]
Betalains	*Opuntia ficus-indica* [L.]	Spectrophotometric analysis	UV-Vis	λ = 536 nm	n/a	n/a	n/a	n/a	n/a	[41]
Betalains	*Rivina humilis* L. fruits, juice	Spectrophotometric analysis	UV-Vis	λ = 535 nm	n/a	n/a	n/a	n/a	n/a	[42]
α-carotene	*Daucus carota* L. roots	HPLC	UV-Vis	450 nm	Supelcosil LC-18 column (15 cm × 4.6 cm, 5 µm)	Isocratic	Methanol/10% (*v/v*) Acetonitrile: H_2O	50 µL	n/a	[43]
Lutein	Commercial/Milk	HPLC	UV-Vis	450 nm	RP C30 YMC (250 × 4.6 mm, 5 µm)	Isocratic	ethanol, tert-butyl-methyl-ether (MTBE) as the mobile phase	50 µL	n/a	[44]

Table 1. *Cont.*

Food Colorant	Food Matrix	Analytical Technique	Detection	Detection Settings (i.e., λ, Ionisation)	Column	Elution	Mobile Phase	Inj. Volume	Figures of Merit (LOD, LOQ, Linear Range)	Ref.
Lutein	Hawaii "T. erecta", Carmen "T. patula".	HPLC	DAD	450 nm	Waters-Spherisorb column SC-04 (125 × 4.0 mm, ODS2, 3.0 μm)	Gradient	(A) Acetonitrile–methanol (9:1 *v/v*): (B) Ethyl acetate	100 μL	n/a	[45]
Astaxanthin	Microalgae and yeasts	HPLC	DAD	470 nm	Chiralcel OD-RH column (5 μm, 150 mm × 4.6 mm)	n/a	(A) Acetonitrile and (B) phosphoric acid (3.5 mM)	n/a	n/a	[46]
Crocetin, Crocin	Grape seed, monascus, gardenia, and red radish	Spectrophotometric analysis	UV-Vis	438 nm; 462 nm	n/a	n/a	n/a	n/a	n/a	[47]
Monascus red pigments	Beetroot red and paprika extract	High Resolution Mass Spectrometry	HPLC-QTOF-MS	ESI (+)	Kinetex c18 column (2.6 μm, 50 mm × 4.6 mm)	Gradient	(A) Acetonitrile; (B) H$_2$O; (C) aqueous HCOOH 1% *v/v*	n/a	n/a	[48]

5.2. Sample Preparation for Natural Colorant Analysis

Several sample preparation protocols are reported in the literature by applying various techniques. The applied protocol is strongly dependent by the type and nature of the food sample. Below in Table 2, a short description of the sample preparation protocols is given, along with their application for the clean-up of food samples, for the quantification of natural food colorants. A hydrolysis step with a deprotonation step (ethanol, HCl solution) is being reported depending on the food matrix, including dilution methods and SFE with supercritical CO_2.

Table 2. Sample preparation techniques for the analysis of natural food colorants in food products.

Food Colorant	Extraction/Sample Preparation	Ref.
3-Deoxyanthocyanidins	Ground sample, with 1% HCl in methanol, centrifugation hydrolysis;	[34]
Anthocyanin-derived extracts	Comparison of different extraction methods (ultrasonic and natural extraction) vs. magnetic stirring	[35]
	Extraction with ethanol: H_2O (1:1 v/v) acidified with 0.01% HCl	[36]
Cyanidin 3-glucoside	Extraction with 0.1% HCl (v/v) in methanol, combination of the extracts, evaporation, and dissolution	[37]
Betacyanins	Mixing with water, filtration, addition of ethanol (precipitation of pectic substances and proteins)	[38]
Betalains	Sample dissolution in ethanol, agitated and homogenized	[39]
Betalains	Filtration of water extract (no pH adjustment)	[40]
Betalains	Lyophilization and macerated with PBS (pH 5.0) in 1:5 w/w ratio, followed by spray-drying	[41]
Betalains	Dilution of the juice; filtration; addition of Se^{4+}, Zn^{2+}, and Cu^{2+}	[42]
α-carotene	Comparison between simple extraction and Supercritical Fluid extraction (CO_2); Simple extraction: Hexane/acetone; SFE: SC-CO_2 (SFE)	[43]
Lutein	Sample dilution in 95% ethanol and extraction with acetone and petroleum ether. Evaporation and reconstitution	[44]
Lutein	Extraction with organic solvent (isopropanol), centrifugation and supernatant extracted with hexane	[45]
Astaxanthin	Extraction with ethyl acetate, filtration	[46]
Crocetin, Crocin	Dilution in DMSO	[47]

5.3. Analytical Techniques in the Use of Synthetic Food Colorants Determinations

The need to determine synthetic colorants in food matrices originating from their known toxicity, renders the analytical task even more challenging as food matrices are ordinarily very complex. Various analytical techniques are used to determine synthetic food colorants in food samples, including spectrophotometry, thin layer chromatography, capillary electrophoresis, high performance liquid chromatography and mass spectrometry (MS).

Certain chemical properties and characteristics of the substances/colorants that influence their separation, such as hydrophilicity/hydrophobicity, existence of acidic or alkaline groups should to be taken into account. Using a Reversed Phase (RP) liquid chromatography separation, more polar compounds are eluting first followed by the less polar. However, their chromatographic separation is normally performed at neutral pH (ca. 7), and thus, any presence of acidic or alkaline groups could affect the elution sequence.

Ordinarily, organic solvents such as methanol, acetonitrile, or their mixture are used for analysis by HPLC. The addition of acetonitrile improves significantly chromatographic peaks' shape (i.e., asymmetry). Nevertheless, the addition of an inorganic electrolyte as a chemical modifier to the mobile phase can be considered as important in order to advance the separation of all the ionizable species [12,28,37,49].

Food colorants are compounds that absorb exceedingly in the visible region. Thus, spectrophotometry is sufficient and appropriate for their quantitative analysis. It is generally preferred as a quite straightforward technique, with respective low instrumental cost (i.e., compared to MS/MS). However, in several cases, its main drawback is the lack of specificity, as in case of mixtures of absorbing species. A solution to overcome the problem of specificity is the application of mass spectrometry (MS). In this case, all spectral interventions or interferences, presented on UV–Vis/DAD detectors, are overpassed. High analytical sensitivity could succeed, even in more difficult food matrices, though after proper clean-up. In addition, tandem MS technique could provide structural information based on the molecular mass/ion and the respective fragmentation pattern. Regarding the ionization mode, in most cases, for synthetic colorants, the electro spray ionization (ESI) is preferred because synthetic food colorants are polar molecules, and their ionization efficiency depends on the existence of matrix interferences, present in sample or in the mobile phase. In general, negative mode (ESI-) is more effective, though in other non-regulated substances (i.e., Sudan I-IV) the positive ionization is preferred. During the MS/MS analysis, chemical modifiers (i.e., $HCOONH_4$ or CH_3COONH_4) are added to the mobile phases, in order to improve and facilitate the better ionization of each target analyte.

Capillary electrophoresis follows in frequency of use the HPLC-DAD/UV-Vis or MS/MS techniques, applied for the quantification of food colorants. These methods present good separation of both small and large molecules, using high voltages. Other reported techniques are FIA (Flow Injection Analysis) and TLC (Thin Layer Chromatography). These could be considered as relatively simple analytical techniques, even for quantification, though in certain cases they could lack specificity and could be affected by matrix interferences.

For synthetic food colorants, all the respective references containing details about the tested matrices, analytical techniques, detection and settings, analytical columns if used, elution, mobile phases, injection volumes, and figures of merit (LOD, LOQ) are presented below in Table 3.

As it could be extrapolated from Table 3, a significant number of LC-MS, LC-MS/MS or LC-UV/Vis methods are available, which are dedicated to simultaneous detection of either a significant or limited number of artificial colorants (whether authorized or delisted), even including illegal Sudan-type dyes. In addition, to Table 3, Figure 3 gives the percentage distribution of the analytical techniques, regarding the analysis of synthetic food colorants. It could be easily concluded that HPLC/U(H)PLC is the most frequently applied technique, followed by capillary electrophoresis and enzyme-linked immunosorbent assay (ELISA) as well as other residual methods. In the case of ELISA, it needs to be highlighted that it cannot be applied for a group of substances/food colorants but only for standalone substances, for which the monoclonal antibodies have been developed.

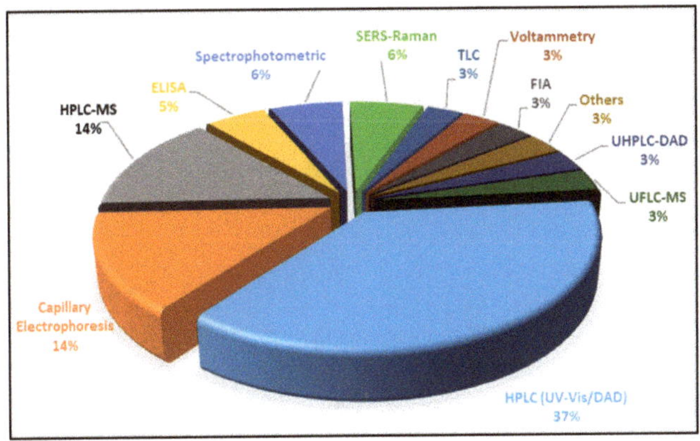

Figure 3. Distribution of techniques for the analysis of synthetic food colorants.

Table 3. Analytical techniques for the determination of natural food colorants in food samples.

Food Colorant	MATRIX	Analytical Technique	Detection	Column	Elution	Mobile Phase	Inj. Volume	Figures of Merit (LOD, LOQ)	Ref.
Brilliant blue	Liquid foods	CE	UV (λ = 220 nm) 36 cm capillary; Separation voltage (8 kV, −8 kV);	Fused-silica cappilaries of 375 od μm and 75 μm i.d	-	30 mM PBS buffer (pH6), with 0.9 mg/mL dASNPs and 2 mM β-cyclodextrin (CD)	Electro kinetic injection	LOD = 0.36 mg/L LOQ = 0.63 mg/L	[2]
Amaranth, ponceau 4R, sunset yellow, tartrazine, brilliant blue		spectrophotometric kinetic method	UV-Vis $\lambda_{Prussian\ blue}$ = 760 nm	n/a	n/a	n/a	n/a	LOD = 0.2–6.0 mg/L	[9]
Sudan I	Non-alcoholic drinks, sweets, jellies	Enzyme-linked Immuno-sorbent assay (ELISA)	n/a	n/a	n/a	n/a	n/a	LOD = 0.07 ng/mL	[29]
Sudan I	Non-alcoholic drinks, jellies	HPLC	UV- Vis (478 nm)	C_{18} (250 × 4.6 mm, 5.0 μm)	Gradient	methanol/2% CH_3COOH	20 μL	LOD = 0.14 ng/mL	[29]
Tartrazine, quinoline Yellow, sunset yellow, Carmoisine, Amaranth, ponceau 4R, Erythrosine, Red 2G, allura Red AC, Patent Blue V, Indigo Carmine, brilliant blue, Green S	Beverages, dairy powders, jellies, candies, condiments, icings, syrups, extracts	HPLC	DAD Various wavelengths	Discovery C18 (250 mm × 4.6 mm 5 μm)	Gradient	CH_3COONH_4 0.13 M (pH = 7.5; NaOH)/methanol: acetonitrile 80:20 v/v	20 μL	LOD = 1.87–22.1 μg/L	[49]
Tartrazine, sunset yellow, brilliant Blue, acid red	Powder	SERS-Raman	confocal microscope Raman spectrometer system	n/a	n/a	n/a	n/a	LOD = 10^{-7} M	[50]
Allura red, sunset yellow, tartrazine	Soft drinks	HPLC	DAD	n/a	Gradient	methanol (HPLC grade) and NaH_2PO_4/ Na_2HPO_4 buffer (0.10 M, pH = 7.0).	n/a	LOD = 0.06–0.30 μg/mL	[51]
Allura red, sunset yellow, tartrazine	Soft drinks	HLA-Go	-	-	-	-	-	-	[51]
Azorubine, amaranth, cochineal red A, red 2G, allura red, azocarmine B (AZO B), azocarmine G (AZO G), ponceau 2R, ponceau 6R, tartrazine, sunset yellow, quinoline yellow, orange II, metanil yellow (MY), patent blue V, indigo carmine, brilliant blue	Solid food/liquid beverages	HPLC	DAD DAD, $\lambda_{quant.}$: −620 nm (blue); −515 nm (red); −420 and 480 nm (yellow).	C8 (150 × 4.6 mm, 3 μm)	Gradient	Acetonitrile/sodium acetate (pH = 7)	20 μL	5–300 mg/kg (solid food samples) 5–100 mg/L (drinks)	[52]

Table 3. *Cont.*

Food Colorant	MATRIX	Analytical Technique	Detection	Column	Elution	Mobile Phase	Inj. Volume	Figures of Merit (LOD, LOQ)	Ref.
Brilliant blue, Tartrazine, amaranth, carmine, sunset yellow, allura red, erythrosine	Wine and soft drinks	UFLC	ESI (−)-MS/MS	ODS II (100 mm × 2.0 mm; 2.2 μm)	Gradient	A: Acetonitrile: CH_3COONH_4 5.0 mM/ (B) H_2O: CH_3COONH_4 5.0 mM	5 μL	LOD = 0.45–1.51 μg/L LOQ = 1.51–5.00 μg/L	[53]
Brilliant blue, tartrazine, amaranth, sunset yellow	Wine and soft drinks	TLC-UV-Vis	UV-Vis	TLC-PET 20 × 20 silica gel	n/a	8 mL 2-propanol and 3 mL NH_4OH	5 μL (standards) and 30 μL sample	n/a	[54]
Allura red, sunset yellow, tartrazine	Solid food/liquid beverages	Spectrophoto-metric BLLS/RBL	Absorbance spectra-pH data Spectral measurements (300–600 nm) at different pH	n.a	n.a	n.a	n.a	LOD = 0.54 mg/L	[55]
Sunset yellow	Beverage	HPLC	ESI (−)-MS	C18-ether column (150 mm × 4.6 mm, 5 μm)	Isocratic	(A) 63% aqueous solution 20 mM CH_3COONH_4: (B) 37% methanol	20 μL	n/a	[56]
Carmoisine, sunset yellow	Beverage	HPLC	ESI (−)-MS	C18 column (250 mm × 2 mm, 4 μm)	Isocratic	(A) Methanol and (B) 10 mM $HCOONH_4$ (45:55, *v/v*)	20 μL	LOD = 10–12 μg/L	[57]
Allura red	Beverage	HPLC	ESI (−)-MS	HSS-T3 column (2.1 mm 100 mm, 1.8 μm)	Gradient	A: H_2O: CH_3COONH_4 1.0 mM/ (B) Methanol: CH_3COONH_4 1.0 mM	20 μL	n/a	[58,59]
Brilliant blue, tartrazine, allura red, amaranth, Azorubine, patent Blue V, ponceau 4R	Various food products	HPLC	DAD Various wavelengths	Xterra RP18 column (250 × 4.6 mm, 5 μm)	Gradient	A) 0.1 M CH_3COONH_4 in water and (B) 0.1 M CH_3COONH_4 in methanol	20 μL	LOD = 0.02–1.49 mg/L	[60]
Brilliant blue, indigo carmine, allura red, carminic acid, ponceau 4R, sunset yellow, tartrazine	Dairy powders, color beverages, jellies, candies, condiments, icings, syrups,	CE	UV (200 nm) Condiiton with 1 M NaOH, H_2O electrode polarity (25kV)	-	-	Running buffer of pH 10 (20 mM NaOH solution to 15 mM disodium tetraborate (borax) to 20 mM NaOH, until the desired pH	Large-volume injection	LODs 0.05–0.40 μg/mL	[61]
Brilliant blue, indigo carmine, allura red, carminic acid, ponceau 4R, sunset yellow, tartrazine, fast green FCF	Liquid foods	CE	UV (λ = 200 nm) Condiiton with 1 M NaOH, H_2O electrode polarity (25kV)	-		Running buffer of pH 10 (20 mM NaOH to 15 mM disodium tetraborate (borax) to 20 mM NaOH, until desired pH	Large-volume injection	LOD = 0.002–0.026 μg/mL	[62]

Table 3. *Cont.*

Food Colorant	MATRIX	Analytical Technique	Detection	Column	Elution	Mobile Phase	Inj. Volume	Figures of Merit (LOD, LOQ)	Ref.
Sunset yellow, carmoisine, amaranth, ponceau 4R, erythrosine, red 2G, allura red	Soft drinks	HPLC	DAD Various wavelengths	Symmetry C18 (Waters, Milford, USA) column (150 mm × 4.6 mm, 5 μm)	Gradient	CH_3COONH_4 buffer (1% *w/v*) (0.13 M) (pH: 7.5) by addition of 0.1 M aq. NH_3 (solvent A), methanol (solvent B) and acetonitrile (solvent C)	n/a	0.5–1.4 μg/mL	[63]
Tartrazine, quinoline yellow, sunset yellow, carmoisine, ponceau 4R, allura red, indigo carmine, brilliant blue	Various foods and medicines	HPLC	DAD Various wavelengths	C18 column (250 mm × 4.6 mm, 5 μm)	Isocratic	A) Triton X-100 (0.25%, *v/v*) and (B) 50 mmol/L PBS (pH 7)	20 μL	LOD = 0.05–0.44 μg/mL LOQ = 0.05–1.12 μg/mL	[64]
Tartrazine, sunset yellow, azorubine, amaranth, cochineal red, red 2G, allura red AC, Brilliant Black BN, brown FK, Brown HT Patent Blue V, brilliant Blue FCF, Green S	Fish roe	HPLC	DAD Various wavelengths	Xterra RP18 column (250 × 4.6 mm, 5 μm)	Gradient	(a) 100 mmol/L CH_3COONa buffer (pH 7.0) and (b) Acetonitrile	20 μL	LOD = 0.02–1.49 mg/L	[65]
Brilliant blue, Indigo carmine, allura red, erythrosine, ponceau 4R, sunset yellow, Lemon yellow	Protein-rich samples	HPLC	DAD Various wavelengths	RP-C18 Column	Gradient	Methanol- 20 mM of CH_3COONH_4	20 μL	LOD = 0.1–0.4 mg/kg	[66]
Brilliant blue, tartrazine, sunset yellow, amaranth, carminic acid, acid red, allura red	Meat products	UHPLC	PDA Various wavelengths	BEH C18 (100 × 2.1 mm, 1.7 μm	Gradient	Acetonitrile/ CH_3COONH_4	5 μL	LOD = 0.01 mg/kg LOQ = 0.05 mg/kg	[67]
Carminic acid, sunset yellow, tartrazine	Non-alcoholic drinks, sweets, jellies	Capillary Electrophoresis (CE)	UV (λ = 220 nm) 36 cm capillary; Separation voltage (8 kV, −8 kV);	Fused-silica capillaries of 375 od μm and 75 μm i.d	-	30 mM PBS buffer (pH6), with 0.9 mg/mL dASNPs and 2 mM β-cyclodextrin (CD)	Electrokinetic injection	LOD = 0.03 -0.072 mg/L LOQ = 0.16–0.31 mg/L	[68]
Amaranth, Ponceau 4R, Sunset yellow, tartrazine, Sudan I-IV	Soft drinks/solid samples	HPLC-ESI (+)-MS	ESI (+)-MS	Spherigel C18 (250 × 4.6 mm, 5 μm)	gradient	aq. methanol 0.1% HCOOH/aqueous methanol 20 mM CH_3COONH_4/1% CH_3COOH	20 μL	LOD = 2.0–3.5 ng LOQ = 5.4–10.5 ng	[69]
Brilliant blue, allura red, amaranth, Erythrosine, ponceau 4R, sunset yellow, tartrazine	Soft drinks and processed meats	HPLC	DAD Various wavelengths	Inertsil ODS-SP column (250 × 4.6 mm, 5 μm)	Gradient	CH_3COONH_4 (0.1 M, pH = 7.2)—methanol-acetonitrile (9:1 *v/v*)	20 μL	LOD: 0.005 μg/mL LOQ: 0.018 μg/mL	[70]

Table 3. *Cont.*

Food Colorant	MATRIX	Analytical Technique	Detection	Column	Elution	Mobile Phase	Inj. Volume	Figures of Merit (LOD, LOQ)	Ref.
Brilliant blue, sunset yellow, tartrazine	Dairy powders, beverages, jellies, candies, syrups, extracts	Flow injection (FIA)	Amperometric detection (boron-doped diamondelectrode)	-	-	$E_{det}.2 = -450$ mV (100 ms duration) vs. Ag/AgCl (3.0 M KCl).		LOD = 0.8–3.5 µM	[71]
Brilliant blue, sunset yellow, tartrazine	Beverages, jellies, candies, condiments, icings, syrups	Differential pulse voltammetry (DPV)	Cathodically pretreated boron-doped diamond (BDD) elctrode	-	-	30 Hz; amplitude(a), 40 mV; for DPV scan rate (v), 0 mVs	-	LOD = 13.1–143 nM	[72]
Erythrosine, carmoisine, amaranth, ponceau 4R, Red 3G	Syrups	Capillary electrophoresis (CE)	Laser-induced fluorescence detection Various $\lambda_{exc.}/\lambda_{emis.}$: (Nd:YAG laser with wavelength 532 nm and power 5 mW)	Fused silica capillary with I.D. 50 µm, O.D. 360 µm, length 30 cm	n/a	$V = +17$ kV (intensity of electrical field was 460 Vcm^{-1}).	n/a	LOD = 0.2–0.4 µg/mL	[73]
Tartrazine, sunset yellow, azorrubine, bordeaux S, ponceau 4R, erytrosine, red no 40, patent blue V, indigo carmin, brilliant blue	Alcoholic beverages	CE	UV/vis PBS 10 mM with sodium dodecyl sulfate 10 mM, pH 11, and +25 kV of voltage	Fused silica capillary (73 cm)	n/a	Phosphate buffer solution of 10 mM with sodium dodecyl sulfate 10 mM, pH 11, and +25 kV voltage	n/a	LOD = 0.4–2.5 µg/mL LOQ = µg/mL	[74]
Tartazine, Amaranth, Sunset yellow, allura red, Lutein, lycopene, β-carotene	Various foodstuff	HPLC	DAD Various wavelengths	C18 column (250 mm × 4.6 mm, 5 µm)	Gradient	1% CH_3COONH_4, methanol and acetone;	20 µL	LOD = 0.2–50 ng/mL	[75]
Tartrazine, Quinoline yellow, Sunset Yellow, Carmoisine, Brilliant Blue	Solid foods	spectrophotometric method	UV-Vis Various wavelengths	n/a	n/a	n/a	n/a	LOQ = 1–5 µg/mL	[76]
Allura red	Liquid foods	spectrophotometric method	UV-Vis (506 nm)	n/a	n/a	n/a	n/a	LOD = 2.35 µg/L	[77]

Table 3. *Cont.*

Food Colorant	MATRIX	Analytical Technique	Detection	Column	Elution	Mobile Phase	Inj. Volume	Figures of Merit (LOD, LOQ)	Ref.
Tartrazine, New red, Amaranth, Ponceau 4R, Sunset yellow, Allura red, Acid red, Brilliant Blue, Acid red, Erythrosine, Acid orange, Basic flavine O, Basic orange, Siperse blue 106, Crystal violet, Leucine malachite green, Leucine crystal violet	Meat	UHPLC	DAD (200–800 nm)	C18 column (2.1 mm × 50 mm, 1.7 μm)	Gradient	(A) 20 mM CH_3COONH_4–0.02% acetic acid (pH 5) and (B) acetonitrile	2 μL	LOD = 0.96–2.16 μg/kg LOQ = 1.61–7.19 μg/kg	[78]
New red, Amaranth, Camine, Sunset yellow, Acid Red G, Allura red, Acid Scarlett GR, Erythrosine, Rhodamine B, Sudan I, Para red, Sudan II, Sudan III, Sudan red 7B, Sudab IV, Sudan Orange G	Hotpot condiment	HPLC	DAD Various wavelengths	C18 column (4.6 mm × 250 mm, 5 μm)	Gradient	(A) Methanol; (B) 0.01 M PBS (pH = 7.5)	20 μL	LOD = 0.001–0.00 3 mg/kg	[79]
Allura Red, Ponceau 4R	Granulated drinks	UV–visible spectrophotometer	UV–visible spectrophotometer (ZCDS)	n/a	n/a	n/a	n/a	LOD = 0.059–0.102 μg/mL LOQ = 0.198–0.341 μg/mL	[80]
Brilliant Blue, Sunset Yellow, Tartrazine	Non-alcoholic drinks, sweets, jellies	HPLC	UV-Vis (630 nm, 480 nm, 430 nm)	Modified C18 column (250 × 4.6 mm, 5 μm) with a 0.25% (*v/v*) Triton X-100 aq. solution at pH 7	Isocratic	0.25 mL of Triton X-100 (Sigma) up to 100 mL with 50 mmol l−1 phosphate buffer solution at pH 7	20 μL	LOD = 0.143–0.080 mg/L	[81]
Tartrazine, Amaranth, Sunset Yellow, Allura red, Ponceau 4R, Erythrosine	Soft drinks, sugar and gelatin based confectionery	HPLC-UV	UV 430 nm, 510 nm	C18 column (250 mm × 4.6 mm, 5 μm	Gradient	(A) 0.1 mol/L ammonium acetate aqueous solution (pH 7.5, adjusted with 10 mol/L NaOH-methanol–acetonitrile (30:70, *v/v*)	20 μL	LOD = 0.015–0.32 ng/mL	[82]
Allurea red, Amaranth, Erythrosine, Ponceau 4R, Sunset Yellow	Beverages, alcoholic drinks and fish foe	SERS-Raman	Radiation of 514.5 nm from an air-cooled argon ion laser was used for SERS excitation	n/a	n/a	n/a	n/a	LOD = 10^{-7}–10^{-5} M	[83]
Sunset yellow	beverage, dried bean curd, braised pork	ELISA	n/a	n/a	n/a	n/a		LOD = 25 pg mL^{-1}	[84]

Table 3. *Cont.*

Food Colorant	MATRIX	Analytical Technique	Detection	Column	Elution	Mobile Phase	Inj. Volume	Figures of Merit (LOD, LOQ)	Ref.
(40 food colorants) Ponceau 6R, Tartrazine, Fast yellow AB, Amaranth, Indigotine, Naphthol yellow S, Chrysoine, Ponceau 4R, Sunset yellow FCF, Red 10B, Orange G, Acid violet 7, Brilliant black PN, Allura red AC, Yellow 2G, Red 2G, Uranine, Fast red E, Green S, Ponceau 2R, Azorubine, Orange I, Quinoline yellow, Martius yellow, Ponceau SX, Ponceau 3R, Fast green FCF, Eosine, Brilliant blue FCF, Orange II, Orange RN, Acid blue 1, Erythrosine, Amido black 10B, Acid red 52, Patent blue V, Acid green 9, Phloxine B, Benzyl violet 4B, Rose bengal	Drinks, syrups and candies	HPLC	HPLC–DAD	C18 column (50 mm × 4.6 mm, 1.8 μm)	Gradient	(A) 0.1 mol/L of CH_3COONH_4 pH 6.7 and (B) was Methanol–Acetonitrile (50:50, *v/v*)	5 μL	LOD = 0.03–0.1 μg/g	[85]
Ponceau 4R, Sunset Yellow, Allura Red, Azophloxine, Ponceauxylidine, Erythrosine, Orange II	Animal feed and meat	LC	ESI (−)-MS	C18 column (2.1 mm × 150 mm, 5 μm)	Gradient	(A) 20 mmol/L CH_3COONH_4: Acetonitrile		LOD = 0.02–21.83 ng/mL	[86]
New Coccine, Indigo Carmine, Erythrosine, Tartrazine, Sunset Yellow FCF, Fast Green FCF, Brilliant Blue FCF, Allura Red AC, Amaranth, Dimethyl Yellow, Fast Garnet GBC, Para Red, Sudan I, Sudan II, Sudan III, Sudan IV, Sudan Orange G, Sudan Red 7B, Sudan Red B, Sudan Red G	Chili powders; commercial syrup preserved fruits	LC	ESI (−) and ESI (+) MS/MS	Acclaim Polar Advantage C16 (3 mm, 4.6 × 150 mm)	Gradient	(A) Acetonitrile and (B) 20 mM CH_3COONH_4 −1.0% CH_3COOH		LOQ = 0.005–1 mg/kg	[87]
Multi-class (53 food colorants)	Spices	UHPLC	QTOF-MS (sequential window acquisition of all theoretical fragment-ion spectra-SWATH)	Acquity UPLC BEH C18 column (2.1 × 100 mm, 1.7 μm,)	Gradient	(A) Acetonitrile and (B) 10 mM CH_3COONH_4 pH = 6.7	n/a	n/a	[88]
Multi-class (34 water soluble synthetic food colorants)	Beverages, syrup, chewing gum	HPLC	DAD-IT-TOF/MS (λ = 200–700 nm)	Atlantis™ dC18 (4.6 mm × 250 mm, 5 μm)	Gradient	(A)20 mM $HCOONH_4$ buffer; (B) methanol/acetonitrile (1:1 *v/v*)	20 μL	LODs = 0.009–0.102 μg/mL; LOQs = 0.045–0.203 μg/mL	[89]

The applied analytical techniques are followed by proper detection approaches. In this framework, simple detector UV-Vis/DAD is mostly applied, followed by MS/MS detectors, UV-Vis spectrometry, and electrochemical detection. The UV-Vis/DAD detection wavelengths depend on the analyte color (i.e., blue, yellow, red) set in any case in the maximum absorbance.

Regarding the MS, listed and EU-approved food colorants could be analyzed in the negative ionization, while for other substances (i.e., Sudan I-IV) positive ionization is applied.

From observation among the available methods of analysis (Table 3 and Figure 3), it could be concluded that traditional TLC methods require a significant sample preparation step and a time-consuming analytical procedure. On the other hand, the HPLC methods need longer analysis time, compared to the respective LC-MS/MS methods, in order to obtain good separation for the same number of analytes [87–89].

As reported recently by Periat et al., full-scan screening methods using HR-MS (High Resolution Mass Spectrometry) have proven to be an alternative to triple quadrupole methods as they could maximize the number of control and analyzed target colorants. Main advantages of the HR-MS can be the reduced sample preparation and the combined targeted analysis with untargeted screening of food colorants with high MS resolving power. Quadropole Time-of-Flight (QTOF) used by Li et al. and by Periat et al. for the detection and identification of coloring compounds in spices provided not only mass accuracy but also MS/MS spectra information and thus increased selectivity. A drawback of the approach could be the high cost of the instrumentation [85,86]. As reported by Li et al., HR-MS accurate mass measurements can detect a large number of target analytes, avoiding isobaric interferences in complex samples [89]. A combination of an ESI (or APCI) ionization with an anion trap analyzer linked to a TOF mass analyzer (ESI/APCI-IT-TOF/MS) provides simultaneously multi tandem MS (up to MS^2) with respective mass accuracy. Currently, there is an increasing interest on the fragmentation mechanism of synthetic food dyes; use of ESI-IT-TOF/MS^n in positive as well as in negative ionization modes [87–89] has been increased.

5.4. Sample Preparation for the Determination of Synthetic Colorants in Foods

Currently, there is no generally accepted/standard method for synthetic colorant extraction in laboratories. Nevertheless, most extraction procedures follow a common approach, which normally involves firstly the release of desired analytes from their matrices, followed afterwards by removal of extraneous matter/interferences by applying an efficient extraction protocol (i.e., solid–liquid or liquid–liquid extraction) [90].

The applied sample preparation protocols are strongly dependent on the type and nature of the food sample. A short description of the sample preparation protocols, along with their application to the clean-up of food samples, for the analysis of synthetic food colorants is given in Table 4.

Membrane filtration involves the permeation of the analyte through a thin layer of material. Explicitly, in case of beverages, when filtration is involved, a degassing step needs to be done in advance, in order to remove CO_2 [90].

Solid phase extraction (SPE) is one of the most commonly used techniques in determination of food colorants, presenting certain advantages such as simplicity. Polyamide resin used for SPE cleanup retains polar compounds with chemical groups that can be protonated. In acidic pH, during SPE, the colorants are adsorbed to the polyamide stationary phase mainly by Van der Waals interactions. Other hydrophilic substances can mask SPE interaction sites by reducing their binding power for the colorants and consequently reducing the capacity of the cartridges. Some substances, such as amaranth, are strongly retained by SPE cartridges, and the ammonia solution used for elution could be insufficient for its release (low recoveries).

Table 4. Sample preparation techniques for the determination of synthetic food colorants in food samples.

Food Colorant	MATRIX	Extraction/Sample Preparation	Ref.
Amaranth, Ponceau 4R, Sunset Yellow, Tartrazine and Brilliant Blue		Reduction of iron (III) in sodium acetate/hydrochloric acid solution (pH 1.71) followed by a chromogenic reaction with potassium hexacyanoferrate (III) to form the Prussian blue species	[9]
Sudan I	Non-alcoholic drinks, sweets, jellies	Extraction, sonication, centrifugation, filtration. Sample extracts or liquid samples were filtrated	[29]
Tartrazine, quinoline Yellow, Sunset Yellow, Carmoisine, Amaranth, Ponceau 4R, Erythrosine, Red 2G, Allura Red AC, Patent Blue V, Indigo Carmine, Brilliant Blue FCF, Green S	Dairy powders, color beverages, jellies, candies, condiments, icings, syrups, extracts	Beverages: degassed by stirring Solid: dissolved in water, ultrasonication, filtration	[49]
Tartrazine, Sunset yellow, Brilliant Blue, Acid red	Powder	Fabrication of flower-like silver nanostructures by adding 10 mL of ultrapure water, 2 mL of PVP solution (1%) and 0.2 mL of silver nitrate solution (1 mol/L)	[50]
Allura Red, Sunset Yellow, and Tartrazine	Soft drinks	Food solutions were prepared with dilution with methanol–water mixture (*v/v*, 50/50)	[51]
Azorubine, amaranth, cochineal red A, red 2G, allura red, azocarmine B (AZO B), azocarmine G (AZO G), ponceau 2R, ponceau 6R, tartrazine, sunset yellow, quinoline yellow, orange II, metanil yellow (MY), patent blue V, indigo carmine and brilliant blue FCF.	Solid food/liquid beverages	4 g solid sample +20 mL ethanol-H_2O (1:1 *v/v*), ultrasound and shaking, centrifugation, separation, and SPE in polyamide (PA) cartridge Beverages: degas (ultrasound), diluted 1.1 with H_2O, centrifuge	[52]
Brilliant Blue FCF, Tartrazine, Amaranth, Carmine, Sunset yellow, Allura red, Erythrosine	Wine and soft drinks	Magnetic dispersive solid-phase extraction (M-dSPE):	[53]
Brilliant Blue FCF, Tartrazine, Amaranth, Sunset yellow,	Wine and soft drinks	Degassing followed by SPE with Sep-Pack C18 and elution with 2-propanol	[54]
Allura red, sunset yellow, tartrazine	Solid food/liquid beverages	Sample dilution (0.5–2.0 g) in 100 mL H_2O	[55]
Sunset yellow	Beverage	No extraction	[56]
Carmoisine, sunset yellow	Beverage	Samples were diluted with water and filtered through 0.2 μm polypropylene membrane	[57]
Allura red	Beverage	filtered through 0.45 μm nylon filter and diluted 1:20 (*v/v*) in ultrapure water	[58,59]
Brilliant blue, tartrazine, allura red Amaranth, Azorubine, Patent Blue V, Ponceau 4R	Various food products	Beverages: sample sonicated, addition of aq. NH_3, filtration; Solid: homogenization, addition of aq. NH_3, sonication, centrifugation	[60]
Brilliant Blue FCF, Indigo carmine, Allura red, carminic acid, Ponceau 4R, Sunset yellow, tartrazine	Dairy powders, color beverages, jellies, candies, condiments, icings, syrups, extracts	Flavored milk samples: diluted with ethanol (1:1 *v/v*), SPE with PA cartridge	[61]
Brilliant Blue FCF, Indigo carmine, Allura red, carminic acid, Ponceau 4R, Sunset yellow, tartrazine, fast green FCF	Liquid foods	Beverages: degas (ultrasound) and directly to CE Milk: diluted with ethanol (1:1 *v/v*), SPE with PA cartridge Jelly: blended with ethanol: H_2O 1:1 *v/v* at 65 °C × 4 h + SPE (Polyamide cartridge)	[62]
Sunset yellow, Carmoisine, Amaranth, Ponceau 4R, Erythrosine., Red 2G, Allura red	Soft drinks	Extraction stage, followed by sonication, centrifugation, and concentration step + clean up via SPE on polyamide (PA) cartridges	[63]
Tartrazine, Quinoline Yellow, Sunset Yellow, Carmoisine, Ponceau, 4R, Allura Red, Indigo Carmine, Brilliant Blue	Various foods and medicines	Homogenization, dissolution, filtration	[64]
Tartrazine, Sunset Yellow FCF, Azorubine Amaranth Cochineal Red, Red 2G), Allura Red AC, Brilliant Black BN, Brown FK and Brown HT, Patent Blue V, Brilliant Blue FCF, and Green S	Fish roe	Extraction with aq. NH_3, centrifugation, pH adjustment, addition of PA sorbent and extraction with methanol	[65]
Brilliant blue, Indigo carmine, allura red, erythrosine, ponceau 4R, sunset yellow, Lemon yellow	Protein-rich samples	Purification/deproteinization with chitosan	[66]

Table 4. *Cont.*

Food Colorant	MATRIX	Extraction/Sample Preparation	Ref.
Brilliant blue, tartrazine, sunset yellow, amaranth, carmininic acid, acid red, allura red	Meat products	ASE (static) with ethanol-H_2O_NH_3 75:24:1 $v/v/v$, 85oC for 10 min	[67]
Carminic acid, sunset yellow, tartrazine, brilliant blue.	Non-alcoholic drinks, sweets, jellies	d-SPME with diamino-moiety functionalized silica nanoparticles (dASNPs) and β-cyclodextrin (β-CD) and pseudo-stationary phases (PSPs). Optimization of pH (2.5), sorbent, amount of dASNPs; ionic strength; extraction time and mode; desorption time; interferences (no interferences identified)	[68]
Amaranth, Ponceau 4R, Sunset yellow, tartrazine, Sudan I-IV	Soft drinks/solid samples	Liquid: filtration, degassing Solid: homogenization, extraction with DMSO, sanitation, centrifugation, and filtration	[69]
Brilliant blue FCF, Allura red, Amaranth, Erythrosine, Ponceau 4R, Sunset Yellow, Tartrazine	Soft drinks and processed meats		[70]
Brilliant blue, sunset yellow, tartrazine	Dairy powders, color beverages, jellies, candies, condiments, icings, syrups, extracts	Homogenization, addition in 0.1 M H_2SO_4 (gelatin dissolution), ultrasonic and dilution with supporting electrolyte	[71]
Brilliant blue, sunset yellow, tartrazine	Dairy powders, color beverages, jellies, candies, condiments, icings, syrups, extracts	-	[72]
Erythrosine, carmoisine, amaranth, ponceau 4R, Red 3G	Syrups	Dilution to PBS (20 mM, pH 11) as a background electrolyte (BGE) in the ratio of 1:10 or 1:2 for CE–LIF	[73]
Tartrazine, sunset yellow, azorrubine, bordeaux S, ponceau 4R, erytrosine, red no 40, patent blue V, indigo carmine, brilliant blue FCF	Alcoholic beverages	degassed by mechanical agitation and filtered	[74]
Tartrazine, Amaranth, Sunset yellow, Allura red, Lutein, Lycopene, β-carotene	Various foodstuff	Ultrasound-assisted solvent extraction: immersion to methanol, sonication, centrifugation, extraction with acetone, evaporation. Liquid: 0.5 mL + 1 mL methanol; Solid: 0.2 g + 1 mL methanol	[75]
Tartrazine, Quinoline yellow, Sunset Yellow, Carmoisine, Brilliant Blue	Solid foods	Solid: a portion of food was diluted in H_2O, centrifuges and diluted with equal volume of CH_3COOH 3 M Liquid: a portion of was diluted in a mixture of NH_3: ethanol 2:73%, mix, and centrifuged.	[76]
Allura red	Liquid foods	Sample filtered, pH adjusted (4.0), extraction with Acetonitrile by SPE (MCI GEL CHP20P resin)	[77]
Tartrazine, New red, Amaranth, Ponceau 4R, Sunset yellow, Allura red, Acid red, Brilliant Blue, Acid red, Erythrosine, Acid orange, Basic flavine O, Basic orange, Siperse blue 106, Crystal violet, Leucine malachite green, Leucine crystal violet	Meat	Microwave assisted extraction: sample with methanol/H_2O (95:5 v/v) followed by SPE (C18 column), evaporation to dryness and reconstitution with methanol	[78]
New red, Amaranth, Carmine, Sunset yellow, Acid Red G, Allura red, Acid Scarlett GR, Erythrosine, Rhodamine B, Sudan I, Para red, Sudan II, Sudan III, Sudan red 7B, Sudab IV, Sudan Orange G	Hotpot condiment	Direct solvent extraction: sample with solvent (acetone-methanol), vortex, centrifugation, evaporation, pH adjustment	[79]
Allura Red, Ponceau 4R	Granulated drinks	Powdered sample dissolved in distilled water,	[80]
Brilliant Blue, Sunset Yellow, Tartrazine	Non-alcoholic drinks, sweets, jellies	Solid: dilution in H_2O and filtration	[81]
Tartrazine, Amaranth, Sunset Yellow, Allura red, Ponceau 4R, Erythrosine	Soft drinks, sugar and gelatin based confectionery	Ionic liquid dispersive liquid phase microextraction. Sample + ionic liquid [1-Octyl-3-methylimidazolium tetrafluoroborate ([C_8MIM][BF_4])] and methanol addition.	[82]

Table 4. *Cont.*

Food Colorant	MATRIX	Extraction/Sample Preparation	Ref.
Allura red, Amaranth, Erythrosine, Ponceau 4R, Sunset Yellow	Beverages, alcoholic drinks and fish foe	Synthesis of the G/Ag nanoparticle composite	[83]
Sunset yellow	Beverage, dried bean curd, braised pork		[84]
(40 food colorants) Ponceau 6R, Tartrazine, Fast yellow AB, Amaranth, Indigotine, Naphthol yellow S, Chrysoine, Ponceau 4R, Sunset yellow FCF, Red 10B, Orange G, Acid violet 7, Brilliant black PN, Allura red AC, Yellow 2G, Red 2G, Uranine, Fast red E, Green S, Ponceau 2R, Azorubine, Orange I, Quinoline yellow, Martius yellow, Ponceau SX, Ponceau 3R, Fast green FCF, Eosine, Brilliant blue FCF, Orange II, Orange RN, Acid blue 1, Erythrosine, Amido black 10B, Acid red 52, Patent blue V, Acid green 9, Phloxine B, Benzyl violet 4B, Rose bengal.	Drinks, syrups, and candies	Drinks: degas, evaporation Solid: grind, mixing with solvent and SPE with PA column with 1% NH_3/ethanol solution	[85]
Ponceau 4R, Sunset Yellow, Allura Red, Azophloxine, Ponceauxylidine, Erythrosine Orange II	Animal feed and meat	Homogenization, extraction with ethanol:NH_3:H_2O (80:1:19 *v/v*), evaporation, reconstitution	[86]
New Coccine, Indigo Carmine, Erythrosine, Tartrazine, Sunset Yellow FCF, Fast Green FCF, Brilliant Blue FCF, Allura Red AC, Amaranth, Dimethyl Yellow, Fast Garnet GBC, Para Red, Sudan I, Sudan II, Sudan III, Sudan IV, Sudan Orange G, Sudan Red 7B, Sudan Red B, Sudan Red G	Chili powders; commercial syrup preserved fruits	Homogenization and extraction with Acetonitrile twice	[87]
Multi-class (53 food colorants)	Spices	Extraction with H_2O/methanol/acetonitrile/THF, 9:1:5:5, *v/v/v/v*)	[88]
Multi-class (34 water soluble synthetic food colorants)	Beverages, syrup, chewing gum	Beverages: degassing, pH adjustment, and dilution. Syrup: dilution, sonication, and pH adjustment. Chewing gum: washing with water, pH adjustment.	[89]

Dispersive solid phase extraction (d-SPE) analysis is a simple sample preparation methodology that is suitable for a wide variety of food and agricultural products, as is also QuEChERS, introduced for pesticides from Anastassiades et al. [91]. In case of synthetic colorants, a modified QuEChERS method has been reported (magnetic-dSPE) using cross-linking magnetic polymer (NH_2-LDC-MP) containing less hydrophilic amino groups and more lipophilic styrene monomer for cleaning up the synthetic food colorants from wine and soft drinks [53].

Liquid–liquid extraction (LLE) deals with the separation of substances based on their relative solubility in two different immiscible liquids. Common solvents for the extraction of synthetic food colorants from food matrices are water, ethanol, methanol, isopropyl alcohol, ammoniac ethanol, ethyl acetate, ammonia, cyclohexane, and tetra-n-butyl ammonium phosphate. Wu et al. has also reported an extraction method based on Ionic liquid dispersive liquid phase microextraction using the ionic liquid (1-Octyl-3-methylimidazolium tetrafluoroborate ((C_8MIM)(BF_4))) [81].

In the literature, a limited number of protocols exists dealing with other types of extraction methods for synthetic food colorants, such as MAE and Ultrasound Assisted Extraction (UAE). These kinds of extractions require special instrumentation and most probably can be beneficial for a laboratory, as extractions with organic solvents are characterized by consumption of high volumes of solvents, are time consuming, and in some cases have low recoveries [90].

6. Conclusions

The use of food colorants in the production of foods leads to the need for the development of accurate, precise, sensitive, and selective analytical methods for their analysis and quantification. Certain interest in the impacts of food colorants is being reported worldwide. There is a plethora of analytical research works that deal with the analytical challenge of the analysis and quantification of either natural or synthetic food colorants. The research community gives more attention to the appropriate analysis, in sufficient concentration or mass fraction levels, mostly to synthetic food colorants rather than natural ones.

Analytical methodologies have much more to offer in this direction and, as it could be concluded from synthetic colorants, HPLC is the most frequently used followed by capillary electrophoresis. In terms of detection methods, the simple UV-Vis/DAD is the predominant one followed by tandem MS. The analytical techniques and sample preparation methodologies presented cover the existing methodologies mainly applied during the last decade.

Regarding sample preparation, this is highly sample dependent. It could involve the application of different extraction techniques, such as membrane filtration, liquid–liquid and solid phase extraction techniques, for cleaning-up the highly complex matrix of food products. Sample preparation is of great importance and must be carefully developed, in order to avoid or eliminate existing matrix interferences aiming to the development of simple, selective, and precise methods of extraction.

In the case of simple liquid samples, dilution and injection are preferred, though in other cases such as high protein content foods, specific steps need to be followed for sufficient sample clean-up.

Author Contributions: Conceptualization, E.T., H.G., and K.N.; methodology, K.N.; formal analysis, E.T. and K.N.; investigation, K.N.; resources, E.T. and K.N.; data curation, E.T.; writing—original draft preparation, K.N.; writing—review and editing, H.G. and E.T.; visualization, E.T.; supervision, E.T.; project administration, E.T. All authors have read and agreed to the published version of the manuscript.

Funding: This research received no external funding.

Acknowledgments: We acknowledge support of this work by the project "FoodOmicsGR Comprehensive Characterisation of Foods" (MIS 5029057) which is implemented under the Action "Reinforcement of the Research and Innovation Infrastructure", funded by the Operational Programme Competitiveness, Entrepreneurship and Innovation (NSRF2014-2020) and co-financed by Greece and the European Union (European Regional Development Fund).

Conflicts of Interest: The authors declare no conflict of interest.

References

1. Codex Alimentarius. Available online: http://www.codexalimentarius.org/standards/gsfa/ (accessed on 18 September 2018).
2. Carocho, M.; Morales, P.; Ferreira, I.C.F.R. Natural food additives: Quo vadis? *Trans Food Sci. Technol.* **2015**, *45*, 284–295. [CrossRef]
3. Amchova, P.; Kotolova, H.; Ruda-Kucerova, J. Health safety issues of synthetic food colorants. *Reg. Toxic. Pharm.* **2015**, *73*, 914–922. [CrossRef]
4. Aberamound, A. A Review Article on Edible Pigments Properties and Sources as Natural Biocolorants in Foodstuff and Food Industry. *World J. Dairy Food Sci.* **2011**, *6*, 71–78.
5. Burrows, A.J.D. Palette of Our Palates: A Brief History of Food Coloring and Its Regulation. *Compreh. Rev. Food Sci. Food Saf.* **2009**, *9*, 394–408. [CrossRef]
6. Martins, N.; Roriz, C.L.; Morales, P.; Barros, L.; Ferreira, I.C. Food colorants: Challenges, opportunities and current desires of agro-industries to en-sure consumer expectations and regulatory practices. *Trends Food Sci. Technol.* **2016**, *52*, 1–15. [CrossRef]
7. Rodriguez-Amaya, D.B. Natural food pigments and colorants. *Curr. Opin. Food Sci.* **2016**, *7*, 20–26. [CrossRef]
8. Tumolo, T.; Lanfer-Marquez, U.M. Copper chlorophyllin: A food colorant with bioactive properties? *Food Res. Int.* **2012**, *46*, 451–459. [CrossRef]
9. Ni, Y.; Wang, Y.; Kokot, S. Simultaneous kinetic spectrophotometric analysis of five synthetic food colorants with the aid of chemometrics. *Talanta* **2009**, *78*, 432–441. [CrossRef]
10. Sigurdson, G.T.; Tang, P.; Giusti, M.M. Natural colorants: food colorants from natural sources. *Annu. Rev. Food Sci. Technol.* **2017**, *8*, 261–280. [CrossRef]
11. Carocho, M.; Barreiro, M.F.; Morales, P.; Ferreira, I.C. Adding molecules to food, pros and cons: A review on synthetic and natural food additives. *Compr. Rev. Food Sci. Food Saf.* **2014**, *13*, 377–399. [CrossRef]

12. Dias, M.I.; Ferreira, I.C.; Barreiro, M.F. Microencapsulation of bioactives for food applications. *Food Funct.* **2015**, *6*, 1035–1052. [CrossRef] [PubMed]

13. Patakova, P. Monascus secondary metabolites: Production and biological activity. *J. Ind. Microbiol. Biotechnol.* **2013**, *40*, 169–181. [CrossRef] [PubMed]

14. Wang, C.; Chen, D.; Chen, M.; Wang, Y.; Li, Z.; Li, F. Stimulatory effects of blue light on the growth, monascin and ankaflavin production in Monascus. *Biotechnol. Lett.* **2015**, *37*, 1043–1048. [CrossRef]

15. Dias, M.G.; Camoes, M.F.G.; Oliveira, L. Carotenoids in traditional Portuguese fruits and vegetables. *Food Chem.* **2009**, *113*, 808–815. [CrossRef]

16. Carocho, M.; Ferreira, I.C. A review on antioxidants, prooxidants and related controversy: Natural and synthetic compounds, screening and analysis methodologies and future perspectives. *Food Chem. Toxicol.* **2013**, *51*, 15–25. [CrossRef]

17. Masone, D.; Chanforan, C. Study on the interaction of artificial and natural food colorants with human serum albumin: A computational point of view. *Comput. Biol. Chem.* **2015**, *56*, 152–158. [CrossRef]

18. Oplatowska-Stachowiak, M.; Elliottt, C.T. Food colors: Existing and emerging food safety concerns. *Crit. Rev. Food Sci. Nutr.* **2017**, *57*, 524–548. [CrossRef]

19. Feng, J.; Cerniglia, C.E.; Chen, H. Toxicological significance of azo dye metabolism by human intestinal microbiota. *Front. Biosci.* **2012**, *4*, 568–586. [CrossRef]

20. Golka, K.; Kopps, S.; Myslak, Z.W. Carcinogenicity of azo colorants: Influence of solubility and bioavailability. *Toxicol. Lett.* **2004**, *151*, 203–210. [CrossRef]

21. Vojdani, A.; Vojdani, C. Immune reactivity to food coloring. *Altern. Ther. Health Med.* **2015**, *21*, 52–62.

22. Mpountoukas, P.; Pantazaki, A.; Kostareli, E.; Christodoulou, P.; Kareli, D.; Poliliou, S.; Mourelatos, C.; Lambropoulou, V.; Lialiaris, T. Cytogenetic evaluation and DNA interaction studies of the food colorants amaranth, erythrosine and tartrazine. *Food Chem. Toxic.* **2010**, *48*, 2934–2944. [CrossRef]

23. Pan, X.; Qin, P.; Liu, R.; Wang, J. Characterizing the Interaction between tartra-zine and two serum albumins by a hybrid spectroscopic approach. *J. Agric. Food Chem.* **2011**, *59*, 6650–6656. [CrossRef] [PubMed]

24. Basu, A.; Kumar, G.S. Study on the interaction of the toxic food additive Carmoisine with serum albumins: A microcalorimetric investigation. *J. Hazard Mater.* **2014**, *273*, 200–206. [CrossRef] [PubMed]

25. Datta, S.; Mahapatra, N.; Halder, M. pH-insensitive electrostatic interaction of carmoisine with two serum proteins: A possible caution on its uses in food and pharmaceutical industry. *J. Photochem. Photobiol. B* **2013**, *124*, 50–62. [CrossRef] [PubMed]

26. Wang, L.; Zhang, G.; Wang, Y. Binding properties of food colorant Allura Red with human serum albumin in vitro. *Mol. Biol. Rep.* **2014**, *41*, 3381–3391. [CrossRef]

27. Wu, D.; Yan, J.; Wang, J.; Wang, Q.; Li, H. Characterisation of interaction between food colourant A Allura red AC and human serum albumin: Multi-spectroscopic analyses and docking simulations. *Food Chem.* **2015**, *170*, 423–429. [CrossRef]

28. Tellier, F.; Steibel, J.; Chabrier, R.; Ble, F.X.; Tubaldo, H.; Rasata, R.; Chambron, J.; Duportail, G.; Simon, H.; Rodier, J.F.; et al. Sentinel lymph nodes fluorescence detection and imaging using Patent Blue V bound to human serum albumin. *Biomed. Opt. Exp.* **2012**, *3*, 2306–2316. [CrossRef]

29. Wang, Y.; Wei, D.; Yang, H.; Yang, Y.; Xing, W.; Li, Y. Development of a highly sensitive and specific monoclonal antibody-based enzyme-linked immunosorbent assay (ELISA) for detection of Sudan I in food samples. *Talanta* **2009**, *77*, 1783–1789. [CrossRef]

30. Moller, P.; Wallin, H. Genotoxic hazards of azo pigments and other colorants related to 1-phenylazo-2-hydroxynaphthalene. *Mutat. Res.* **2000**, *462*, 13–30. [CrossRef]

31. Stiborova, M.; Martinek, V.; Rydlova, H.; Hodek, P.; Frei, E. Sudan I is a Potential Carcinogen for Humans: Evidence for Its Metabolic Activation and Detoxication by Human Recombinant Cytochrome P450 1A1 and Liver Microsomes. *Cancer Res.* **2002**, *62*, 5678–5684.

32. European Food Safety Authority. Available online: http://www.efsa.europa.eu (accessed on 15 September 2018).

33. Lehto, S.; Buchweitz, M.; Klimm, A.; Straßburger, R.; Bechtold, C.; Ulberth, F. Comparison of food colour regulations in the EU and the US: A review of cur-rent provisions. *Food Addit. Contam.* **2017**, *34*, 335–355. [CrossRef] [PubMed]

34. Dykes, L.; Rooney, W.L.; Rooney, L.W. Evaluation of phenolics and antioxidant activity of black sorghum hybrids. *J. Cereal Sci.* **2013**, *58*, 278–283. [CrossRef]

35. Sivakumar, V.; Vijaeeswarri, J.; Anna, J.L. Effective natural dye extraction from different plant materials using ultrasound. *Ind. Crop. Prod.* **2011**, *33*, 116–122. [CrossRef]

36. Sagdic, O.; Ekici, L.; Ozturk, I.; Tekinay, T.; Polat, B.; Tastemur, B.; Senturk, B. Cytotoxic and bioactive properties of different color tulip flowers and degradation kinetic of tulip flower anthocyanins. *Food Chem. Toxicol.* **2013**, *58*, 432–439. [CrossRef] [PubMed]

37. Longo, L.; Scardino, A.; Vasapollo, G. Identification and quantification of anthocyanins in the berries of *Pistacia lentiscus* L., *Phillyrea latifolia* L. and *Rubia peregrina* L. *Innov. Food Sci. Emerg. Technol.* **2007**, *8*, 360–364. [CrossRef]

38. Stintzing, F.C.; Schieber, A.; Carle, R. Betacyanins in fruits from red-purple pitaya, Hylocereus polyrhizus (Weber) Britton & Rose. *Food Chem.* **2002**, *77*, 101–106.

39. Ravichandran, K.; Saw, N.M.M.T.; Mohdaly, A.A.; Gabr, A.M.; Kastell, A.; Riedel, H.; Cai, Z.-Z.; Knorr, D.; Smetanska, I. Impact of processing of red beet on betalain content and antioxidant activity. *Food Res. Int.* **2013**, *50*, 670–675. [CrossRef]

40. Cassano, A.; Conidi, C.; Drioli, E. Physico-chemical parameters of cactus pear (Opuntia ficus-indica) juice clarified by microfiltration and ultrafiltration processes. *Desalination* **2010**, *250*, 1101–1104. [CrossRef]

41. Otalora, M.C.; Carriazo, J.G.; Iturriaga, L.; Nazareno, M.A.; Osorio, C. Micro-encapsulation of betalains obtained from cactus fruit (Opuntia ficus- indica) by spray drying using cactus cladode mucilage and maltodextrin as encapsulating agents. *Food Chem.* **2015**, *187*, 174–181. [CrossRef]

42. Khan, M.I.; Giridhar, P. Enhanced chemical stability, chromatic properties and regeneration of betalains in Rivina humilis L. berry juice. *LWT—Food Sci. Technol.* **2014**, *58*, 649–657. [CrossRef]

43. Sun, M.; Temelli, F. Supercritical carbon dioxide extraction of carotenoids from carrot using canola oil as a continuous co-solvent. *J. Supercrit. Fluids* **2006**, *37*, 397–408. [CrossRef]

44. Sobral, D.; Costa, R.G.B.; Machado, G.M.; Paula, J.C.J.; Teodoro, V.A.M.; Nunes, N.M.; Pinto, M.S. Can lutein replace annatto in the manufacture of Prato cheese? *LWT—Food Sci. Technol.* **2016**, *68*, 349–355. [CrossRef]

45. Khalil, M.; Raila, J.; Ali, M.; Islam, K.M.S.; Schenk, R.; Krause, J.-P.; Schweigert, F.H.; Rawel, H. Stability and bioavailability of lutein ester supplements from Tagetes flower prepared under food processing conditions. *J. Funct. Foods* **2012**, *4*, 602–610. [CrossRef]

46. Grewe, C.; Menge, S.; Griehl, C. Enantioselective separation of all-E- astaxanthin and its determination in microbial sources. *J. Chromatogr. A* **2007**, *1166*, 97–100. [CrossRef] [PubMed]

47. Wada, M.; Kido, H.; Ohyama, K.; Ichibangase, T.; Kishikawa, N.; Ohba, Y.; Nakashima, M.N.; Kuroda, N.; Nakashima, K. Chemiluminescent screening of quenching effects of natural colorants against reactive oxygen species: Evaluation of grape seed, monascus, gardenia and red radish extracts as multi-functional food additives. *Food Chem.* **2007**, *101*, 980–986. [CrossRef]

48. Thalhamer, B.; Buchberger, W. Adulteration of beetroot red and paprika extract based food colorant with Monascus red pigments and their detection by HPLC-QTof MS analyses. *Food Control* **2019**, *105*, 58–63. [CrossRef]

49. Minioti, K.S.; Sakellariou, C.F.; Thomaidis, N.S. Determination of 13 synthetic food colorants in water-soluble foods by reversed-phase high- performance liquid chromatography coupled with diode-array detector. *Anal. Chim. Acta* **2007**, *583*, 103–110. [CrossRef]

50. Ai, Y.-J.; Wu, Y.-X.; Dong, Q.-M.; Li, J.-B.; Xu, B.-J.; Yu, Z.; Ni, D. Rapid qualitative and quantitative determination of food colorants by both T Raman spectra and Surface-enhanced Raman Scattering (SERS). *Food Chem.* **2018**, *241*, 427–433. [CrossRef]

51. Al-Degs, Y.S. Determination of three dyes in commercial soft drinks using HLA/GO and liquid chromatography. *Food Chem.* **2009**, *117*, 485–490. [CrossRef]

52. Bonan, S.; Fedrizzi, G.; Menotta, S.; Elisabetta, C. Simultaneous determination of synthetic dyes in foodstuffs and beverages by high-performance liquid chromatography coupled with diode-array detector. *Dyes Pigment.* **2013**, *99*, 36–40. [CrossRef]

53. Chen, X.H.; Zhao, Y.G.; Shen, H.Y.; Zhou, L.X.; Pan, S.D.; Jin, M.C. Fast determination of seven synthetic pigments from wine and soft drinks using magnetic dispersive solid-phase extraction followed by liquid chromatography–tandem mass spectrometry. *J. Chromatogr. A* **2014**, *1346*, 123–128. [CrossRef]

54. De Andrade, F.I.; Guedes, M.I.F.; Vieira, Í.G.; Mendes, F.N.P.; Rodrigues, P.A.S.; Maia, C.S.C.; Ávila, M.M.M.; Ribeiro, L.d.M. Determination of synthetic food dyes in commercial soft drinks by TLC and ion-pair HPLC. *Food Chem.* **2014**, *157*, 193–198. [CrossRef] [PubMed]

55. El-Sheikh, A.H.; Al-Degs, Y.S. Spectrophotometric determination of food dyes in soft drinks by second order multivariate calibration of the absorbance spectra-pH data matrices. *Dyes Pigments* **2013**, *97*, 330–339. [CrossRef]

56. Gosetti, F.; Gianotti, V.; Polati, S.; Gennaro, M.C. HPLC–MS degradation study of E110 Sunset Yellow FCF in a commercial beverage. *J. Chromatogr. A* **2005**, *1090*, 107–115. [CrossRef] [PubMed]

57. Gosetti, F.; Frascarolo, P.; Mazzucco, E.; Gianotti, V.; Bottaro, M.; Gennaro, M.C. Photodegradation of E110 and E122 dyes in a commercial aperitif: A high performance liquid chromatography–diode array–tandem mass spectrometry study. *J. Chromatogr. A* **2008**, *1202*, 58–63. [CrossRef]

58. Gosetti, F.; Chiuminatto, U.; Mazzucco, E.; Calabrese, G.; Gennaro, M.C.; Marengo, E. Identification of photodegradation products of Allura Red AC (E129) in a beverage by ultra-high performance liquid chromatography–quadrupole-time-of-flight mass spectrometry. *Anal. Chim. Acta* **2012**, *746*, 84–89. [CrossRef]

59. Gosetti, F.; Chiuminatto, U.; Mazzucco, E.; Calabrese, G.; Gennaro, M.C.; Marengo, E. Non-target screening of Allura Red AC photodegradation products in a beverage through ultra-high performance liquid chromatography coupled with hybrid triple quadrupole/linear ion trap mass spectrometry. *Food Chem.* **2013**, *136*, 617–623. [CrossRef]

60. Harp, B.P.; Miranda-Bermudez, E.; Barrows, J.N. Determination of seven certified color additives in food products using liquid chromatography. *J. Agric. Food Chem.* **2013**, *61*, 3726–3736. [CrossRef]

61. Huang, H.-Y.; Shih, Y.-C.; Chen, Y.-C. Determining eight colorants in milk beverages by capillary electrophoresis. *J. Chromatogr. A* **2002**, *959*, 317–325. [CrossRef]

62. Huang, H.-Y.; Chiu, C.-W.; Sue, S.-L.; Cheng, C.-F. Analysis of food colorants by capillary electrophoresis with large- volume sample stacking. *J. Chromatogr. A* **2003**, *995*, 29–36. [CrossRef]

63. Karanikolopoulos, G.; Gerakis, A.; Papadopoulou, K.; Mastrantoni, I. Determination of synthetic food colorants in fish products by an HPLC-DAD method. *Food Chem.* **2015**, *177*, 197–203. [CrossRef] [PubMed]

64. Khanavi, M.; Hajimahmoodi, M.; Ranjbar, A.M.; Oveisi, M.R.; Ardekani, M.R.S.; Mogaddam, G. Development of a green chromatographic method for simultaneous determination of food colorants. *Food Anal. Methods* **2012**, *5*, 408–415. [CrossRef]

65. Kirschbaum, J.; Krause, C.; Brückner, H. Liquid chromatographic quantification of synthetic colorants in fish roe and caviar. *Eur. Food Res. Technol.* **2006**, *222*, 572–579. [CrossRef]

66. Kong, C.; Fodjo, E.K.; Li, D.; Cai, Y.; Huang, D.; Wang, Y.; Shen, X. Chitosan-based ad-sorption and freeze deproteinization: Improved extraction and purifica-tion of synthetic colorants from protein-rich food samples. *Food Chem.* **2015**, *188*, 240–247. [CrossRef]

67. Liao, Q.G.; Li, W.H.; Luo, L.G. Applicability of accelerated solvent extraction for synthetic colorants analysis in meat products with ultrahigh performance liquid chromatography–photodiode array detection. *Anal. Chim. Acta* **2012**, *716*, 128–132. [CrossRef]

68. Liu, F.-J.; Liu, C.-T.; Li, W.; Tang, A.-N. Dispersive solid-phase micro- extraction and capillary electrophoresis separation of food colorants in beverages using diamino moiety functionalized silica nanoparticles as both extractant and pseudostationary phase. *Talanta* **2015**, *132*, 366–372. [CrossRef]

69. Ma, M.; Luo, X.; Chen, B.; Su, S.; Yao, S. Simultaneous determination of water-soluble and fat-soluble synthetic colorants in foodstuff by high- perfo-mance liquid chromatography-diode array detection-electrospray mass spectrometry. *J. Chromatogr. A* **2006**, *1103*, 170–176. [CrossRef]

70. Ma, K.; Yang, Y.N.; Jiang, X.X.; Zhao, M.; Cai, Y.Q. Simultaneous determination of 20 food additives by high performance liquid chromatography with photo-diode array detector. *Chin. Chem. Lett.* **2012**, *23*, 492–495. [CrossRef]

71. Medeiros, R.A.; Lourencao, B.C.; Rocha-Filho, R.C.; Fatibello-Filho, O. Flow injection simultaneous determination of synthetic colorants in food using multiple pulse amperometric detection with a boron-doped diamond electrode. *Talanta* **2012**, *99*, 883–889. [CrossRef]

72. Medeiros, R.A.; Lourencao, B.C.; Rocha-Filho, R.C.; Fatibello-Filho, O. Simultaneous voltammetric determination of synthetic colorants in food using a cathodically pretreated boron-doped diamond electrode. *Talanta* **2012**, *97*, 291–297. [CrossRef]

73. Ryvolova, M.; Taborský, P.; Vrabel, P.; Krasenský, P.; Preisler, J. Sensitive determination of erythrosine and other red food colorants using capillary electrophoresis with laser-induced fluorescence detection. *J. Chromatogr. A* **2007**, *1141*, 206–211. [CrossRef] [PubMed]

74. Prado, M.A.; Boas, L.F.V.; Bronze, M.R.; Godoy, H.T. Validation of methodology for simultaneous determination of synthetic dyes in alcoholic beverages by capillary electrophoresis. *J. Chromatogr. A* **2006**, *1136*, 231–236. [CrossRef] [PubMed]

75. Shen, Y.; Zhang, X.; Prinyawiwatkul, W.; Xu, Z. Simultaneous determination of red and yellow artificial food colourants and carotenoid pigments in food products. *Food Chem.* **2014**, *157*, 553–558. [CrossRef] [PubMed]

76. Sorouraddin, M.H.; Rostami, A.; Saadati, M.A. simple and portable multi-colour light emitting diode based photocolourimeter for the analysis of mixtures of five common food dyes. *Food Chem.* **2011**, *127*, 308–313. [CrossRef]

77. Soylak, M.; Unsal, Y.E.; Tuzen, M. Spectrophotometric determination of trace levels of allura red in water samples after separation and pre-concentration. *Food Chem. Toxicol.* **2011**, *49*, 1183–1187. [CrossRef]

78. Sun, H.; Sun, N.; Li, H.; Zhang, J.; Yang, Y. Development of multiresidue analysis for 21 synthetic colorants in meat by microwave-assisted extraction–solid-phase extraction–reversed-phase ultrahigh performance liquid chromatography. *Food Anal. Methods* **2013**, *6*, 1291–1299. [CrossRef]

79. Tang, B.; Xi, C.; Zou, Y.; Wang, G.; Li, X.; Zhang, L.; Chen, D.; Zhang, J. Simultaneous determination of 16 synthetic colorants in hotpot condiment by high performance liquid chromatography. *J. Chromatogr. B* **2014**, *960*, 87–91. [CrossRef]

80. Turak, F.; Ozgur, M.U. Simultaneous determination of allura red and ponceau 4R in drinks with the use of four derivative spectrophotometric methods and comparison with high-performance liquid chromatography. *J. AOAC Int.* **2013**, *96*, 1377–1386. [CrossRef]

81. Vidotti, E.C.; Costa, W.F.; Oliveira, C.C. Development of a green chromatographic method for determination of colorants in food samples. *Talanta* **2006**, *68*, 516–521. [CrossRef]

82. Wu, H.; Guo, J.B.; Du, L.M.; Tian, H.; Hao, C.X.; Wang, Z.F.; Wang, J.Y. A rapid shaking-based ionic liquid dispersive liquid phase micro extraction for the simultaneous determination of six synthetic food colourants in soft drinks, sugar-and gelatin-based confectionery by high-performance liquid chromatography. *Food Chem.* **2013**, *141*, 182–186. [CrossRef]

83. Xie, Y.; Li, Y.; Niu, L.; Wang, H.; Qian, H.; Yao, W. A novel surface-enhanced Raman scattering sensor to detect prohibited colorants in food by graphene/silver nanocomposite. *Talanta* **2012**, *100*, 32–37. [CrossRef] [PubMed]

84. Xing, Y.; Meng, M.; Xue, H.; Zhang, T.; Yin, Y.; Xi, R. Development of a polyclonal antibody-based enzyme-linked immunosorbent assay (ELISA) for detection of sunset yellow FCF in food samples. *Talanta* **2012**, *99*, 125–131. [CrossRef] [PubMed]

85. Yoshioka, N.; Ichihashi, K. Determination of 40 synthetic food colors in drinks and candies by high-performance liquid chromatography using a short column with photodiode array detection. *Talanta* **2008**, *74*, 1408–1413. [CrossRef] [PubMed]

86. Zou, T.; He, P.; Yasen, A.; Li, Z. Determination of seven synthetic dyes in animal feeds and meat by high performance liquid chromatography with diode array and tandem mass detectors. *Food Chem.* **2013**, *138*, 1742–1748. [CrossRef] [PubMed]

87. Tsai, C.-F.; Kuo, C.-H.; Shih, D.Y.-C. Determination of 20 synthetic dyes in chili powders and syrup-preserved fruits by liquid chromatography/tandem mass spectrometry. *J. Food Drug Anal.* **2015**, *23*, 453–462. [CrossRef] [PubMed]

88. Périat, A.; Bieri, S.; Mottier, N. SWATH-MS screening strategy for the determination of food dyes in spices by UHPLC-HRMS. *Food Chem. X* **2019**, *1*, 100009. [CrossRef]

89. Li, X.Q.; Zhang, Q.H.; Ma, K.; Li, H.M.; Guo, Z. Identification and determination of 34 water-soluble synthetic dyes in foodstuff by high performance liquid chromatography–diode array detection–ion trap time-of-flight tandem mass spectrometry. *Food Chem.* **2015**, *182*, 316–326. [CrossRef]

90. Yamjala, K.; Nainar, M.S.; Ramisetti, N.R. Methods for the analysis of azo dyes employed in food industry—A review. *Food Chem.* **2016**, *192*, 813–824. [CrossRef]

91. Anastassiades, M.; Lehotay, S.J.; Stajnbaher, D.; Schenck, F.J. Fast and easy multiresidue method employing acetonitrile extraction/partitioning and "dispersive solid-phase extraction" for the determination of pesticide residues in produce. *J. AOAC Int.* **2003**, *86*, 412–431.

Review

Rapid Solid-Liquid Dynamic Extraction (RSLDE): A Powerful and Greener Alternative to the Latest Solid-Liquid Extraction Techniques

Daniele Naviglio [1], Pierpaolo Scarano [2], Martina Ciaravolo [1] and Monica Gallo [3,*]

[1] Department of Chemical Sciences, University of Naples Federico II, via Cintia; Monte S. Angelo Complex, 80126 Naples, Italy
[2] Department of Science and Technology, University of Sannio, Via Port'Arsa 11, 82100 Benevento, Italy
[3] Department of Molecular Medicine and Medical Biotechnology, University of Naples Federico II, via Pansini 5, 80131 Naples, Italy
* Correspondence: mongallo@unina.it; Tel.: +39-081-746-3117

Received: 26 May 2019; Accepted: 4 July 2019; Published: 5 July 2019

Abstract: Traditionally, solid-liquid extractions are performed using organic and/or inorganic liquids and their mixtures as extractant solvents in contact with an insoluble solid matrix (e.g., the Soxhlet method) or using sequential atmospheric pressure systems that require long procedures, such as maceration or percolation. The objective of this procedure is the extraction of any compounds that can be carried out from the inner solid material to the outlet, resulting in a solution containing colorants, bioactive compounds, odorous substances, etc. Over the years, in the extraction techniques sector, there have been many important changes from the points of view of production, quality, and human and environmental safety due to improvements in technology. In more recent times, the interest of the scientific community has been aimed at the study of sustainable processes for the valorization of extracts from vegetables and food by-products, through the use of non-conventional (innovative) technologies that represent a valid alternative to conventional methods, generally through saving time and energy and the formation of fewer by-products. Therefore, with the development of principles based on the prevention of pollution, on a lower risk for human health, and on a low environmental impact, new systems have been implemented to reduce extraction times and solvent consumption, to improve efficiency, and to increase the productivity of the extracts. From this point of view, rapid solid-liquid dynamic extraction (RSLDE), performed using the Naviglio extractor, compared to traditional applications, is a technique that is able to reduce extraction times, generally leads to higher yields, does not require heating of the system, allows one to extract the active ingredients, and avoids their degradation. This technique is based on a new solid-liquid extraction principle named Naviglio's principle. In this review, after reviewing the latest extraction techniques, an overview of RSLDE applications in various research and production sectors over the past two decades is provided.

Keywords: solid-liquid extraction; green extraction; RSLDE; bioactive compounds; Naviglio extractor; Naviglio's principle

---

1. Introduction

Solid-liquid extraction processes, both traditional ones (maceration and percolation) and those introduced more recently (e.g., supercritical fluid extraction (SFE) and accelerated solvent extraction (ASE)), are based on two fundamental principles: diffusion and/or osmosis. On the basis of these principles, it is possible to make some general forecasts in relation to the extractive system, and it is possible to roughly hypothesize the extraction times and yields with respect to a generic solid matrix (generally vegetable) [1]. Three variables are to be optimized to achieve the best extractive conditions:

by decreasing the "granulometry" of the solid, the extractive yield increases because of an increased surface area of contact between solid and liquid; the raising of the "temperature" of the system reduces the time of extraction due to the increase in diffusion phenomena (Fick's law); the increase of the "affinity" of the extraction liquid towards the compounds to be extracted increases the effectiveness of the extraction process (*similis similia solvuntur*). However, the extractive principles which these techniques are based on have no active effect on the characteristics of the process, such as extraction times, yield, and efficiency. In fact, once the conditions have been set, the system reaches an equilibrium condition that can change only by modifying some parameters, such as the temperature or the addition of other extraction liquid [2]. For this reason, it is suggested that the extractive batch must be mixed during extraction to avoid a partial extraction due to the slow diffusion of compounds extracted.

Though solid-liquid extraction is a technique that has been known for a long time and is still widely used, there are still many unknown aspects that require further investigation to fully understand the mechanism. In the field of solid-liquid extraction techniques, it is possible to distinguish conventional extraction techniques, including maceration, percolation, squeezing, counter-current extraction, extraction through Soxhlet, and distillation, from unconventional (or innovative) ones. Conventional extractions have been used for many years, although they have many drawbacks: they require the use of high quantities of expensive and pure solvents, since during the process they consume a high amount; they have a low selectivity of extraction; they have a high solvent evaporation rate during the process; and they are generally characterized by long extraction times and by the thermal decomposition of thermolabile compounds [3,4]. To overcome all these limitations, new and promising solid-liquid extraction techniques, which are defined as non-conventional, have been introduced, mainly in the industrial field, such as ultrasound-assisted extraction (UAE) [5], supercritical fluid extraction (SFE) [6], microwave-assisted extraction (MAE) [7], extraction with accelerated solvent [8], solid phase microextraction [9], enzyme-assisted extraction [10], and rapid solid-liquid extraction dynamic (RSLDE) via the Naviglio extractor [11]. On the other hand, the interest of the scientific community has recently been aimed at the study of sustainable processes, so all these extraction techniques have common objectives, including the extraction of active ingredients (bioactive compounds) from the vegetable matrices, as well as their by-products for the valorization of waste, to improve the selectivity of the processes, to isolate the bioactive compounds in more suitable forms for detection and separation, and to provide an effective and reproducible method that is independent of the variability of the sample matrices; furthermore, high extraction yields are preferable to promote the economy of the process [12–14].

However, the extraction procedure generally takes place in a single solution (a single-step process), and it is difficult to set two or more extraction stages, because of the rise in extractant volume and time. Only the Soxhlet extractor limits the solvent volume, because it uses the distillation of the solvent, and the process works with fresh solvent. This can be considered a multi-step extractive process. Vice versa, RSLDE is based on a different principle. In fact, "the generation, with a suitable solvent, of a negative pressure gradient between the outside and the inside of a solid matrix containing extractable material, followed by a sudden restoration of the initial equilibrium conditions, induces forced extraction of the compounds not chemically linked to the main structure of which the solid is made" [11]. RSLDE changes the philosophy of solid-liquid extraction; the extraction happens thanks to a negative gradient of pressure between the inner material and the outside of the solid matrix (high pressure inside and low pressure outside; Naviglio's principle). When the gradient of pressure is removed, the liquid flows out of the solid in a very fast manner and carries out all substances not chemically bonded to the main structure of the solid. This means that in this case the extraction is an "active" process because the gradient of pressure forces out the molecules, while techniques based on diffusion and osmosis are "passive" processes because the molecules are not forced out of the matrix.

According to this principle, the solid-liquid extraction process is first of all independent of the affinity that the compounds to be extracted from the solid matrix have towards the extracting solvent: they are, in fact, extracted by a difference of pressure between the liquid inside the matrix and the liquid on the outside of it. They are extracted out of the solid with a suction effect and can therefore also be extracted in solvents with opposite or different polarity. Furthermore, the pressure effect on the solid matrix and following the de-pressure leads to an active action with respect to the extraction process, as a small quantity of material is extracted at each pressure and depression cycle (the "active" solid-liquid extractive process), the extent of which is closely correlated with the pressure difference generated between the inside and the outside of the solid matrix and to the features of the solid matrix. Based on this new and innovative extractive principle, it has been made possible, in many cases, to use water as an extraction solvent, a condition that cannot be achieved with traditional techniques, such as maceration and percolation; in this case, the fermentative process is slower due to the movement of liquid around the solid, and this prevents the microorganisms from growing [15].

This review aims to give a brief overview of the various extraction techniques, focusing mainly on RSLDE and its various fields of application thanks to the introduction of an innovative solid-liquid principle of extraction.

2. State of the Art of Solid-Liquid Extraction Techniques

Solid-liquid extraction techniques are the basis of many analytical procedures for the preparation of samples and are reported in the official methods of analysis [16]. On the other hand, they are applied to the production of small quantities of homemade extracts such as alcoholic beverages and herbal teas [17,18]. These extraction procedures are also applied to industrial production. In fact, in many industrial processes, the initial phase of the preparation of a product requires the application of a solid-liquid extraction technique to isolate the extractable material contained in the most varied solid matrices, mainly vegetables. An important example is represented by medicinal plants, from which active ingredients with pharmacological properties are obtained; related fields are those of herbal medicine, cosmetics, and perfumery, which are the most ancient applications. In other industrial sectors such as the beverage industry, a solid-liquid extraction is used to obtain alcoholic extracts of fruit peels, flowers, leaves, etc., which are then mixed with water and sugar to obtain the finished product. The list could continue by referring to multiple industrial applications that are very similar.

The solid-liquid extraction is based on a simple phenomenon: if a solid matrix containing extractable compounds is immersed in a liquid, the latter begins to enrich itself with certain chemically related substances that move from the inside to the surface of the solid and then from the surface into the liquid. This principle is based on diffusion and osmosis and is performed by maceration, which is the simplest and most economical extraction technique and is therefore widely used. The maceration process requires only a closable glass or stainless steel container in which extractable solid is covered with liquid. To overcome the rapid saturation of liquid strictly around the solid, desultory agitation is required. Unfortunately, it is not always applicable, because it requires long contact times between the solid and the liquid; for example, plants cannot be macerated in water at room temperature for a very long time due to rotting phenomena. The production needs of the industry, which require the obtaining of large volumes of extracts in a short amount of time, have found an application in percolation extraction; in this case it is possible to process large quantities of solid material with large volumes of liquid and obtain the extract quite quickly, albeit sacrificing the efficiency of the extraction, which remains low due to the limited contact between the solid and the extracting liquid [19]. In this case, the solid matrix is not completely exhausted and could be re-extracted with another technique.

For special applications, such as the production of essential oils and, in general, compounds with low vapor pressure, it is possible to resort to steam distillation [20]. This solid-liquid extraction technique is particular in that it requires the transport of volatile compounds through a steam flow; since the isolated product is an essential oil, it can be considered a solid-liquid extraction technique. In any case, the extraction system is subjected to strong heating; therefore, the thermolabile compounds undergo transformations and consequently are not kept intact. As a result of this, steam distillation is not often applicable.

These examples serve to indicate that each of the solid-liquid extraction techniques that are currently used are not universally applicable since they are limited. Moreover, the extractive principle on which they are based is essentially linked to the phenomena of diffusion and osmosis of the substances contained in the solid, which tend to occupy the entire volume of the extracting liquid, after extraction. Therefore, desultory agitation of the extraction batch is necessary. To increase the efficiency of these extraction systems and to reduce the time of extraction, a temperature increase is used, which affects the increase in diffusion (Fick's law), in order to reduce extraction times and increase yields. Generally, this expedient is not often applicable (over 40 °C) to vegetable matrices, because they contain substances that degrade due to heat, especially active principles [21,22].

The use of ultrasounds for the extraction of active ingredients from medicinal plants leads to the same results as extraction by pressing (squeezing). Furthermore, the system heats up due to the prolonged treatment, the solid matrix is completely crushed, and a mixture that is very difficult to separate into its constituents is obtained. Among other things, the use of ultrasound energy of more than 20 kHz may have an effect on the active phytochemicals through the formation of free radicals [23]. However, due to its speed, its economic advantage, and the relatively low-cost technology involved, UAE is one of the techniques used in the industry for bioactive compound extractions. As a result, in many cases, ultrasounds can be a good alternative to pressing because it simplifies the extractive system [24].

An alternative extraction technique is based on the use of supercritical fluids, mainly based on the use of carbon dioxide. In the supercritical phase, carbon dioxide assumes the characteristics of a non-polar solvent and is comparable to liquid n-hexane; with this method, it is therefore possible to extract non-polar compounds from solid matrices. The advantage of this technique is that, at the end of the extraction, the solvent, the carbon dioxide, is removed in the form of gas, enabling the possibility of recovering the concentrated extracted compounds with a very low environmental impact (green extraction). This technique finds applications at an industrial level, such as the extraction of oil from seeds, caffeine from coffee, nicotine from tobacco, etc. [25], but it is still very expensive and not universally applicable due to the difficulty of changing the polarity of carbon dioxide and for the interference of water contained in solids [26].

Another extraction technique is Soxhlet extraction, which is reported as an official extraction method [27] for numerous analytical methods in which an initial preparation of a solid sample extract is expected. The Soxhlet method also uses system heating, since it is based on the principles of diffusion and osmosis, so it cannot be used for substances that degrade due to heat [28]. Soxhlet extraction is a good method for the extraction of high boiling substances such as polycyclic aromatic hydrocarbons (PAH), polychlorobiphenyls (PCBs), dioxine, triglycerides, and so on. Nowadays, an improved method to perform Soxhlet extraction is named Soxtec [29]; this process is based on the same principles; however, thanks to pressure control, it is possible to accelerate the recirculation of the extractant solvent. In this way, the process is about 10 times faster [30].

To increase extraction yields and reduce time, accelerated solvent extraction (ASE) can be used [31]. This technique is based on an increase in diffusion because it is possible to extract solids by using liquids operating above their boiling temperature while being maintained in a liquid state by the increase in pressure. The material to be extracted is placed in a cylindrical steel container, and the extracting solvent is introduced; the temperature of the system is raised above the boiling point of the solvent, which is maintained in the liquid state thanks to a simultaneous increase in pressure (the vial is sealed to resist high pressure values: 100–200 bar). After a short contact period, the solid matrix is completely extracted. With this technique, it is not possible to extract thermally unstable compounds [32].

In this paper, a review of innovative solid-liquid extraction technology is presented, which can be used as a valid alternative to the existing ones, RSLDE, which can be considered a green means of extraction. The application of green technology aims to preserve the natural environment and its resources, and to limit the negative influence of human involvement [33]. The philosophy of green chemistry is to develop and encourage the utilization of procedures that reduce and/or eliminate the use or production of hazardous substances. The extraction takes place for the generation of a negative pressure gradient from the inside towards the outside of the solid matrix, so it can be carried out at room temperature, or even sub-environmental temperatures [11]. The functioning of this innovative system is based on a new solid-liquid extractive principle, as it is not equivalent to others reported in literature. The patent of the instrument named the Naviglio extractor was released in 2000 [34] and registered in 1998. An extractive cycle consists of both static and dynamic phases. During the static phase, the liquid is maintained under pressure at about 10 bar on the solid to be extracted and is left long enough to let the liquid penetrate inside the solid and to balance the pressure between the inside and the outside of the solid (about 1–3 min). After this, at the beginning of the dynamic phase, the pressure immediately drops to atmospheric pressure, causing a rapid flowing of liquid from the inside to the outlet of the solid matrix. At this moment, there is a suction effect of the liquid from the inside towards the outside of the solid. This rapid displacement of the extracting solvent transports the extractable material (compounds not chemically linked) outwards. The cycles can be repeated until the solid runs out. Experimental tests carried out to date on more than 200 vegetables have shown that, working at a pressure of about 10 bar, most solid matrices, regardless of the degree of crumbling, can be extracted using about 30 extractive cycles (two-minute static phase; two-minute dynamic phase) that are completed in two hours. Furthermore, the reproducibility of the extraction on the same matrix in terms of yield was proven, and experiments were carried out to compare this method with other extraction techniques, which showed that RSLDE had a higher recovery and a higher quality of extract, and in no case was the alteration of thermolabile substances induced [11] (Figure 1).

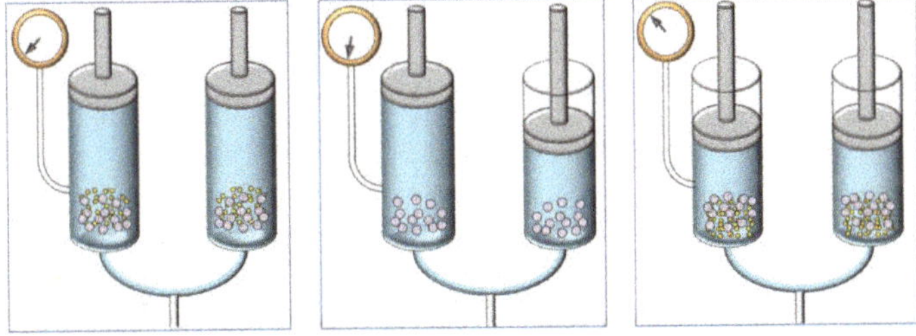

Figure 1. Schematic representation of the Naviglio extractor consisting of two extraction chambers connected via a conduit: the first two images show the dynamic phase, while the third image the static phase.

3. Comparison between the Various Solid-Liquid Extraction Techniques: Pros and Cons

The choice of methods and technologies related to an extraction process based on solid-liquid contact is not simple and this depends largely on the structural complexity and composition of the solid matrix; therefore it is not easy to find universal methods suitable for every type of Solid-Liquid extraction. In choosing the most appropriate techniques, operating conditions, solvents, etc., knowledge of the chemical properties of the compounds to be extracted and their behavior in the presence of different solvents is of fundamental importance. Due to the large extent of vegetables, operating conditions (granulometry of solid, different extractant liquids and their mix, temperature etc.) to the date, numerical and/or mathematical models that could anticipate the time and yield of extraction starting from precise conditions (solid type, solvent, temperature and so on) are not available. Alongside the aforementioned classical techniques, over the years others have been added; more complex and efficient and based on innovative extraction principles, such as extraction with supercritical fluids (SFE), ultrasound extraction (UAE), microwave extraction (MAE), accelerated solvent extraction (ASE) and finally the rapid solid liquid dynamic extraction (RSLDE) that uses the Naviglio extractor, which due to its characteristics of efficiency and improvement compared to other extraction techniques, was the subject of this review. RSLDE is an interesting new and innovative Solid-Liquid technology because it changes the philosophy of extraction; diffusion and osmosis are negligible in respect to the extraction based on difference in pressure between the inner material and the outlet of the solid matrix; this makes the extractive process "active" because it forces molecules out of solid. Below are the positive and negative aspects, pros and cons, of the main Solid-Liquid extraction techniques, nowadays existing, and briefly summarized in Table 1.

Table 1. Comparison and main characteristics of Solid-Liquid extraction techniques are herein presented.

Extraction Technique	Solvent	Granulometry	Time	Yield	Quality Extracted	Extract Stability	References
Squeezing	Indifferent	Not important	Minimum	Exhaustive	Poor	Poor	[35–38]
Maceration	Fundamental	Important	Long	Exhaustive	Great	Great	[39,40]
Decotion	Fundamental	Important	Long	Exhaustive	Great	Great	[41,42]
Percolation	Fundamental	Important	Middle	Partial	Good	Good	[43]
Soxhlet	Fundamental	Important	Long	Exhaustive	Poor	Poor	[28,44]
SCD	Indifferent	Not important	Middle	Partial	Poor	Poor	[45]
MAE	Fundamental	Not important	Middle	Partial	Poor	Poor	[46,47]
UAE	Fundamental	Not important	Middle	Partial	Great	Great	[48,49]
SFE	Indifferent	Not Important	Middle	Exhaustive	Poor	Poor	[50,51]
ASE	Fundamental	Not important	Minimum	Exhaustive	Poor	Poor	[8,52]
RSLDE	Indifferent	Not important	Minimum	Exhaustive	Great	Great	[53]

Abbreviations: SCD: steam current distillation; MAE: microwave-assisted extraction; UAE: ultrasound-assisted extraction; SFE: supercritical fluid extraction; ASE: accelerated Solid-Liquid extraction; RSLDE: rapid Solid-Liquid dynamic extraction.

3.1. Squeezing

Squeezing is an ancient extraction technique based on extraction from vegetables of substances such as dyes, perfumes, poisons, and even substances with marked healing properties [35]. The technique is very simple, as it consists of a pressing system based on the application of pressure, by pestles, mortars, mullers, presses, etc., on the mass of a plant material; this mechanical action serves to obtain the exudates of vegetables in which important substances are contained [36]. This technique is unusual because liquid is not used to extract molecules from the inner material of a solid, but in spite of this it is counted among Solid-Liquid extraction techniques for the final effect of obtaining extracts. Squeezing finds its greatest use in the food industry, particularly in the extraction of oil from seeds and oleaginous fruits (olive oil, sunflower oil, etc.) and for the extraction of essential oils from the fruits of the genus *Citrus*. The advantage of squeezing is that it does not use any thermal gradient, which can induce the peroxidation of the extracted oils. Crushing solids releases elements contained in a vegetable contaminated by a series of undesired compounds. The product resulting from this process is rarely used as it is; in most cases, it is necessary to resort to a sophisticated separation processes in order to isolate the desired compounds. For this reason, despite being an ancient technique, applications are limited in number. Because the application of squeezing to other vegetables does not produce useful extracts, the need to discover new technology became a priority. In a paper by Vongsak et al. (2013), varying extraction methods, such as squeezing, decoction, maceration, percolation, and Soxhlet extraction have been used to extract phenolics and flavonoids from fresh and dried leaves of *M. oleifera*. The results show that maceration was more advantageous than other methods for the extraction of phenolics and flavonoids with the highest antioxidant activity [37].

Another particular and specific Solid-Liquid extraction technique is enfleurage, which uses extractant liquid that is not in contact with any solid. It is based on the dissolution of aromatic compounds in a liquid located above a vegetable. In this way, volatile compounds that fill a closed environment solubilize in the liquid (generally n-hexane). This extraction process has been widely used in the cosmetic industry in ancient times. This process was based on the observation that some fats had the peculiarity of absorbing odors from delicate parts of vegetables, generally flowers, which were then used as perfumes. With enfleurage, excellent quality oils are obtained; however, being an extremely expensive method, it is used today for demonstration purposes only [38]. Compared to squeezing, enfleurage is an unusual Solid-Liquid extraction technique because there is no contact between components. Briefly, squeezing and enfleurage are two simple and ancient techniques of Solid-Liquid extraction that have few and limited applications; both techniques do not use Solid-Liquid contact and are not often applicable.

3.2. Maceration

Another simple and economic separation technique is maceration, which is carried out in steel containers that can have both small and large capacities (starting from a few liters to industrial-level amounts) and inert material both towards the solid matrix and the extracting solvent. This Solid-Liquid extraction technique is the first and the oldest that is based on diffusion and osmosis and, for this reason, is counted as the reference technique for many applications involving the extraction of active principles from officinal plants [39]. The solid to be extracted is introduced into the inert container and completely covered by the solvent. In order to obtain the most complete extraction possible, the container must be hermetically closed, and agitation of the batch is required in order to enable the diffusion of compounds extracted in the liquid and thus to avoid the equilibrium of extracted substances. The extraction process is generally quite long and requires days or even weeks to complete. In this extraction process, both diffusion and osmosis phenomena, strongly dependent on temperature, occur. The extraction process is sped up with the increase in temperature (Fick's law) or, more recently, with the use of ultrasounds or microwaves that increase the kinetic energy of the molecules that are found within the solid matrix and potentially extractables. This technique is recommended for extracting soluble or thermolabile active ingredients and for those matrices that, when hot, can lose substances of therapeutic interest (active principles). Maceration requires only occasional agitation for the diffusion of substances that are extracted in the mass of the extracting liquid. It is useful to underline that it is necessary to carry out maceration with limited quantities of solvent in several cycles and to subject the extracted material to squeezing, so as to avoid a strong loss of active ingredients. In fact, a dried and ground vegetable matrix absorbs a certain amount of solvent and, depending on its absorbent capacity, retains a more or less high portion in which the active ingredients are dissolved. A study by Ćujić et al. (2016) indicated that maceration was an effective and simple technique for the extraction of bioactive compounds from chokeberry fruit, even if it requires a long extraction time [40]. The maceration technique, with regard to extraction in aqueous phase, presents some variations, as it is not possible to use water at room temperature to extract vegetables because microbiological processes of fermentation take place more rapidly than extractive processes. To remedy this inconvenience, the infusion can be used, which can be represented as maceration for very short periods of time (1–2 min or until it has cooled), and is obtained by immersing medicinal plants or solid foods in water boiling to extract the active ingredients. In this case, the extraction is certainly faster, but the degradation of thermally labile substances also becomes faster. Briefly, maceration is a good technique of Solid-Liquid extraction and is applied in many cases for the production of extracts from officinal plants [39]. It is easy to apply, but extraction times are too long. The loss of liquid in the solid matrix is also relevant.

3.3. Decoction

Another variation of the classic maceration is decoction, which is carried out by contacting the matrix with the solvent operating at boiling temperature, for a variable time up to 30 min. At the end of extraction, the liquid is filtered, and the squeezed liquid of the extracted matrix impregnated with the solvent is added to it. This technique is therefore reserved for compact materials that have thermoresistant active principles, in accordance with which extraction requires the intervention of heat. The metabolic and antioxidant profiles of 10 herbal preparations have been assessed, and data have shown that the infusion procedure positively affected the extractability of the phenolic compounds compared to decoctions [41]. However, decoctions, like infusions, are easily altered and have limited validity; in fact, their shelf life is very short, and for this reason the extracts (generally named tisanes) must be consumed immediately after production. The last variant of classical maceration is digestion, which consists of heating the matrix in contact with the solvent from 35 to 60 °C. This technique is used when moderate heat is allowed in order to increase the extraction power of the solvent: if the solvent used is very volatile, it is necessary that the container in which the digestion is carried out is provided with a suitable refluxing refrigerant system for the recovery and recycling of the solvent itself. Digestion yields are higher than those obtained in maceration even if, with cooling and resting, cloudiness and the formation of precipitates occur [42]. Briefly, the decoction is a valid method to extract the active ingredients or aromas from the parts of medicinal plants or foods that are harder, such as roots, seeds, bark or wood, but it is not suitable for thermolabile compounds.

3.4. Percolation

3.4.1. Simple Percolation

In simple percolation, a particular container (cylindric percolator) filled with the matrix is used. In this matrix, the extracting liquid is recirculated by the means of a pump. Percolators can be made of glass, enameled iron, porcelain, or steel, and the shape depends on the nature of the matrix to be extracted. The percolation involves a series of fundamental operations: A good grinding is required; in fact, the degree of pulverization greatly influences the efficiency and the extraction time. Preliminary humidification is necessary, as the particles of the matrix to be extracted in contact with the solvent tend to swell. In the absence of this operation, the interstitial spaces diminish, and the regular outflow of the liquid is thus prevented. It is necessary to fill the percolator after a layer of cotton and sand is placed on the bottom, with the aim to block the solid matrix and as a filtration element. The matrix is added in a uniform and compact manner, minimizing and humidifying the contents of the chromatographic column. No spaces are left in the solid matrix. Preventive maceration occurs, and this works to soften the tissues and facilitates extraction. Finally, the extracting liquid is added to the percolator head and comes into contact with the solid matrix, and this liquid exerts a dynamic solubilization action on the matrix. In this technique, diffusion and osmosis occur as they do in maceration; the difference is in the continuous movement of extractant liquid through the solid, and this constitutes the driving force of percolation. The leachate thus forms, after filtration on cotton or sand, comes out from the percolator and is collected.

3.4.2. Continuous Percolation

In this extraction technique, a series of percolators are used, and in these the matrix to be extracted is placed and continuously fed in a counter-current by leachates that are less rich in extracted substances and that come from successive percolators, where the matrix is in a more exhausted state. Passing from one diffuser to the other, the solvent will be increasingly enriched by the extractable components while the matrix becomes increasingly impoverished in solutes. In this way, a concentration gradient is guaranteed in each diffuser. By optimizing the soaking and distribution ratios, the percolation operation can take place using a series of a few diffusers (5–10). One work by Chanda et al. (2012) reports a comparison between three different methods for extracting antioxidants from *Syzygium cumini* L. leaves: sequential cold percolation, decoction, and maceration. The results show that the sequential cold percolation extraction method is the best method of extracting leaf antioxidants from this plant [43]. In general, percolation does not require trained personnel to perform extraction operations. Furthermore, temperature and/or ultrasounds and microwaves can accelerate the extraction process; however, as for maceration, it is necessary to make the same considerations regarding thermolabile substances. Moreover, it is worth noting that, if the Solid-Liquid contact is very fast, the yield of extraction is not high, and generally about 50% (*w/w*) of extractable substances are lost in the matrix by the end of the process. Briefly, percolation is a very fast way to extract active principles from vegetables, but the process is not exhaustive.

3.5. Soxhlet Extraction

The Soxhlet method was introduced to determine whether it was possible to extract with the same Solid-Liquid ratio without using great quantities of extractant liquid and using "fresh" liquid [28]. Through the Soxhlet apparatus, this was possible because the liquid in contact with the solid is consistently "fresh" because of its distillation from the boiling liquid in the flask.

The Soxhlet method is used to extract compounds with high thermic stability due to a high temperature (the boiling point of the solvent). These substances are concentrated during the extractive process. The main advantage of this device is the use of a minimum quantity of solvent, thanks to its continuous purification and distillation after each passage through the matrix. The material to be extracted is placed in a porous thimble placed in the extraction chamber, which is placed on a distillation flask in which the solvent to be heated is placed. As the liquid boils, its vapors rise along a side tube up to the refrigerant mounted on the extractor. The liquid obtained from the condensation of the vapors falls into the extraction chamber passing through the material contained in the porous thimble, filling it until it reaches the elbow of the lateral siphon. At this point, due to its weight, the percolated liquid is sucked into the underlying flask, from which it is distilled again. The cycle described above is repeated several times until the extraction is considered complete: in this way, it is possible to extract all the soluble material from the matrix always using the same volume of solvent previously loaded in the boiler, renewed continuously by the distillation process. A review by De Castro and Priego-Capote (2010) describes the advantages and shortcomings of this centenary technique as well as the attempts to improve its performance and the achievements reached. In addition, currently, automation of Soxhlet procedures opened the door to the commercialization of a number of different approaches [44]. However, the Soxhlet apparatus and similar equipment cannot be used for the extraction of matrices that contain thermolabile active principles; moreover, this apparatus is not scalable to an industrial dimension. Briefly, Soxhlet extraction is a good tool for the extraction of many classes of compounds at a laboratory level; in fact, it is reported in official methods of analysis for the extraction of fats from foods, for the extraction of IPAs and PCBs from soil, etc. The limit of this type of extraction is related to the high temperature of the solvent, which does not allow for the extraction of thermo-sensible substances.

3.6. Steam Current Distillation (SCD)

SCD is a preferred method applied to the extraction of essential oils from vegetables. The solid matrix is placed in a distillation flask in which steam is forced to pass through; volatile compounds are moved in a condenser where they pass from gas to liquid form, and at the bottom of the condenser a container is placed. The principle of this technique is based on the fact that the vapor pressure of volatile substances allows them to be removed from the vegetable. For this reason, this technique of Solid-Liquid extraction is different because volatile compounds are not extracted in a liquid. However, like maceration and enfleurage, this technique is considered Solid-Liquid extraction for the final effect of extraction.

Distillation can be simple (e.g., the traditional distillation of wine waste) or in a steam current; in the latter case, the process becomes faster and the yield is higher in comparison with simple distillation. The auxiliary fluid that is used to assist distillation is generally represented by water in the form of steam, as it is very simple to generate steam (steam generators), has a high latent heat value (for this reason this process is expensive), and it is also particularly suitable for extractions of essential oils from aromatic plants. In a paper by Wei et al. (2012), steam distillation extraction and one-step high-speed counter-current chromatography were applied to separate and purify some bioactive compounds from the essential oil of *Flaveria bidentis* (L.) Kuntze, and good yields and high purity of the compounds (96.8% on average) were obtained [45]. However, despite its many advantages, steam current distillation cannot be used for all classes of organic compounds since the temperatures reached in the treatment can still be critical for the integrity of some of the molecules involved. Briefly, steam current distillation is an ancient technique for the extraction of thermally stable volatile compounds contained in vegetables, particularly for the extraction of essential oils. It is the most common technique for the production of distillates and similar products worldwide.

3.7. Microwave-Assisted Extraction (MAE)

MAE is a fast and efficient extraction technique based on the use of microwaves to heat the sample/solvent mixture in order to facilitate and speed up the extraction of the analyte. It is essentially very similar to the maceration process, but the introduction of a source of microwaves contributes to accelerating the extractive process. Unlike traditional heat sources, which act on a surface, from which heat diffuses towards the inner layers of the matrix by conduction and convection, a microwave heat source acts on the entire volume (if the medium is homogeneous) or on localized heating centers, consisting of the polar molecules present in the product. Therefore, whereas with conventional heating some time is required to heat the container before the heat is transferred to the solution, the microwaves directly heat the solution and the solid matrix, and the temperature gradient is kept to a minimum. Currently, MAE is already widely used in the laboratory for the extraction of organic pollutants from different matrices and for the isolation of natural products. It allows for a considerable reduction in the process time and in the solvent volumes used with respect to the classical extraction conducted through Soxhlet extractors. Recently, the MAE of phenolics from pomegranate peels was studied by Kaderides et al. (2019) and the extraction efficiency was compared with that of ultrasound extraction. The obtained extracts presented a high antioxidant activity [46]. However, MAE has the disadvantages that the tested samples must be thermostable, and a filtration phase is necessary, which in some cases can be very complex. Briefly, the introduction of microwaves in the maceration batch is used to perform MAE. This process is more accelerated than maceration because microwaves immediately heat both the solid and the liquid. The increasing temperature accelerates the extractive process, but at the same time if the energy of microwave is too high it is possible to damage the solid matrix and transform the active principles [47].

3.8. Ultrasonic Assisted Extraction (UAE)

The ultrasonication technique consists of passing a series of ultrasound pulses with increasing intensity through titanium probes immersed in a liquid medium. The probes convert the pulsed electrical energy applied to their heads into a vibrational impulse, which in a gaseous medium is transformed into ultrasound, while in a liquid medium, due to its incompressibility, it becomes an implosion. The waves generated by the pressure impulse in these particular vibrational conditions can cause cavitation, a phenomenon that consists in the formation of millions of small bubbles, during the negative pressure phase, which can implode in one of the subsequent compression phases. The implosion of each bubble causes a sudden change in temperature and pressure within the latter. The collapse of the cavity near a liquid-solid interface, however, differs considerably from the cavitation in a homogeneous liquid. In fact, passing through a liquid, the ultrasound expansion cycles exert a negative pressure on the liquid, with the molecules moving away from each other: if the ultrasound is sufficiently intense, the negative pressure exceeds the tensile strength of the liquid molecules generating a cavity. Cavitation bubbles form in the pre-existing weak points of the liquid and inside the solid spaces. Both were filled with gas in the powdered matter and suspended. Micro-bubbles, prior to cavitation were suspended in the irradiated liquid. Thus, there were devastating effects on the cellular structure. At high intensities, then, due to an inertial effect, a small cavity can quickly develop during the expansion half-cycle and will not have time to re-compress during the compression half-cycle. The bubble thus formed in the following cycle will suffer the same effect and increase in size, and the phenomenon will be repeated in the subsequent cycles until the bubble reaches a critical size such that it collapses with an increase in thermal energy. Instead, at lower acoustic intensities, cavity development may occur with a slower process. Under these conditions, a cavity will oscillate in size until it reaches the critical dimension, defined as a resonant dimension, where it can efficiently absorb the energy coming from the ultrasonic irradiation. The frequency range of use of the ultrasound is outside the perception limit of the human ear. Sonication is a technique that is used in many fields: the most widespread laboratory applications are in the field of biomedical and pharmaceutical research to lyse bacteria or cells in culture, in the field of environmental analysis for the extraction of various molecules, in the cosmetic and pharmaceutical industry for the preparation of creams and emulsions, and in biotechnology for the homogenization of immiscible liquids and the solubilization of difficult compounds. In the extractive field, this technique uses ultrasound frequencies to break up the cellular structure and facilitate diffusion processes. Goula et al. (2017) have carried out comparative studies between ultrasound-assisted and conventional solvent extraction in terms of processing procedure and total carotenoid content extracted from pomegranate wastes. The efficiency of the technique made it possible to produce an oil enriched with antioxidants [48]. The use of ultrasound in solvent extraction is a good remedy for the inconveniences linked to diffusion, but it is not always efficient. Due to the high energy developed inside the extraction batch, the breaking of the cellular structure results in extracts very similar to that obtained with the squeezing technique; in fact, vegetables are finely dispersed in the extractant liquid, and the resulting mixture is complex, so filtration and separation of active principles are required. Moreover, extracted compounds suffer from the direct bombing of cavitation generated by ultrasounds and can undergo transformations resulting in the loss of their beneficial activity. Briefly, for UAE, there are many variables to consider in order to obtain a good yield, so often the development of the various parameters lengthens the experimentation time. In the extraction of active ingredients from plants, for example, the results are comparable to extraction by squeezing, if not worse, due to the heating of the system for a prolonged time. The solid matrix is completely crushed, and a mixture is impossible to separate in its constituents is obtained, which makes this technique difficult to apply on an industrial level. This technique is often used in a laboratory procedure for sample preparation [49].

3.9. Supercritical Fluid Extraction (SFE)

SFE is a recent and very complex Solid-Liquid extraction technology [50]. The technique is based on the possibility of being able to use an extraction solvent that is a fluid (usually carbon dioxide) with properties that are intermediate between those of gases and liquids, named a supercritical fluid. Through modest variations in temperature and/or pressure, it is possible to modulate the properties of gases in a wide range and use their criticality to control phase behavior in extraction/separation processes. In practice, above the critical temperature, it is possible to continuously regulate the solubility of the fluid over a wide range, either with a small change in isothermal pressure or with a small isobaric temperature change. This ability to regulate the solvent power of a supercritical fluid is the main feature on which the SFE systems are based. These solvents can be used to extract and then efficiently recover the selected products. Since supercritical fluids have density, viscosity, and other properties that are intermediate between those of the substance in the gaseous state and those of the substance in the liquid state, the first and obvious advantage of this technique is that at the end of the extraction process the carbon dioxide is brought to ambient temperature and pressure, and gasification consequently leaves the substances extracted from the solid matrix. This fact makes SFE a green technique for Solid-Liquid extraction. A second advantage is represented by the best transport speeds: although the densities of supercritical fluids approximate those of conventional liquids, their transport properties are closer to those of gases. For example, the viscosity is many orders of magnitude lower than that of liquids, and the same diffusion coefficients are 100 times larger than typical ones observed in conventional liquids. The choice of CO_2 as a supercritical fluid offers the following advantages: it spreads through extractive matrices faster than typical solvents that have a larger molecular size; it is cheap and can be obtained easily; it has higher diffusion coefficients and lower viscosities than the liquid solvent; it has a strong permeability, so the extraction time can be considerably shorter than that required by extraction with a common solvent; it is odorless, non-toxic, does not burn, does not explode, and does not damage the ozone layer; the working temperature is close to room temperature (31.1 °C), particularly suitable for heat-sensitive material, which would be decomposed by heat treatment; recovery is simple and convenient and can be recycled without treatment; extraction and removal are combined in a single technique, significantly shortening processing times in a simple and convenient way; and it has a variable solvent power, depending on the selected operating conditions (pressure and temperature). The application of supercritical fluids in the extraction of bioactive compounds and their operative extraction conditions has been reported in a review by da Silva et al. (2016) [51]. Supercritical fluid technology offers features that overcome many limitations of conventional extraction methods. However, the limitation of this technique consists in the need for specialized equipment as well as a lower solubilizing capacity for water-soluble compounds, which can be solved in part by adding traces of polar liquids (methanol and acetone), and the request for specialized personnel for its use. Briefly, SFE is the best choice for the extraction of non-polar substances, such as nicotine from tobacco, caffeine from coffee, and oil from seeds, at an industrial scale; it is completely green extraction technology. Unfortunately, this technique is very expensive, requires specialized personnel, and is not versatile.

3.10. Accelerated Solvent Extraction (ASE)

ASE represents a useful and innovative approach for the extraction of a wide class of compounds from matrices of complex chemical-physical entity. The extraction of the analytes from the matrices takes place using a solvent kept in liquid phase at temperatures above the boiling temperature thanks to the application of high pressure. This means that this technique requires a stainless steel container that resists the high pressures generated by raising temperatures beyond the boiling point of the solvent. The increase in temperature, in fact, accelerates the desorption of the analytes (Fick's law) from the sample and their solubilization in the solvent, allowing for an effective extraction in a short period of time. In the use of solvents under high temperature and pressure, it is possible to influence the extraction process by modifying some chemical-physical parameters of the solvent–matrix system. The greatest effect on extraction is given by the temperature, as it influences the physical properties of the solvent and the interaction between the liquid phase and the material raising the molecular diffusion. Less significant is the effect of pressure that, even at low values, facilitates the penetration of the solvent into the pores of the sample. The essential function for ASE is that of keeping the solvent in a liquid state during the process. An advantage in the use of liquids at high temperature under pressure, with respect to supercritical fluids, is the fact that the former has a greater solvent strength and that, being used in methods that involve extractions at atmospheric pressure, no modifications or preliminary tests are required for the evaluation of their extractive efficiency. The other advantage is represented by the fact that, using liquid solvents, there are no phase changes in the return of the system to atmospheric conditions and therefore there is no need for liquid or packed restrictors or traps for the recovery of the analytes from the extract [52]. In their study, Cai et al. (2016) investigated the extraction efficiency of anthocyanins from purple sweet potatoes using conventional extraction, UAE, and ASE. The results show that extraction efficiencies were opposite for anthocyanins, and phenolics/flavonoids for the three methods [8]. On the other hand, the limits of the ASE method are represented by a partial extraction because of the static system and a possible degradation of the active ingredients due to operative conditions. Briefly, ASE is a good technique for the extraction of thermally very stable substances because raising the temperature behind the boiling point of the solvent remained in contact with the solid and the liquid at high temperatures for the entire experimental period. Due to the high pressure generated in the system, this technique is only used for sample preparation at a laboratory scale (10–20 mL).

3.11. Rapid Solid-Liquid Extraction (RSLDE)

Introduced in 2000 [34], RSLDE, through the use of the Naviglio extractor, represents a valid alternative to all existing Solid-Liquid extraction techniques and brings significant advantages in obtaining high quality extracts. First of all, it is not necessary to heat the extractive system, as the action performed is mechanical. Current extraction techniques (percolation, Soxhlet, steam current distillation and ultrasound), based on the principles of osmosis and diffusion, require an increase in temperature to increase the extraction efficiency. In the case of thermolabile compounds, however, the increase in temperature contributes to their degradation, as reported below. RSLDE requires a few extraction cycles, about 30 (depending on the matrix, but always within hours), to bring a large number of vegetable matrices to complete exhaustion. Compared to maceration, the official extraction method for many processes, RSLDE has been proven to be both quick and comprehensive. Moreover, it is possible to use water as an extracting liquid for many applications thanks to the reduced extraction times, while the prolonged contact of the plant solid matrices with water is unthinkable for maceration. RSLDE is an inexpensive technique and requires minimal energy expenditure when compared with SFE or ASE, both of which, among other things, require the use of high temperatures. In a recent work by Posadino et al. (2018), RSLDE was used to obtain polyphenolic antioxidants from the Cagnulari grape marc. The results indicate Naviglio extraction, as a green technology process, can be used to exploit wine waste to obtain antioxidants that can be used to produce enriched foods and nutraceuticals high in antioxidants [53].

In summary, the main advantages of RSLDE are as follows: exhaustion in a short period of time, with solid matrices containing extractable substances, at low operating temperatures (environment or sub-environment); reproducibility of the extraction since there is a real possibility of standardizing extracts for their active ingredients, with a guarantee of the production of high-quality extracts. From a careful comparison between the main characteristics of each of the Solid-Liquid extraction techniques described above, it is possible to state that, at present, no technique simultaneously provides all of the advantages offered by RSLDE, in terms of granulometry of the solid material, of the solvent type, or of the yield, time, quality, and stability of the extract (Table 1). On the other hand, due to its ease of use, its low-energy consumption, and the speed of the extraction process, RSLDE can be used as an exploratory and research technique for solid matrices that are not yet known and can be used to deal with materials that must undergo processes of washing (this means that the substances extracted from the solid are not important and for this reason will be discharged), such as polymers with clathrates, cork, etc. RSLDE has been advantageously used in processes in which it is important to fix substances inside the solid, as in leather tanning operations where both chrome solutions and solutions containing natural tannins are used (data not yet published). Moreover, RSLDE finds important applications in the beverage industry for the preparation of many alcoholic drinks; extracts from the ethyl alcohol of citrus peels (lemon, mandarin, orange, etc.) and from tonics and bitters from herbal extracts have been derived. In the perfume industry, it is possible to obtain fragrant and aromatic plant extracts by replacing maceration with RSLDE; in the same way, formulations of preparations in cosmetics and in herbal medicine have been improved compared to classical techniques, and alterations in the extracts obtained can be determined and made less active. Briefly, the application of this technology is very simple. It requires little energy for functioning, the time of extraction is low (two hours is the reference time), and the active principles are not degraded. Moreover, it is possible to apply RSLDE at temperatures below room temperature.

Figure 2 shows a pie chart showing the application percentages of each technique mentioned in the work obtained from a qualitative analysis of the literature data.

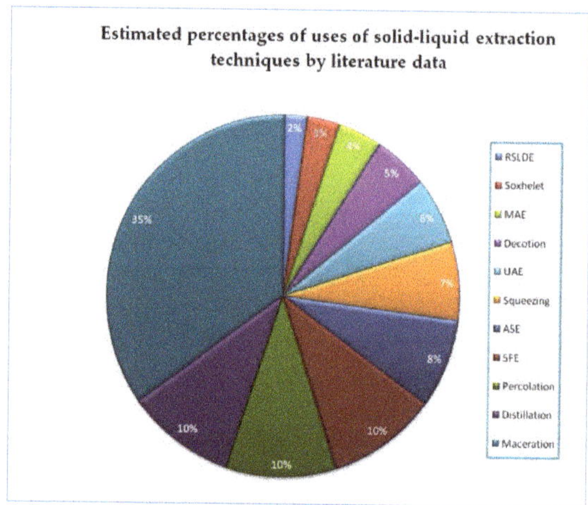

Figure 2. Pie chart showing the percentage of applications of each cited extraction technique.

4. RSLDE Applications in Various Industries

4.1. The Pharmaceutical Sector

In the pharmaceutical sector, RSLDE has been used in the preparation of high-quality standardized extracts, including medicinal plant extracts and herbal extracts, fluid extracts, mother tinctures, glycerinated extracts, glyceric macerates, liposoluble extracts, bitter medicines, etc., all of which were obtained in a much shorter period of time (4–8 h) compared to maceration, which took 21 days (maceration data provided by the Official Pharmacopoeia) [39]. The speed of the process, the extraction at low temperature, and the high efficiency guarantee the total recovery of non-degraded active ingredients contained in medicinal plants [11].

Paullinia cupana seeds, commonly called guarana, are natural sources of phenolic antioxidants and antimicrobial compounds, and the use of guarana extract is interesting for the food, pharmaceutical, and cosmetic industries, where such natural additives are required [54]. A work by Basile et al. (2005) has reported extraction from *Paullinia cupana* var. *sorbilis* Mart. (*Sapindaceae*) seeds via RSLDE. Moreover, the antibacterial and antioxidant activity of the ethanol extract was assessed towards selected bacteria and in different antioxidant models [55].

Cardiospermum halicacabum is a herbaceous plant belonging to the *Sapindaceae* family, widely used in traditional medicine for its therapeutic properties. Menichini et al. (2014) analyzed the chemical composition of extracts from aerial parts and seeds, the inhibitory properties against some enzymes, and the antioxidant effects obtained using RSLDE and the Soxhlet method. The findings suggested the potential of both seeds and aerial parts of *C. halicacabum* for the treatment of neurological disorders [56]. Moreover, RSLDE was used to extract the flowering aerial parts of *Schizogyne sericea*, a halophytic shrub that is widespread on the coastal rocks of Tenerife (Canary Islands). The extracts obtained were assayed for in vitro biological activities. Results showed that aqueous extracts, rich in phenolic acids, were endowed with relevant radical scavenging activity [57].

Therefore, among the green extraction techniques used to improve the sensitivity and the selectivity of analytical methods, RSLDE represents a sustainable alternative to classical sample-preparation procedures used in the past [58]. In a study by Cozzolino et al. (2016), the extraction of curcuminoids by RSLDE was performed from *Curcuma longa* roots, focusing the interest on curcumin, the major phenolic component of the root that has been shown to have high antioxidant activity [59]. On the other hand, some studies have shown that curcumin exerts anti-tumor effects for its ability to induce apoptosis in cancer cells without cytotoxic effects on healthy cells. Moreover, some research has demonstrated an absence of toxicity in humans when dosing this active principle for short periods of time. Therefore, for its beneficial and healing properties, curcumin obtained by the described extraction method may be used as a natural dietary supplement [60]. A comparison between three extraction processes, including traditional maceration in n-hexane and ethyl alcohol, supercritical fluid extraction (SFE), and cyclically pressurized extraction (CPE), also known as RSLDE, has been carried out for the extraction of pyrethrins, predominantly nonpolar natural compounds with insecticidal properties found in pyrethrum, an extract of certain species of chrysanthemums [61].

4.2. The Cosmetic Sector

In cosmetics and perfumery, both in production and research, it is possible to produce extracts from vegetable matrices that contain pigments and odorous substances for the production and formulation of creams and perfumes. Officinal plants are raw, cosmetic materials that have been used in numerous formulations since ancient times. Plant extraction methods are carried out to obtain active phytocomplexes, both lipo and hydrosoluble [62]. The active ingredients of plants can be obtained from the plant complex or can be taken with drugs, a term that indicates the part or parts of the plant in which the active ingredients are present. Plant drugs are essentially whole plants (fragmented or cut), parts of plants, algae, fungi, or lichens in an untreated state, generally in dried form, but sometimes fresh. Phytocosmetic plants include vasal reinforcers (e.g., root ruscus, blueberry fruits, and gingko leaves), emollients (e.g., mallow, altea, and borage), stimulants (e.g., lavender, thyme, sage, juniper, and rosemary), bioactivators (e.g., calendula and carrot). They can be used as such or through their fluid extracts, and, with the addition of natural excipients, they can be used for natural functional cosmetics. The excipients are products that support and convey active and functional plant extracts. Essential oils or essences are an important part of phytocosmetics; they are obtained by the distillation of medicinal aromatic plants, obtaining a separation of the volatile component distillable from the non-volatile. These essential oils are diluted in appropriate solvents and applied in aesthetics according to their properties [63].

A review by Barbulova et al. (2015) reported some examples of the most important applications of agricultural food by-products in cosmetics and their performance as efficacy and safety [64]. In another review, Zappelli et al. (2016) showed examples of active cosmetic ingredients developed through biotechnological systems, whose activity on the skin has been scientifically proven through in vitro and clinical studies [65]. More recently, the reasons and the characteristics as well as the challenges of plant cell culture-based productions for the cosmetic and food industries are discussed in a review by Eibl et al. (2018) [66].

In the case of RSLDE, an active action is carried out towards the substances to be extracted; in fact, the compounds not chemically bound to the solid matrix are extracted in small quantities at each extraction cycle (active process) until the matrix is completely exhausted. The advantage is that the whole process takes place in the order of hours. The important consequences of the use of this technique are the possibility of extracting vegetable matrices with water. Therefore, it is possible to extract substances at temperatures even lower than room temperature for any thermolabile compounds. Moreover, these applications can be implemented on industrial, domestic, and lab scales [11].

4.3. The Herbal Sector

In the herbal and phytotherapy sector, both in production and research, RSLDE can be used for the extraction of plants and medicinal herbs for the production of fluid extracts. Since it is not necessary to heat the extraction system, it is possible to produce teas and/or infusions at room temperature, keeping the active ingredients unaltered.

Fresh plants of *Malva silvestris* were extracted with water using RSLDE, and the effects of terpenoids and phenol isolated from this plant on the germination and growth of dicotyledon *Lactuca sativa* L. (lettuce) were studied [67].

In a study by Ferrara et al. (2014), a conventional extraction technique (UAE) and a cyclically pressurized Solid-Liquid extraction (RSLDE) were compared, in order to obtain qualitative and quantitative data related to the bioactive compounds of saffron. The results obtained showed that extracts via RSLDE had significant advantages in terms of extraction efficiency and the quality of the extract [68].

4.4. The Food and Beverage Sector

In the food sector, both in production and research, RSLDE has been applied in various ways. Lycopene, the carotenoid responsible for the red color of many fruits and vegetables, is considered fundamental for its antioxidant action. Therefore, its extraction is of great interest in various sectors. In fact, it can be used both for the formulation of functional foods and in cosmetics. In addition, lycopene can be extracted from tomato processing waste using only water as an extract liquid. The use of water as an extracting phase considerably reduces the cost of the entire process when compared with the commonly used solvent-based procedure or with the newer supercritical extraction process of lycopene from tomato waste. Lycopene, not soluble in water, was recovered in a quasi-crystalline solid form and purified by solid-phase extraction using a small amount of organic solvent [69]. Lycopene can be used as a dye and/or a natural antioxidant. Moreover, through RSLDE, it is also possible to produce limoncello, a lemon liqueur, in just two hours, avoiding the long traditional maceration that takes 7–14 days [70]. Nowadays the industrial process for the production of lemon liquor is performed via maceration, as home-made products are made, and the process requires at least 48 h of lemon peel infusion in alcohol.

In a paper by Formato et al. (2013), two Solid-Liquid extraction techniques, supercritical fluid extraction (SFE) with and without modifiers and cyclically pressurized Solid-Liquid extraction or RSLDE, were compared on the basis of the extraction of acidic compounds contained in hops. The results showed that both techniques were valid for the extraction of α and β acids from hops. By suitably varying the parameters of the two extractive procedures, it was possible to obtain extracts for use in the production of beer and of dietary supplements and drugs [71].

In order to obtain the alcoholic extracts of some herbal mixtures, the traditional maceration procedure was compared to RSLDE. Three different mixtures of various parts of plants were extracted with both methods, and results were compared. Organoleptic tests performed on alcoholic bitters obtained from different extracts have been used to determine the optimum extraction time for the two different methods used. The results showed that the bitters produced with RLSDE were more appreciated than bitters prepared by maceration [72]. In another work by Naviglio et al., (2014), the extraction process for the production of *Cinchona calisaya* elixir starting from the same vegetable mixture was performed by a conventional maceration and RSLDE. The results show that, compared to the conventional method, RSLDE allowed for extraction at room temperature using cyclical extraction pressurization. In this way, it was possible to avoid thermal stress on thermolabile substances while simultaneously reducing extraction time [73].

Based on the Naviglio extractor, a system of extraction has been devised that can be used for student laboratory experiments to illustrate RSLDE (two-syringe system). In a paper by Naviglio et al. (2015), students compare two extraction techniques for the preparation of limoncello: maceration and the two-syringe system. The development of the two-syringe system for simple manual operations reduced the risks of the procedure, allowing students to evaluate the extraction efficiency of the two methods [74]. In another work, two extraction processes for the production of limoncello—the traditional maceration of lemon peels and RSLDE—were compared. Alcoholic extracts were analyzed by gas chromatography, and alcoholic extracts were analyzed by UV spectrophotometry to identify the more abundant chemical species, while the organoleptic tests performed on the final product (limoncello) provided an indication of the taste of the final product. Results showed that the RSLDE process was 120 times faster than maceration and had a greater efficiency in a short period of time [75].

The results of a recent study by Gigliarelli et al. (2017) indicated that RSLDE is an effective technique for the extraction of piperine from fruits of *Piper longum* compared to other techniques, such as Soxhlet extraction, decotion (International Organization for Standardization), and conventional maceration [76].

Stevioside and rebaudioside A are the main diterpene glycosides present in the leaves of the *Stevia rebaudiana* plant, which is used in the production of foods and low-calorie beverages. In this context, RSLDE constitutes a valid alternative method to conventional extraction by reducing the extraction time and the consumption of toxic solvents and favoring the use of extracted metabolites as food additives and/or nutraceuticals [77,78].

Portulaca oleracea, commonly known as purslane, is a wild plant pest of orchards and gardens, but is also an edible vegetable rich in beneficial nutrients. The purpose of a work by Gallo et al. (2017) was to compare various Solid-Liquid extraction techniques to determine the most efficient technique for the extraction of biomolecules from leaves of purslane. Therefore, extraction reproducibility was tested on the same matrix in terms of weighting, and comparison experiments were performed. RSLDE showed a higher recovery and a higher quality of extract, and in no cases did it induce the alteration of heat-sensitive substances [79].

RSLDE can be used for applications other than extraction, such as the quick hydration of legumes and the de-structuring of vegetables at room temperature (heat-free cooking) for a better preservation of nutritional elements. Hydrating beans before cooking reduces cooking time, increases their tenderness and weight, and improves their appearance after cooking. Naviglio et al. (2013) described a process of cyclically pressurized soaking for the rapid hydration of cannellini beans at room temperature. The hydration process was approximately 10-fold faster than the traditional soaking procedure, and the microbial load developed by the end of this process was much lower compared to that obtained using the traditional process [80].

A large amount of food waste and by-products is produced from farm to plate and represents valuable sources for the production of compounds with a high added value. Consequently, the application of innovative approaches is necessary due to the limitation of conventional processes. In this context, RSLDE can be a useful tool to increase extract yield and quality, reducing extraction time, temperature, and toxic solvents. In a study by Bilo et al. (2018), RSLDE was used for the synthesis of a new bioplastic produced from rice straw, an agricultural waste that is generally not recovered. The results show that, depending on the environmental humidity, the material shows a dual mechanical behavior that can be exploited to obtain shrink films and sheets or to drive a shape memory effect. Therefore, rice straw bioplastics could represent a new potential eco-material for different application fields [81]. In addition, RSLDE could provide an innovative approach to increase the production of specific compounds from food waste for use as nutraceuticals or as ingredients in the design of functional foods. As it is currently known, grape pomace is a by-product of winemaking that can be conveniently reused in many different ways, including agronomic use as well as cosmetic industry applications. Moreover, the by-products can also be used in the energy field as biomass for the production of biogas or for food plants used for the production of energy. As an added value, grape pomace resulting from the production of wine also contains numerous bioactive compounds. In a very recent work by Gallo et al. (2019), to extract polyphenols, grape peels were processed via RSLDE, which does not require the use of any organic solvent, nor does it include heating or cooling processes that can cause the loss of substances of interest [82]. Still, within the framework of the guiding principles for eco-innovation, which aims at a zero waste economy, many residues have the potential to be reused as raw material for new products and applications and in other production systems, such as those in the nutraceutical and pharmaceutical sectors. Another recent work by Gallo et al. (2019) shows an alternative process for the extraction of lycopene from tomato waste through RSLDE. The high purity of lycopene obtained using this procedure make the process very attractive, and the pure product obtained could be used in various applications [83].

5. Conclusions

The efficiencies of conventional and non-conventional extraction methods mostly depend on the critical input parameters, on understanding the nature of the matrices to be extracted, on the chemistry of the compounds, and on scientific expertise. In particular, RSLDE introduces a new Solid-Liquid extractive technology that is based on the difference in pressure between the inner material and the outlet of the solid matrix, which generates a suction effect (matter transfer) that causes the compounds that are not chemically linked to the solid matrix to be extracted. Therefore, RSLDE makes it possible to optimally replace most of the current Solid-Liquid extraction techniques and brings about considerable new features and advantages in obtaining high-quality extracts. It changed the concept of Solid-Liquid extraction based only on diffusion and osmosis phenomena (passive process of extraction) using a new philosophy that bases on the generation of pressure gradient between the inner material and the outlet of the solid (active process of extraction). In addition, the areas of application of RSLDE are numerous and include the pharmaceutical, cosmetic, herbal, and food and beverage industries. Furthermore, RSLDE can be used for the extraction of food waste, a by-product of various industrial, agricultural, domestic, and other food sectors, which is currently increasing due to the increase in these activities. These by-products can be used as a potential source of bioactive and nutraceutical compounds that have important applications in the treatment of various disorders.

Author Contributions: Conceptualization, D.N. and M.G.; investigation, D.N., P.S., M.C., and M.G.; data curation, P.S., M.C.; writing—original draft preparation, M.G.; writing—review and editing, D.N. and M.G.; visualization, D.N.; supervision, M.G.

Funding: This research received no external funding.

Conflicts of Interest: The authors declare no conflict of interest. They do not derive any financial benefit from the publication of this review, in fact none of the authors is part of the company that produces the extractor, nor does it receive any gain from the sale of the instruments.

References

1. Azmir, J.; Zaidul, I.S.M.; Rahman, M.M.; Sharif, K.M.; Mohamed, A.; Sahena, F.; Omar, A.K.M. Techniques for extraction of bioactive compounds from plant materials: A review. *J. Food Eng.* **2013**, *117*, 426–436. [CrossRef]

2. Aguilera, J.M. Solid-liquid extraction. In *Extraction Optimization in Food Engineering*; CRC Press: Boca Raton, FL, USA, 2003; pp. 51–70.

3. Wang, L.; Weller, C.L. Recent advances in extraction of nutraceuticals from plants. *Trends Food Sci. Technol.* **2006**, *17*, 300–312. [CrossRef]

4. Galanakis, C.M. Recovery of high added-value components from food wastes: Conventional, emerging technologies and commercialized applications. *Trends Food Sci. Technol.* **2012**, *26*, 68–87. [CrossRef]

5. Chemat, F.; Rombaut, N.; Sicaire, A.G.; Meullemiestre, A.; Fabiano-Tixier, A.S.; Abert-Vian, M. Ultrasound assisted extraction of food and natural products. Mechanisms, techniques, combinations, protocols and applications. A review. *Ultrason. Sonochem.* **2017**, *34*, 540–560. [CrossRef] [PubMed]

6. Khaw, K.Y.; Parat, M.O.; Shaw, P.N.; Falconer, J.R. Solvent supercritical fluid technologies to extract bioactive compounds from natural sources: A review. *Molecules* **2017**, *22*, 1186. [CrossRef] [PubMed]

7. Ekezie, F.G.C.; Sun, D.W.; Cheng, J.H. Acceleration of microwave-assisted extraction processes of food components by integrating technologies and applying emerging solvents: A review of latest developments. *Trends Food Sci. Technol.* **2017**, *67*, 160–172. [CrossRef]

8. Cai, Z.; Qu, Z.; Lan, Y.; Zhao, S.; Ma, X.; Wan, Q.; Li, P. Conventional, ultrasound-assisted, and accelerated-solvent extractions of anthocyanins from purple sweet potatoes. *Food Chem.* **2016**, *197*, 266–272. [CrossRef]

9. Souza-Silva, É.A.; Jiang, R.; Rodríguez-Lafuente, A.; Gionfriddo, E.; Pawliszyn, J. A critical review of the state of the art of solid-phase microextraction of complex matrices I. Environmental analysis. *Trends Anal. Chem.* **2015**, *71*, 224–235. [CrossRef]

10. Kumar, S.J.; Kumar, G.V.; Dash, A.; Scholz, P.; Banerjee, R. Sustainable green solvents and techniques for lipid extraction from microalgae: A review. *Algal Res.* **2017**, *21*, 138–147. [CrossRef]

11. Naviglio, D. Naviglio's principle and presentation of an innovative solid–liquid extraction technology: Extractor Naviglio®. *Anal. Lett.* **2003**, *36*, 1647–1659. [CrossRef]

12. Barba, F.J.; Zhu, Z.; Koubaa, M.; Sant'Ana, A.S.; Orlien, V. Green alternative methods for the extraction of antioxidant bioactive compounds from winery wastes and by-products: A review. *Trends Food Sci. Technol.* **2016**, *49*, 96–109. [CrossRef]

13. Chemat, F.; Rombaut, N.; Meullemiestre, A.; Turk, M.; Perino, S.; Fabiano-Tixier, A.S.; Abert-Vian, M. Review of green food processing techniques. Preservation, transformation, and extraction. *Innov. Food Sci. Emerg. Technol.* **2017**, *41*, 357–377. [CrossRef]

14. Al Jitan, S.; Alkhoori, S.A.; Yousef, L.F. Phenolic acids from plants: Extraction and application to human health. In *Studies in Natural Products Chemistry*; Elsevier: Amsterdam, The Netherlands, 2018; Volume 58, pp. 389–417.

15. Azwanida, N.N. A review on the extraction methods use in medicinal plants, principle, strength and limitation. *Med. Aromat. Plants* **2015**, *4*. [CrossRef]

16. AOAC Method 43.290. *Official Methods of Analysis of the AOAC*, 15th ed.; Association of Official Analytical Chemists: Washington, DC, USA, 1990.

17. Willson, K.C.; Clifford, M.N. *Tea: Cultivation to Consumption*; Springer Science & Business Media: Berlin, Germany, 2012.

18. Liguori, L.; Russo, P.; Albanese, D.; Di Matteo, M. Production of low-alcohol beverages: Current status and perspectives. In *Food Processing for Increased Quality and Consumption*; Academic Press: Cambridge, MA, USA, 2018; pp. 347–382.

19. Aspé, E.; Fernández, K. The effect of different extraction techniques on extraction yield, total phenolic, and anti-radical capacity of extracts from *Pinus radiata* Bark. *Ind. Crop. Prod.* **2011**, *34*, 838–844. [CrossRef]

20. Božović, M.; Navarra, A.; Garzoli, S.; Pepi, F.; Ragno, R. Essential oils extraction: A 24-hour steam distillation systematic methodology. *Nat. Prod. Res.* **2017**, *31*, 2387–2396. [CrossRef] [PubMed]

21. Joana Gil-Chávez, G.; Villa, J.A.; Fernando Ayala-Zavala, J.; Basilio Heredia, J.; Sepulveda, D.; Yahia, E.M.; González-Aguilar, G.A. Technologies for extraction and production of bioactive compounds to be used as nutraceuticals and food ingredients: An overview. *Compr. Rev. Food Sci. Food Saf.* **2013**, *12*, 5–23. [CrossRef]

22. Rostagno, M.A.; Prado, J.M. *Natural Product Extraction: Principles and Applications*; No. 21; Royal Society of Chemistry: London, UK, 2013.

23. Esclapez, M.D.; García-Pérez, J.V.; Mulet, A.; Cárcel, J.A. Ultrasound-assisted extraction of natural products. *Food Eng. Rev.* **2011**, *3*, 108. [CrossRef]

24. Gallo, M.; Ferrara, L.; Naviglio, D. Application of ultrasound in food science and technology: A perspective. *Foods* **2018**, *7*, 164. [CrossRef] [PubMed]

25. Jesus, S.P.; Meireles, M.A.A. Supercritical fluid extraction: A global perspective of the fundamental concepts of this eco-friendly extraction technique. In *Alternative Solvents for Natural Products Extraction*; Springer: Berlin, Germany, 2014; pp. 39–72.

26. Sánchez-Camargo, A.D.P.; Parada-Alonso, F.; Ibáñez, E.; Cifuentes, A. Recent applications of on-line supercritical fluid extraction coupled to advanced analytical techniques for compounds extraction and identification. *J. Sep. Sci.* **2019**, *42*, 243–257. [CrossRef] [PubMed]

27. AOAC Method 963.15. *Agricultural Chemicals, Contaminants, Drugs*, 15th ed.; Association of Official Analytical Chemists: Arlington, VA, USA, 1990.

28. Jensen, W.B. The origin of the Soxhlet extractor. *J. Chem. Educ.* **2007**, *84*, 1913. [CrossRef]

29. Anderson, S. Soxtec: Its principles and applications. In *Oil Extraction and Analysis. Critical Issues and Competitive Studies*; AOCS Publishing: Champaign, IL, USA, 2004; pp. 11–25.

30. Carro, N.; Cobas, J.; García, I.; Ignacio, M.; Mouteira, A.; Silva, B. Development of a method for the determination of SCCPs (short-chain chlorinated paraffins) in bivalve mollusk using Soxtec device followed by gas chromatography-triple quadrupole tandem mass spectrometry. *J. Anal. Sci. Technol.* **2018**, *9*, 8. [CrossRef]

31. Molino, A.; Rimauro, J.; Casella, P.; Cerbone, A.; Larocca, V.; Chianese, S.; Musmarra, D. Extraction of astaxanthin from microalga *Haematococcus pluvialis* in red phase by using generally recognized as safe solvents and accelerated extraction. *J. Biotechnol.* **2018**, *283*, 51–61. [CrossRef] [PubMed]

32. He, Q.; Du, B.; Xu, B. Extraction optimization of phenolics and antioxidants from black goji berry by accelerated solvent extractor using response surface methodology. *Appl. Sci.* **2018**, *8*, 1905. [CrossRef]

33. Hilali, S.; Fabiano-Tixier, A.S.; Ruiz, K.; Hejjaj, A.; Nouh, F.A.; Idlimam, A.; Chemat, F. Green extraction of essential oils, polyphenols and pectins from orange peel employing solar energy. Towards a Zero-Waste Biorefinery. *ACS Sustain. Chem. Eng.* **2019**. [CrossRef]

34. Naviglio, D. Rapid and Dynamic Solid–Liquid Extractor Working at High Pressures and Low Temperatures for Obtaining in Short Times Solutions Containing Substances that Initially Were in Solid Matrixes Insoluble in Extracting Liquid. Italian Patent 1,303,417, 6 November 2000.

35. Baldwin, E.A.; Bai, J.; Plotto, A.; Cameron, R.; Luzio, G.; Narciso, J.; Ford, B.L. Effect of extraction method on quality of orange juice: Hand-squeezed, commercial-fresh squeezed and processed. *J. Sci. Food Agric.* **2012**, *92*, 2029–2042. [CrossRef] [PubMed]

36. Armenta, S.; Garrigues, S.; de la Guardia, M. The role of green extraction techniques in Green Analytical Chemistry. *TrAC Trends Anal. Chem.* **2015**, *71*, 2–8. [CrossRef]

37. Vongsak, B.; Sithisarn, P.; Mangmool, S.; Thongpraditchote, S.; Wongkrajang, Y.; Gritsanapan, W. Maximizing total phenolics, total flavonoids contents and antioxidant activity of Moringa oleifera leaf extract by the appropriate extraction method. *Ind. Crop. Prod.* **2013**, *44*, 566–571. [CrossRef]

38. Stratakos, A.C.; Koidis, A. Methods for extracting essential oils. In *Essential Oils in Food Preservation, Flavor and Safety*; Preedy, V.R., Ed.; Academic Press: Cambridge, MA, USA, 2016; pp. 31–38.

39. European Pharmacopoeia Commission. *European Pharmacopoeia*, 9th ed.; European Directorate for the Quality of Medicines (EDQM): Strasbourg, France, 2014.

40. Ćujić, N.; Šavikin, K.; Janković, T.; Pljevljakušić, D.; Zdunić, G.; Ibrić, S. Optimization of polyphenols extraction from dried chokeberry using maceration as traditional technique. *Food Chem.* **2016**, *194*, 135–142. [CrossRef]

41. Fotakis, C.; Tsigrimani, D.; Tsiaka, T.; Lantzouraki, D.Z.; Strati, I.F.; Makris, C.; Zoumpoulakis, P. Metabolic and antioxidant profiles of herbal infusions and decoctions. *Food Chem.* **2016**, *211*, 963–971. [CrossRef]

42. Manousi, N.; Sarakatsianos, I.; Samanidou, V. Extraction techniques of phenolic compounds and other bioactive compounds from medicinal and aromatic plants. In *Engineering Tools in the Beverage Industry*; Woodhead Publishing: Sawston, UK, 2019; pp. 283–314.

43. Chanda, S.V.; Kaneria, M.J. Optimization of conditions for the extraction of antioxidants from leaves of *Syzygium cumini* L. using different solvents. *Food Anal. Methods* **2012**, *5*, 332–338. [CrossRef]

44. De Castro, M.L.; Priego-Capote, F. Soxhlet extraction: Past and present panacea. *J. Chromatogr. A* **2010**, *1217*, 2383–2389. [CrossRef] [PubMed]

45. Wei, Y.; Du, J.; Lu, Y. Preparative separation of bioactive compounds from essential oil of *Flaveria bidentis* (L.) Kuntze using steam distillation extraction and one step high-speed counter-current chromatography. *J. Sep. Sci.* **2012**, *35*, 2608–2614. [CrossRef] [PubMed]

46. Kaderides, K.; Papaoikonomou, L.; Serafim, M.; Goula, A.M. Microwave-assisted extraction of phenolics from pomegranate peels: Optimization, kinetics, and comparison with ultrasounds extraction. *Chem. Eng. Process. Process Intensif.* **2019**, *137*, 1–11. [CrossRef]

47. Chan, C.H.; Yusoff, R.; Ngoh, G.C.; Kung, F.W.L. Microwave-assisted extractions of active ingredients from plants. *J. Chromatogr. A* **2011**, *1218*, 6213–6225. [CrossRef] [PubMed]

48. Goula, A.M.; Ververi, M.; Adamopoulou, A.; Kaderides, K. Green ultrasound-assisted extraction of carotenoids from pomegranate wastes using vegetable oils. *Ultrason. Sonochem.* **2017**, *34*, 821–830. [CrossRef] [PubMed]

49. Tiwari, B.K. Ultrasound: A clean, green extraction technology. *Trends Anal. Chem.* **2015**, *71*, 100–109. [CrossRef]

50. Sharif, K.M.; Rahman, M.M.; Azmir, J.; Mohamed, A.; Jahurul, M.H.A.; Sahena, F.; Zaidul, I.S.M. Experimental design of supercritical fluid extraction–A review. *J. Food Eng.* **2014**, *124*, 105–116. [CrossRef]

51. Da Silva, R.P.; Rocha-Santos, T.A.; Duarte, A.C. Supercritical fluid extraction of bioactive compounds. *TrAC Trends Anal. Chem.* **2016**, *76*, 40–51. [CrossRef]

52. Nayak, B.; Dahmoune, F.; Moussi, K.; Remini, H.; Dairi, S.; Aoun, O.; Khodir, M. Comparison of microwave, ultrasound and accelerated-assisted solvent extraction for recovery of polyphenols from *Citrus sinensis* peels. *Food Chem.* **2015**, *187*, 507–516. [CrossRef]

53. Posadino, A.; Biosa, G.; Zayed, H.; Abou-Saleh, H.; Cossu, A.; Nasrallah, G.; Pintus, G. Protective effect of cyclically pressurized solid–liquid extraction polyphenols from Cagnulari grape pomace on oxidative endothelial cell death. *Molecules* **2018**, *23*, 2105. [CrossRef]

54. Santana, A.L.; Macedo, G.A. Health and technological aspects of methylxanthines and polyphenols from guarana: A review. *J. Funct. Foods* **2018**, *47*, 457–468. [CrossRef]

55. Basile, A.; Ferrara, L.; Del Pezzo, M.; Mele, G.; Sorbo, S.; Bassi, P.; Montesano, D. Antibacterial and antioxidant activities of ethanol extract from *Paullinia cupana* Mart. *J. Ethnopharmacol.* **2005**, *102*, 32–36. [CrossRef] [PubMed]

56. Menichini, F.; Losi, L.; Bonesi, M.; Pugliese, A.; Loizzo, M.R.; Tundis, R. Chemical profiling and in vitro biological effects of *Cardiospermum halicacabum* L. (*Sapindaceae*) aerial parts and seeds for applications in neurodegenerative disorders. *J. Enzym. Inhib. Med. Chem.* **2014**, *29*, 677–685. [CrossRef] [PubMed]

57. Caprioli, G.; Iannarelli, R.; Sagratini, G.; Vittori, S.; Zorzetto, C.; Sánchez-Mateo, C.C.; Petrelli, D. Phenolic acids, antioxidant and antiproliferative activities of Naviglio® extracts from *Schizogyne sericea* (*Asteraceae*). *Nat. Prod. Res.* **2017**, *31*, 515–522. [CrossRef] [PubMed]

58. Bandar, H.; Hijazi, A.; Rammal, H.; Hachem, A.; Saad, Z.; Badran, B. Techniques for the extraction of bioactive compounds from Lebanese *Urtica Dioica*. *Am. J. Phytomed. Clin. Ther.* **2013**, *1*, 507–513.

59. Cozzolino, I.; Vitulano, M.; Conte, E.; D'Onofrio, F.; Aletta, L.; Ferrara, L.; Gallo, M. Extraction and curcuminoids activity from the roots of *Curcuma longa* by RSLDE using the Naviglio extractor. *ESJ* **2016**, *12*. [CrossRef]

60. Pulido-Moran, M.; Moreno-Fernandez, J.; Ramirez-Tortosa, C.; Ramirez-Tortosa, M. Curcumin and health. *Molecules* **2016**, *21*, 264. [CrossRef]

61. Gallo, M.; Formato, A.; Ianniello, D.; Andolfi, A.; Conte, E.; Ciaravolo, M.; Naviglio, D. Supercritical fluid extraction of pyrethrins from pyrethrum flowers (*Chrysanthemum cinerariifolium*) compared to traditional maceration and cyclic pressurization extraction. *J. Supercrit. Fluids* **2017**, *119*, 104–112. [CrossRef]

62. Cabaleiro, N.; De La Calle, I.; Bendicho, C.; Lavilla, I. Current trends in liquid–liquid and solid–liquid extraction for cosmetic analysis: A review. *Anal. Methods* **2013**, *5*, 323–340. [CrossRef]

63. Ali, B.; Al-Wabel, N.A.; Shams, S.; Ahamad, A.; Khan, S.A.; Anwar, F. Essential oils used in aromatherapy: A systemic review. *Asian Pac. J. Trop. Biomed.* **2015**, *5*, 601–611. [CrossRef]

64. Barbulova, A.; Colucci, G.; Apone, F. New trends in cosmetics: By-products of plant origin and their potential use as cosmetic active ingredients. *Cosmetics* **2015**, *2*, 82–92. [CrossRef]

65. Zappelli, C.; Barbulova, A.; Apone, F.; Colucci, G. Effective active ingredients obtained through Biotechnology. *Cosmetics* **2016**, *3*, 39. [CrossRef]

66. Eibl, R.; Meier, P.; Stutz, I.; Schildberger, D.; Hühn, T.; Eibl, D. Plant cell culture technology in the cosmetics and food industries: Current state and future trends. *Appl. Microbiol. Biotechnol.* **2018**, *102*, 8661–8675. [CrossRef] [PubMed]

67. Cutillo, F.; D'Abrosca, B.; DellaGreca, M.; Fiorentino, A.; Zarrelli, A. Terpenoids and phenol derivatives from *Malva silvestris*. *Phytochemistry* **2006**, *67*, 481–485. [CrossRef] [PubMed]

68. Ferrara, L.; Naviglio, D.; Gallo, M. Extraction of bioactive compounds of saffron (*Crocus sativus* L.) by Ultrasound Assisted Extraction (UAE) and by Rapid Solid-Liquid Dynamic Extraction (RSLDE). *ESJ* **2014**, *10*. [CrossRef]

69. Naviglio, D.; Pizzolongo, F.; Ferrara, L.; Naviglio, B.; Santini, A. Extraction of pure lycopene from industrial tomato waste in water using the extractor Naviglio. *Afr. J. Food Sci.* **2008**, *2*, 037–044.

70. Naviglio, D.; Pizzolongo, F.; Romano, R.; Ferrara, L.; Naviglio, B.; Santini, A. An innovative solid-liquid extraction technology: Use of the Naviglio Extractor for the production of lemon liquor. *Afr. J. Food Sci.* **2007**, *1*, 42–50.

71. Formato, A.; Gallo, M.; Ianniello, D.; Montesano, D.; Naviglio, D. Supercritical fluid extraction of α-and β-acids from hops compared to cyclically pressurized solid–liquid extraction. *J. Supercrit. Fluids* **2013**, *84*, 113–120. [CrossRef]

72. Naviglio, D.; Ferrara, L.; Formato, A.; Gallo, M. Efficiency of conventional extraction technique compared to rapid solid-liquid dynamic extraction (RSLDE) in the preparation of bitter liquors and elixirs. *IOSR J. Pharm.* **2014**, *4*, 14–22.

73. Naviglio, D.; Formato, A.; Gallo, M. Comparison between 2 methods of solid–liquid extraction for the production of *Cinchona calisaya* elixir: An experimental kinetics and numerical modeling approach. *J. Food Sci.* **2014**, *79*, E1704–E1712. [CrossRef]

74. Naviglio, D.; Montesano, D.; Gallo, M. Laboratory production of lemon liqueur (Limoncello) by conventional maceration and a two-syringe system to illustrate rapid solid–liquid dynamic extraction. *J. Chem. Educ.* **2015**, *92*, 911–915. [CrossRef]

75. Naviglio, D.; Formato, A.; Vitulano, M.; Cozzolino, I.; Ferrara, L.; Zanoelo, E.F.; Gallo, M. Comparison between the kinetics of conventional maceration and a cyclic pressurization extraction process for the production of lemon liqueur using a numerical model. *J. Food Process Eng.* **2017**, *40*, e12350. [CrossRef]

76. Gigliarelli, G.; Pagiotti, R.; Persia, D.; Marcotullio, M.C. Optimisation of a Naviglio-assisted extraction followed by determination of piperine content in *Piper longum* extracts. *Nat. Prod. Res.* **2017**, *31*, 214–217. [CrossRef] [PubMed]

77. Gallo, M.; Vitulano, M.; Andolfi, A.; DellaGreca, M.; Conte, E.; Ciaravolo, M.; Naviglio, D. Rapid Solid-Liquid Dynamic Extraction (RSLDE): A new rapid and greener method for extracting two steviol glycosides (stevioside and rebaudioside A) from stevia leaves. *Plant Foods Hum. Nutr.* **2017**, *72*, 141–148. [CrossRef] [PubMed]

78. Gallo, M.; Formato, A.; Formato, G.; Naviglio, D. Comparison between two solid-liquid extraction methods for the recovery of steviol glycosides from dried stevia leaves applying a numerical approach. *Processes* **2018**, *6*, 105. [CrossRef]

79. Gallo, M.; Conte, E.; Naviglio, D. Analysis and comparison of the antioxidant component of *Portulaca oleracea* leaves obtained by different solid-liquid extraction techniques. *Antioxidants* **2017**, *6*, 64. [CrossRef]

80. Naviglio, D.; Formato, A.; Pucillo, G.P.; Gallo, M. A cyclically pressurised soaking process for the hydration and aromatisation of cannellini beans. *J. Food Eng.* **2013**, *116*, 765–774. [CrossRef]

81. Bilo, F.; Pandini, S.; Sartore, L.; Depero, L.E.; Gargiulo, G.; Bonassi, A.; Bontempi, E. A sustainable bioplastic obtained from rice straw. *J. Clean. Prod.* **2018**, *200*, 357–368. [CrossRef]

82. Gallo, M.; Formato, A.; Giacco, R.; Riccardi, G.; Lungo, D.; Formato, G.; Amoresano, A.; Naviglio, D. Mathematical optimization of the green extraction of polyphenols from grape peels through a cyclic pressurization process. *Heliyon* **2019**, e01526. [CrossRef]

83. Gallo, M.; Formato, A.; Ciaravolo, M.; Langella, C.; Cataldo, R.; Naviglio, D. A water extraction process for lycopene from tomato waste using a pressurized method: An application of a numerical simulation. *Eur. Food Res. Technol.* **2019**. [CrossRef]

MDPI

St. Alban-Anlage 66

4052 Basel

Switzerland

Tel. +41 61 683 77 34

Fax +41 61 302 89 18

www.mdpi.com

Foods Editorial Office

E-mail: foods@mdpi.com

www.mdpi.com/journal/foods